The Routledge Handbook to Regional Development in Central and Eastern Europe

Twenty-five years into transformation, Central and Eastern European regions have undergone substantial socio-economic restructuring, integrating into European and global networks and producing new patterns of regional differentiation and development. Yet post-socialist modernisation has not been without its contradictions, manifesting in increasing social and territorial inequalities. Recent studies also suggest there are apparent limits to post-socialist growth models, accompanying a new set of challenges within an increasingly uncertain world.

Aiming to deliver a new synthesis of regional development issues at the crossroads between 'post-socialism' and 'post-transition', this book identifies the main driving forces of spatial restructuring in Central and Eastern Europe, and charts the different regional development paths which take shape against the backdrop of post-crisis Europe. A comparative approach is used to highlight common development challenges and the underlying patterns of socio-economic differentiation alike. The issues investigated within the Handbook extend to a discussion of the varied economic consequences of transition, the social structures and institutional systems which underpin development processes, and the broadly understood sustainability of Central and Eastern Europe's current development model.

This book will be of interest to academics and policymakers working in the fields of regional studies, economic geography, development studies and policy.

Gábor Lux is Senior Research Fellow at the CERS Institute for Regional Studies, the Hungarian Academy of Sciences, Pécs, Hungary.

Gyula Horváth was Research Advisor and former Director-General of the CERS Institute for Regional Studies, the Hungarian Academy of Sciences, Pécs, Hungary.

The Routledge Handbook to Regional Development in Central and Eastern Europe

Edited by Gábor Lux and Gyula Horváth

LONDON AND NEW YORK

First published 2018
by Routledge
2 Park Square, Milton Park, Abingdon, Oxon OX14 4RN

and by Routledge
711 Third Avenue, New York, NY 10017

Routledge is an imprint of the Taylor & Francis Group, an informa business

© 2018 selection and editorial matter, Gábor Lux and Gyula Horváth; individual chapters, the contributors

The right of Gábor Lux and Gyula Horváth to be identified as the authors of the editorial material, and of the authors for their individual chapters, has been asserted in accordance with sections 77 and 78 of the Copyright, Designs and Patents Act 1988.

All rights reserved. No part of this book may be reprinted or reproduced or utilised in any form or by any electronic, mechanical, or other means, now known or hereafter invented, including photocopying and recording, or in any information storage or retrieval system, without permission in writing from the publishers.

Trademark notice: Product or corporate names may be trademarks or registered trademarks, and are used only for identification and explanation without intent to infringe.

British Library Cataloguing-in-Publication Data
A catalogue record for this book is available from the British Library

Library of Congress Cataloging-in-Publication Data
A catalog record for this book has been requested

ISBN: 978-1-4724-8571-7 (hbk)
ISBN: 978-1-315-58613-7 (ebk)

Typeset in Bembo
by Keystroke, Neville Lodge, Tettenhall, Wolverhampton

Printed and bound in Great Britain by
TJ International Ltd, Padstow, Cornwall

Contents

List of figures — vii
List of tables — ix
List of contributors — xi

1 Regional development paths in Central and Eastern Europe and the driving forces of restructuring: an introduction — 1
Gábor Lux

PART I
Economic transformation processes before and beyond the crisis — 13

2 Reintegrating economic space: the metropolitan–provincial divide — 15
Gábor Lux

3 Industrial competitiveness: beyond path-dependence — 29
Gábor Lux

4 The role of business and finance services in Central and Eastern Europe — 47
Zoltán Gál and Sándor Zsolt Kovács

5 The transformation of rural areas in Central and Eastern Europe — 66
Péter Póla

6 Factors influencing regional entrepreneurial activity in Central and Eastern Europe — 87
Balázs Páger

7 Creativity and culture in reproducing uneven development across Central and Eastern Europe — 106
Márton Czirfusz

v

Contents

PART II
Social structures and governance — **121**

8 Changing settlement networks in Central and Eastern Europe with special regard to urban networks — 123
   Zoltán Hajdú, Réka Horeczki and Szilárd Rácz

9 The development of regional governance in Central and Eastern Europe: trends and perspectives — 141
   Ilona Pálné Kovács

10 Managing regional disparities — 159
   László Faragó and Cecília Mezei

11 Rebordering Central and Eastern Europe: Cohesion Policy, cross-border cooperation and 'differential Europeanisation' — 175
   James Wesley Scott

12 (Ethno-)regional endeavours in Central and Eastern Europe — 188
   Nóra Baranyai

PART III
Challenges in sustainable development — **209**

13 The regional dimension of migration and labour markets in Central and Eastern Europe — 211
   Jan Sucháček and Mariola Pytliková

14 The changing role of universities and the innovation performance of regions in Central and Eastern Europe — 225
   Zoltán Gál and Balázs Páger

15 Transition and resilience in Central and Eastern European regions — 240
   Adam Drobniak, Adam Polko and Jan Sucháček

16 New trends in Central and Eastern European transport space — 261
   Ferenc Erdősi

17 Nature and spatial planning in Central and Eastern European countries — 284
   Andrea Suvák

18 Spatial researches in Central and Eastern Europe — 295
   Gyula Horváth

19 Conclusion: an evolutionary look at new development paths — 309
   Gábor Lux and László Faragó

Index — *321*

Figures

2.1	Changes in the share of industrial employment in CEE countries, 1990–2008	18
2.2	The spatial structures of the CEE macro-region, 2013	22
2.3	The transformation and reintegration of post-socialist space	25
3.1	The spatial structures of Central European industry, 2013	35
3.2	The dynamics of industrial employment in the accession and the crisis periods	40
4.1	Share of service activities in employment, 2013	49
4.2	Exports of offshorable services and the sectoral composition of such, 2002, 2007 and 2012	51
4.3	Geographical and sectoral breakdown of the major services offshoring sites, 2010	53
4.4	Foreign ownership as a percentage of total banking sector assets in CEE, 1995–2014	56
4.5	FDI stock in financial services, 2006–2011	58
4.6	Dependencies on foreign bank resources (net external liabilities and debt), 2011	61
5.1	Urban–rural typology of NUTS-3 regions	67
5.2	Population density in CEE regions	68
5.3	The proportion of agriculture in GDP and employment in the CEE macro-region	69
5.4	Development of agricultural output since EU accession	77
6.1	The concept of the study	90
6.2	Regional entrepreneurial activity in the CEE regions	96
6.3	The situation of groups from the aspects of 'Function 1' and 'Function 2'	99
7.1	Cultural-commercial redevelopment of a former industrial site: Manufaktura in Łódź	109
7.2	Budgets of European Capitals of Culture in CEE	113
8.1	Towns with population over 10,000 in the Visegrad countries according to their size, 2011	126
8.2	Distribution of towns and cities with population over 10,000 in South-Eastern Europe, 2011	131
8.3	Urban population in South-Eastern Europe, 1990–2015	132
10.1	Proportion of regional disparities between 2000 and 2013	164
10.2	Cohesion policy target regions within the macro-region	166
13.1	Foreign population stocks from EU8 member states residing in 26 OECD countries, 1995 and 2010	213
13.2	Foreign population stocks from Bulgaria and Romania residing in 26 OECD countries, 1995 and 2010	214

Figures

13.3	Immigration to Central and Eastern European countries, 1995–2014	216
14.1	Regional innovation performance in Europe, 2007 and 2013	234
15.1	(a) Economic resilience of Central and Eastern European countries, 2001–2011; (b) Central and Eastern European countries' economic portfolio	245
15.2	(a) Economic resilience of the best and worst performing regions of Central and Eastern Europe, 2000–2011; (b) Economic portfolio of the best and the worst performing regions of Central and Eastern Europe	247
15.3	(a) Technological resilience of Central and Eastern European countries, 2000–2012; (b) Central and Eastern European countries' technological portfolio	249
15.4	Technological resilience of the Central and Eastern European regions with the best and worst performers, 2001–2011	250
16.1	Reloading stations/belts between two types of gauge networks between East Central Europe and the CIS region, 2014	264
16.2	The sum traffic of airports and specific passenger traffic in the countries of Central and Eastern Europe, 2014	275
16.3	Traffic in Central and Eastern Europe's port regions, 2013	277
18.1	Spatial research workshops in Central and Eastern Europe, 2012	302

Tables

2.1	Foreign Direct Investment (FDI) in post-socialist countries, 1995–2011, and the sectoral breakdown of gross value added (GVA), 2013	20
2.2	Correlation between sectoral employment and nominal GDP per capita	21
3.1	Evaluation of industrial location factors in Central and Eastern Europe	39
4.1	The distribution of Foreign Direct Investment (FDI) by economic activity, 1990–2000	48
4.2	Branch density indicators, 2012	55
5.1	The proportion of cultivated land, 2013	69
5.2	The proportion of cooperative and state ownership of agricultural land before 1989	72
5.3	Average holding size, 2000	74
5.4	Changes in the significance of agriculture in the first years of the transition, 1990–1995	75
5.5	Changes in agricultural output since EU accession	77
5.6	Changes in ownership structure	79
5.7	Areas categorised as predominantly rural and the proportion of population living in them	80
5.8	The SWOT analysis of the macro-region's agriculture and rural areas	82
6.1	Descriptive statistics of the variables related to regional entrepreneurial activity	92
6.2	Descriptive statistics of the regional socioeconomic factors	93
6.3	ANOVA results in the cluster analysis for the CEE regions	93
6.4	Clusters within the CEE regions	95
6.5	The main characteristics of discriminant functions	97
6.6	Canonical discriminant function coefficients	98
6A.1	Structure matrix in the all-sector analysis	103
6A.2	Variables involved in the all-sector analysis	103
6A.3	The functions at group centroids in the all-sector analysis	104
6A.4	Structure matrix in the industrial sector analysis	104
6A.5	The Standardised Canonical – Discriminant Function Coefficients in the industrial sector analysis	104
6A.6	The functions at group centroids in the industrial sector analysis	105
8.1	Major settlement network characteristics in the Visegrad countries, 1990 and 2011	125
8.2	Towns according to administrative functions, 1990 and 2011	127
8.3	The difference between functional urban areas (FUA) and morphological urban areas (MUA) population	129

Tables

8.4	The effects of cyclical development on the urban network and space structure in South-Eastern Europe	132
8.5	Share of capital city regions	137
10.1	Regional disparities in Central and Eastern Europe based on GDP per capita, 2013	167
10.2	Regional policy toolkit and institutional systems of the Central and Eastern European countries	168
12.1	(Ethno-)regional organisations and endeavours in the individual countries	190
12.2	Size of population having regional or ethnic identity per areas	197
12.3	Regional support of (ethno-)regional organisations at the last elections	198
13.1	Overview of dates of lifting restrictions on the free movement of labour for workers from CEE countries	212
14.1	Number of students and HE institutes in CEE countries and capital city regions	228
14.2	The HE and business sector R&D expenditures as a percentage of GDP	231
14.3	Innovation performance of the CEE regions, 2007	236
14.4	Innovation performance of the CEE regions, 2013	236
16.1	Changes in freighting modal split in the countries of CEE, 1990–2013	268
16.2	The distribution of Russian sea foreign trade across Russian and foreign ports	280
17.1	Analysed national-level spatial planning documents of the Visegrad Four countries	289
18.1	The number of regional science researchers in Eastern and Central European countries, 2012	301
18.2	Some characteristics of the major regional scientific journals	304
18.3	The development level of the disciplinary criteria of regional science	306

Contributors

Gábor Lux (co-editor) is Senior Research Fellow at the CERS Institute for Regional Studies of the Hungarian Academy of Sciences. He graduated in 2004 as an economist from the University of Pécs, Faculty of Business and Economics, and earned his PhD in 2009 at the Doctoral School for Regional Policy and Economics. His main areas of research are industrial restructuring and industrial competitiveness in Central and Eastern Europe, industrial policy, urban economic governance and evolutionary economic geography. He is the author, co-author or editor of over 60 publications in Hungarian, English and Russian. He is a recipient of the Hungarian Academy of Sciences Academy Youth Award (2011) and the János Bolyai Research Fellowship (2014–2017). He is a member of the Regional Studies Association, the European Regional Science Association, the Hungarian Regional Science Association and former member of AESOP Young Academics.

Gyula Horváth (co-editor) was Professor in Regional Economics and Policy at the University of Pécs and Research Advisor of the CERS Institute for Regional Studies of the Hungarian Academy of Sciences. From 1997 until 2012, he was Director-General of the Centre for Regional Studies of the Hungarian Academy of Sciences. From 2002 until 2011 he served as president of the Hungarian Regional Science Association, and from 2011 until 2015 as president of the Committee of Regional Science of the Hungarian Academy of Sciences. He was a member of the Academia Europaea (London), and member of the Presidency of the International Academy of Regional Development and Cooperation (Moscow). His research interests included European regional policy, restructuring and regional transformation in Eastern and Central Europe, and the regional development of science and higher education. He undertook extensive advisory and consultancy work within Hungary and Europe. He was visiting professor in Romania, Italy, Ireland, Russia and Denmark. He was author, editor and co-editor of 30 books and over 300 papers in Hungarian and in foreign languages. His latest book entitled *Spaces and places in Central and Eastern Europe: historical trends and perspectives* was published by Routledge in 2015. He passed away on 23 September 2015.

Nóra Baranyai is Research Fellow at the CERS Institute for Regional Studies of the Hungarian Academy of Sciences. She obtained her PhD in political sciences in 2013 at the University of Pécs. Her professional interest is regionalism and ethno-regionalism in Central and Eastern Europe.

Márton Czirfusz is Research Fellow at the CERS Institute for Regional Studies of the Hungarian Academy of Sciences, and Senior Lecturer at Eötvös Loránd University. He is also a member of two Budapest-based independent groups, the 'Helyzet' Working Group for Public Sociology and the Collective for Critical Urban Research. He graduated as a geographer in 2007 and holds a PhD in Geography (2012) from the Eötvös Loránd University. His main research fields are transformation of the Hungarian space-economy after 1990, labour geographies, and critical urban studies, primarily in the Budapest context. He has authored, co-authored or edited more than 40 publications in Hungarian, English and German. He is recipient of the Hungarian Academy of Sciences Academy Youth Award (2016). He serves as managing editor of *Tér és Társadalom*, the most prestigious journal of regional studies in Hungary.

Adam Drobniak is Professor at the University of Economics in Katowice and the University of Opole. He is a member of the Committee of Spatial Economy of the Polish Academy of Sciences and the Regional Studies Association. His research area mainly covers issues referring to local and regional strategic projects, urban resilience and vulnerability, trajectories of transition of post-industrial cities and regions, creative industries and local development design, comparative analysis of cities and regions in the CEE countries and Western Europe. He is the author or co-author of 10 books and over 100 publications in urban and regional studies. He serves as an active expert and moderator of local and regional self-government institutions in the field of development programming for cities (Bytom, Jaworzno, Katowice, Krakow, Mysłowice, Sosnowiec) and regions (Slaskie and Opolskie Voivodeship). He was coordinator of the RSA international Research Network on Transition and Resilience of the Post-Industrial Cities in Europe.

Ferenc Erdősi is Scientific Advisor at the CERS Institute for Regional Studies of the Hungarian Academy of Sciences. He holds the largest body of work in contemporary Hungarian geography as an expert on regional transport and telematics, and a leading specialist in applied regional science. He has been involved in scientific life since 1959, joining the Transdanubian Research Institute of the Hungarian Academy of Sciences and its successors in 1967. In 1993, he was appointed to the position of Professor at the University of Pécs; since 2011, he has been Professor Emeritus. His scientific results have been published in 17 individual monographs, 3 co-authored books, 3 university course books, 93 journal articles, 160 book chapters and several other research papers. He holds the Officer's Cross of the Order of Merit of Hungary (2014). He has played a leading role in the transport network development concepts of multiple Hungarian regions, for which he was awarded the Gábor Baross Award (2004).

László Faragó is Senior Research Fellow at the CERS Institute for Regional Studies of the Hungarian Academy of Sciences. He graduated as an economist at the University of Pécs, and holds a PhD in political sciences. He teaches three different PhD programmes in regional planning and practice. His professional interests are planning theory and practice, space theory and regional policy. He contributed to several territorial plans in practice from the local to the national level. He is the author of 119 scientific publications.

Zoltán Gál is Associate Professor in Regional Economics at the University of Kaposvár and Senior Research Fellow at the CERS Institute for Regional Studies of the Hungarian Academy of Sciences. His research fields include regional innovation, EU innovation policy, regional banking, financial and economic geography. He designed, implemented and managed several national and EU research projects as project coordinator, and was an invited expert at EC DG Research in regional foresight policy. He conducted research at Oxford University. He received the East-European Publication Prize from the SASE (USA) for his study on banking transition. He was one of the organisers of the RSA Research Network on Geographies of Finance and Post-Socialist Transformations. He is a member of the editorial board of the *Journal of Innovation and Entrepreneurship*.

Contributors

Zoltán Hajdú is Scientific Advisor at the CERS Institute for Regional Studies of the Hungarian Academy of Sciences, and was Professor of Geography at the University of Pécs. He graduated in 1976 at the Lajos Kossuth University of Debrecen with a teaching degree, becoming Candidate of Sciences in 1986 and Doctor of Sciences in Geography in 2002. His research interests include political geography, the history of regional geography, the geographic and historical analysis of long-term territorial processes, and public administration. He is the author of 375 publications, including 12 books and 84 journal articles. He is recipient of the Pro Regio Award (2010) and the Officer's Cross of the Order of Merit of Hungary (2013).

Réka Horeczki is PhD Aspirant at the Doctoral School for Regional Policy and Economics at the University of Pécs and Junior Research Fellow at the CERS Institute for Regional Studies of the Hungarian Academy of Sciences. Her research is focused on the socio-economic development of small towns in Central and Eastern Europe.

Sándor Zsolt Kovács is PhD Aspirant at the Doctoral School for Regional Policy and Economics at the University of Pécs and Junior Research Fellow at the CERS Institute for Regional Studies of the Hungarian Academy of Sciences. His research fields include the different factors of local and regional development, financial geography and economic modelling.

Ilona Pálné Kovács is Corresponding Member of the Hungarian Academy of Sciences, Director of the CERS Institute for Regional Studies, and Professor at the Department for Political and International Studies of the University of Pécs, where she is also leader of the PhD programme in political science. She graduated as a lawyer at the University of Pécs, defended her PhD in political science, and was elected as a corresponding member of the HAS in 2013. Her research interests are regional governance and local governments, the management of regional policy teaching, public administration and constitutional law. She was a leader and participant of several domestic and international research projects, including EU framework, ESF, INTERREG, TEMPUS, ESPON, UNESCO, EC DG Regio programmes. She has authored or co-authored over 350 publications. She is a member of several national and international professional organisations (ESF, IGU, RSA and NISPACee).

Cecília Mezei is Research Fellow at the Hungarian Academy of Sciences CERS Institute for Regional Studies and Associate Professor at the University of Kaposvár. She is a local development expert with experience in several applied and basic research projects. Her research areas have focused on local and regional development approaches and practices, local and territorial governance solutions, and the regional development processes, institutions and tools of European countries. She has authored over 50 publications. She is recipient of the Outstanding Young Regionalist Award of the Hungarian Regional Science Association (2011) and the János Bolyai Research Fellowship (2011–2014).

Balázs Páger is Junior Research Fellow at the CERS Institute for Regional Studies of the Hungarian Academy of Sciences. He graduated as economist at the University of Pécs in 2010 and has been doing his doctoral studies at the Doctoral School for Regional Policy and Economics. In his research, he has focused on innovation, as well as measuring regional and national entrepreneurship.

Péter Póla is Research Fellow at the CERS Institute for Regional Studies of the Hungarian Academy of Sciences and Associate Professor at the Eötvös József College of Baja, where he is director of the Economic Institute. His research area has been rural development for nearly two decades, but he has also dealt with local development and social capital, as well as the institutions of local economic development. He is a member of the Hungarian Regional Science Association. He participated in several international research projects on Eastern Europe as participant or project lead. He is the author of over 50 publications.

Adam Polko is Assistant Professor at the University of Economics in Katowice. His research interests focus on the economics of urban public spaces and urban commons, urban regeneration and resilience. He is the author and co-author of books on creative industries, urban resilience, and the economics of urban public services in shrinking regions. He cooperates with several local and regional governments in Poland as an expert in URBACT II working groups, is a member of the Investment Board of the JESSICA Fund in Silesia Region (Poland), and has authored or co-authored several urban regeneration programmes implemented in Polish cities. He was a member of ARCA team which won the first prize in the international urban planning competition 'Nowa Huta of the Future' organised by the city of Kraków. He is a board member of the European Regional Science Association – Polish Section.

Mariola Pytliková is Assistant Professor at CERGE-EI in Prague and Senior Researcher at the VŠB-Technical University of Ostrava, Czechia. She has previously worked at Aarhus University and at the Danish Institute of Governmental Research (KORA) in Copenhagen. She received her PhD in economics from Aarhus University in 2006. Mariola is research fellow at IZA, CReAM CCP and CELSI. In the past she has held visiting posts at Princeton University, the University of Illinois at Chicago, USA, and the University of Modena and Reggio Emilia, Italy, among others. Her research interests are in the field of labour economics and industrial relations; in particular she has worked on topics dealing with international migration, ethnic workforce diversity, gender pay gap and income inequality. She has published in outlets such as the *Economic Journal, European Economic Review, Journal of Population Economics, International Migration Review, European Journal of Population, Journal of Environmental Economics and Management* and the *Economics of Transition*, among others. She is a recipient of Kateřina Šmídková's Award for the best Czech female economist in 2015 and the 2005 Young Economist Award from the Czech Economic Society.

Szilárd Rácz is Scientific Secretary of the CERS Institute for Regional Studies of the Hungarian Academy of Sciences, and since 2008, Secretary of the Hungarian Regional Science Association. His main research fields are the spatial structure and urban network of the Balkans, and regional and urban development in Central Europe. He is the author of 38 publications.

James Wesley Scott is Professor of Regional and Border Studies, Karelian Institute at the University of Eastern Finland in Joensuu. He is a former Research Fellow of the Leibniz-Institute for Regional Development and Structural Planning in Erkner (Germany) and Associate Professor (Privatdozent) of Geography at the Free University of Berlin. He obtained his Habilitation (2006), his PhD (1990) as well as his MA (1986) degrees at the Free University of Berlin and his BSc at the University of California Berkeley (1979). His principal fields of research include urban and regional development, border regions, regional and urban governance, metropolitan area problems, and European and North American geography. He has coordinated numerous EU-funded research projects on cross-border cooperation such as EXLINEA and EUDIMENSIONS within the EU's Fifth and Sixth Framework Programmes, respectively, and he has recently concluded the EUBORDERSCAPES project, financed by the EU's Seventh Framework Programme.

Jan Sucháček is the Head of the Department of Regional and Environmental Economics at the VŠB-Technical University of Ostrava, Czechia, where he has been a Senior Lecturer and Associate Professor since 2003. His research interests focus mainly on urban and regional development, the spatial aspects of European integration, and globalisation. He is the author or co-author of six books and more than 90 articles. He has been active in 20 international research projects and is a member of multiple editorial boards. He was one of the organisers of the RSA Research Network on Transition and Resilience for Post-Industrial Agglomerations in Central Europe.

Andrea Suvák is Junior Research Fellow at the CERS Institute for Regional Studies of the Hungarian Academy of Sciences. She graduated as an economist at the University of Pécs, Faculty of Business and Economics, and shortly after started PhD studies at the Doctoral School for Regional Policy and Economics. Since then her main field of interest has been local economic development, the role of the environment in spatial planning, and environmental ethics. She is the author, co-author or editor of 25 publications in Hungarian and English.

1
Regional development paths in Central and Eastern Europe and the driving forces of restructuring

An introduction

Gábor Lux

Introduction

The transformation of Central and Eastern Europe (CEE) has been the subject of considerable interest in social sciences. With systemic change, EU integration and accession, followed by the years and aftermath of the global financial and economic crisis, the CEE group of countries has undergone deep socio-economic restructuring, leading to new patterns of regional differentiation and development.

Influenced by a combination of inherited and newly emerging factors, territorial disparities have been on the rise. Examples of catching-up with western EU member states in the capital cities and a handful of successful regions are contrasted by the re-emergence of deep socio-economic problems in traditionally underdeveloped peripheries and newly hollowed-out regions still struggling with the legacies of industrial decline and the loss of economic functions. These differences, reflected in several spheres (e.g. competitiveness, social cohesion, governance and sustainability), shape the new national and subnational dividing lines of post-crisis Europe.

This volume, collecting the results of a comparative research project on the driving forces of spatial restructuring and regional development paths, aims to deliver a comprehensive view on the complex system of regional development within CEE. In its chapters, focused on the different aspects of restructuring, the authors identify the common features of spatial restructuring, as well as the underlying patterns of socio-economic differentiation, showing the CEE group to be just as heterogeneous as the EU15.

The global financial and economic crisis serves as a common lynchpin for many contributions, as its far-ranging effects can be said to represent the start of something new – a 'post-transition' period where the inherited problems of post-socialism slowly give way to new dilemmas. The dilemmas of post-transition evidently continue to be influenced by historical and institutional legacies, but the specificity of 'post-socialism' will be weaker, one among a set of influences dominated by deepening European integration, and against the backdrop of a new era of global uncertainties.

Researching the regional development of Central and Eastern Europe

Over the transformation of Central and Eastern Europe, studies on the macro-region have mostly been dominated by macro-level analyses and thematic studies. Research on regional development has been a relatively smaller field of enquiry, going through multiple phases of interest. Pickles (2008), as summarised by Czirfusz (2011), put forward four waves of transition studies: a first wave centred around the broad policy issues of economic reforms, a second wave concerning increasing socio-economic inequalities, a third wave on spatially and socially uneven regional development, and a fourth wave 'when logics and theories of transition studies became objects of scrutiny' (21). In spite of constant academic interest, research has also had its 'blind spots' where much fewer works have been completed. While the EU accession period saw the proliferation of research programmes on structural and cohesion policy, most thematic volumes have either focused on the regional transformation of individual countries, or on specific topics (e.g. declining industrial regions, the spatial distribution of Foreign Direct Investment (FDI), the transformation of rural spaces, or new directions in urban development). In contrast, relatively few works have presented a comprehensive view of the macro-region in a monographic format, and most of these were written more than 15 years ago (Gorzelak, 1996; Heenan and Lamontage, 1999; Bachtler, Downes and Gorzelak, 2000; Turnock, 2001; and Hughes, Sasse and Gordon, 2004). Recent works with a broad outlook include Herrschel (2007) and co-authors with a systemic review of post-socialism and transition from a global perspective, Gorzelak, Bachtler and Smetkowski (2010) on development dynamics and policy responses, Gorzelak and Goh (2010) on the early consequences of the financial crisis, and Lang et al. (2015) on polarisation and peripherisation. However, the new development directions of CEE regions and the questions of the post-crisis period have not yet been adequately explored. More research is needed on the subject, and this book can only hope to contribute to the discussion.

The identities and interests of researchers who deal with CEE subjects are themselves worth noting. In a detailed study of publication patterns across 15 leading international journals in regional studies and human geography, extending to the period between 1995 and 2011, the author of this chapter found 485 articles dealing with various development issues of the CEE macro-region (Lux, 2012). Interest in CEE was more or less stable across this period, with most attention dedicated to the subject of urban development (19 per cent of all papers), followed by public administration (11 per cent), then regional policy, manufacturing, regional differences and rural issues (10 per cent each). The specific problems of post-socialism (8 per cent), border issues (6 per cent) and services (5 per cent) received comparatively less attention. There were two features of particular interest in publication patterns: first, *the clear dominance of individual case studies fitted into western theories, with a much weaker representation of comparative, synthetic or theoretical papers* with a CEE connection. Indeed, where the latter three were found, they were usually written by western primary authors, although sometimes with local co-authorship. Second, *the examined articles demonstrated proof of what is commonly referred to as the 'Anglo-Saxon hegemony in human geography'* (for the broader debate, see Gutiérrez and López-Nieva, 2001; Rodríguez-Pose, 2006; Aalbers and Rossi, 2007; and Paasi, 2013; and for the specific question of CEE, Stenning, 2005, as well as Timár 2004a, 2004b). There was promising evidence of growing internationalisation: the share of papers with CEE primary or co-authors had risen from 52 per cent between 1995 and 2004, to 61 per cent between 2005 and 2011. Nonetheless, for good or ill, the production of regional studies was clearly tied to the dominant research hubs of the United Kingdom, the Netherlands and North America, and reflected the curiosity and research agendas of western academia.

Rationale for the book: emerging doubts about the CEE regional development model

Regional development in Central and Eastern Europe has long been framed by notions of *transition* from one model to another: central planning to market economy, authoritarian rule to democracy and top-down to bottom-up social organisation. While most components of transition have been articulated on a general, society-wide level, some are anchored in space: eastern vs. western orientation, hierarchies vs. networks, and centralisation vs. decentralisation. The spatial embeddedness of transition was not generally made explicit at the time of systematic change, and did not emerge as a cohesive agenda of *spatial justice* (c.f. Soja, 2009), but some of its elements were implicitly present in specific goals, particularly concerning the autonomy of local communities, and non-discrimination in development funding.[1]

The notion of transition, often conceptualised in the form of dichotomies, suggests instability along with fast-paced, substantial and lasting change towards a new stable system. However, the early hopes of transition failed to materialise: rapid change in some regions was counterbalanced by lagging development or decline in others; improvement in varied spheres by new crises and vulnerabilities. There is, perhaps most prominently in Hungary, also a certain sense of missed opportunities: the CEE macro-region has remained a periphery within Europe and on the global stage, and did not produce the standout growth rates, globally competitive firms, brands and narratives of the iconic post-war examples of successful modernisation. Politically, CEE has little influence beyond its borders and own affairs, and does not play on the European, let alone global, scale. It has little in the way of cultural exports, and its outside representation is deeply problematic whether we consider its public image or media coverage. There are success stories and hopeful signs, but hardly a reason for celebration. The roots of this disillusionment run deep. Crucial empirical evidence of increasing regional differences and limited modernisation under post-socialism was already presented in a comprehensive manner by Sokol (2001) and Dunford (2005), and troubling extrapolations until 2030 and 2050 have been provided more recently by the long-term development scenarios of the ESPON ET2050 programme (2015). Twenty-five years represent a rather long period in the development of countries and macro-regions, and yet expected benefits have failed to materialise, while many socio-economic problems thought to be short-lived seem to have become permanent features of the post-socialist condition. The explanations and responses to this dilemma have given rise to two major interpretations of post-socialist development.

One interpretation places emphasis on the slow-paced nature of regional change. Contradicting scenarios that calculate with fast-paced transformation, this approach posits that the malaises of state socialism are deeply rooted, and change must take place over decades, perhaps even generations. These explanations can draw especially relevant lessons from the results of the institutional turn in economic geography (Martin, 2000; Amin, 2001), as well as the emerging field of evolutionary economic geography (Boschma and Frenken, 2006; MacKinnon et al., 2009), which offer an array of useful concepts for discussion. Path-dependent development, lock-ins, institutional rigidities or problems associated with the accumulation of financial, social and knowledge capital may be seen to hold the key to explain 'historically embedded' growth processes. Similarly, institutions act as carriers of history – except in CEE, they are often considered the 'wrong' kind of institutions, either oriented towards reproducing undesirable results (similar to the vicious cycles characterising 'Old Industrial Regions'), or insufficiently prepared to accommodate and realise modern policies. Indeed, the chapters in this volume offer a wealth of evidence that hint at the *evolutionary* nature of CEE regional development paths, and to the outstanding relevance of institutions in shaping them. As Lengyel and Bajmócy (2013, p. 6) propose, '[u]nderstanding the changes in institutions or the behaviour of individual actors, recognising

3

enduring behavioural patterns or bounded rationality arising from centralised decision-making are all crucial components in understanding post-socialist transition'.

Another group of responses seems to lead to the conclusion that *in a sense, 'transition' may no longer be accurately described as an intermediate development phase, but a new, stable system with its own operating logic.* These explanations can trace their origins to early criticism of neo-liberal development policies in CEE, particularly two in-depth papers by Gowan (1995, 1996), but they have taken full form in the 'varieties of capitalism' debate (Bohle and Greskovits, 2004, 2006; Peck and Theodore, 2007; Rugraff, 2008; Nölke and Vliegenthart, 2009) which proposes the existence of a specific 'dependent market economy' (DME) model of capitalism to describe Central and Eastern European countries. The DME is 'neither fish nor fowl', differing from both the 'liberal market economies' (LMEs) exemplified by the Anglo-Saxon countries, and the more regulated 'coordinated market economies' (CMEs) mainly found in continental Europe.[2,3] In particular, Nölke and Vliegenthart (2009) make a persuasive case in charting the modern socio-economic dependencies of the macro-region, and calling attention to its inherently low upgrading potential. Most notably, this train of thought has seen further elaboration by Farkas (2011), who provides empirical proof of this distinct developmental model; by Drahokoupil and Myant (2015), who outline how upgrading processes may take place within a system of external dependencies; and by Medve-Bálint (2014), who places emphasis on the EU's role in providing substantial policy support to establishing and maintaining this development model through its FDI policies.

While they are both useful in understanding the nature of regional development in CEE, and the truth may indeed lie somewhere in the middle, neither of these interpretations are without serious problems or contradictions. The first explanation can be contrasted by the experiences of geographic peripheries which have managed to chart an impressive development trajectory in recent decades, contradicting the notion that development is an a priori sluggish process. These examples include Ireland's rapid modernisation from a rural, low-wage economy through a predominately FDI-based development path (Horváth, 1998), Scandinavia's rise through the welfare state's heavy investments into human capital and the knowledge economy (Pogátsa, 2016), and the developmental states of East Asia (Gereffi, 1995; Rugraff, 2008; Cimoli, Dosi and Stiglitz, 2015). While there is precious little to unite these cases (except perhaps a general attention to developing human capital), and their own problems must not be neglected, they seem to have demonstrated a modernisation performance surpassing that of post-socialist Central and Eastern Europe.

The second explanation – while it is internally self-consistent, and correctly identifies that many of CEE's problems are not merely symptomatic, but systemic – has an uncomfortable tinge of geographic (and in some cases, cultural) determinism, and robs the varied actors of the macro-region of both their *agency* and their *responsibility*. Defeatist narratives neglect to consider the possibilities of individual and collective action, or even to appreciate the instances where they have achieved lasting, meaningful change (Domański, 2004; Pogátsa, 2014). The second explanation also tends to discount, or at least undervalue, the significance of rising inequalities and lagging growth within global/European core regions, and how they influence the growth prospects of CEE countries and regions. If CEE development is characterised by externally dependent relationships, then these dependencies also serve to transmit the ongoing crisis of post-industrial society, projecting its consequences on the admittedly more vulnerable socio-economic fabric of the CEE macro-region (Chapter 4 in this book provides a clear-cut example). All in all, *'fixing' the problems of CEE regional development cannot be discussed independently of 'fixing' the problems of Europe itself*; and in a sense, this is an encouraging reaffirmation of the positive results of European integration.

Aims and scope of the book

In the original research project that served as the foundation of this work,[4] then over the course of planning and realising the book itself, *we have strived to provide a new synthesis of CEE regional development issues from a comparative, theoretically informed, empirically sound, historically embedded, but forward-looking perspective.* Our aim, explored through the book's 19 chapters, was to highlight the common patterns of regional development across the CEE group of countries, but also bring attention to the internal fault lines and differences which divide the macro-region and result in *increasingly divergent paths of regional development.*

These divides, ranging from differences between country groups (often the Visegrad countries – Czechia, Slovakia, Poland and Hungary – South-Eastern Europe, and the special case of Slovenia, although these demarcations are neither exact nor constantly applicable), to centre–periphery relationships (mainly between national capitals and a few more metropolitan areas, and non-metropolitan space) and finer patterns of functional differentiation, have produced a territory which is increasingly heterogeneous, and which is increasingly ill-served by 'one size fits all' development recipes. Indeed, one of our main conclusions is the reaffirmation of the importance of distinction, and, with it, a certain vision of spatial justice and a need for increasing subsidiarity on both the national and European level. There is a moral responsibility to recognise the different interests and values of different territories, and their right to be a part of the future on their own terms, and I believe the contributions in this book reflect this notion.

The effort to strike a balance between synthesis and comparison (detail) has invariably coloured the contents of this book. Although our research project involved significant regional-level groundwork and data collection, whose results were published separately (Horváth, 2015b), and the partial results have resulted in individual research monographs (including Horváth, 2015a) as well as numerous research papers, this book focuses on our main findings. This means *a deliberate emphasis on qualitative analysis.* Data are mainly used either to illustrate salient points, or to offer a macro-regional comparison; with in-depth quantitative methods only being used in chapters 6 and 15. The chapters are 'theoretically informed' by discussing their results within the context of the appropriate contemporary regional development theories, but it is empirical discussion that dominates. Finally, while the book explores the relationship between policy and socio-economic processes, its focus is mainly on the latter. What we *do* investigate with considerable interest is why certain policies have succeeded or failed across CEE countries. The record is not very convincing, and chapter after chapter finds that *policy transfer which disregards its territorial context, and cannot adapt to local needs and capabilities is setting itself up for disappointment.* This is true for both national and EU-financed initiatives. While warnings about 'new regionalist' policies are hardly new (see, for instance, Lovering, 1999's particularly insightful critique), further risks of policy implementation lie in the macro-region's specific, historically inherited institutional weaknesses and rigidities, another theme explored through the book (with a particular focus in chapters 9 and 10). Building efficient, democratic public planning is an unsolved puzzle for CEE societies.

The geographic scope of the original research did not encompass the entirety of CEE space, and this limitation is reflected in the following chapters. The main focus was on the Visegrad group of countries (Czechia, Slovakia, Poland and Hungary), as well as Romania, Bulgaria, and the more developed successor states of Yugoslavia (Slovenia, Croatia, Serbia, Montenegro and Macedonia), with this order of emphasis. Missing are the Baltic states of Estonia, Latvia and Lithuania, whose unique trajectory of development – from a particularly deep post-Soviet crisis to an interesting mixture of Scandinavian knowledge economies and Anglo-Saxon openness – was outside the expertise of our research team. Likewise, while the territory of the former German Democratic Republic offers a host of interesting parallels to our research (c.f. Heimpold and Titze, 2014; Bartoli, Rotondi and Tommasi, 2014; or Horváth, 2012, on the problem of 'the German Mezzogiorno'), its integration into Germany also comes with a host of differences which would have taken a significant effort to resolve within our research effort. Finally, our analysis did not extend to South-Eastern Europe's internal periphery (Albania, Bosnia and Herzegovina and Kosovo), or the post-Soviet countries, whose development paths would merit a separate volume of their own. These omissions are acknowledged as our work's limitations, although we believe that when it comes to offering a synthesis, many of our conclusions remain relevant even in the case of countries we could not investigate in detail.

Structure of contributions

With the preceding comments in mind, the book is arranged into three main thematic units, followed by a conclusion. Each of the chapters examines a specific aspect of regional development in Central and Eastern Europe, discussing the most relevant socio-economic transformation processes which underpin them, and drawing attention to current problems as well as emerging challenges.

Six chapters, forming the first part of the book, introduce the reader to the varied economic consequences of transition. Chapter 2, serving to introduce the following chapters, provides an overview of structural shifts, drawing attention to how they are embedded within processes of globalisation and rescaling in post-Fordist economies. Different economic sectors not only create different spatial patterns, they also offer different development opportunities in different regions. Indeed, the territorially uneven processes of tertiarisation produce markedly superior results in metropolitan areas than outside them, and CEE's urbanisation deficit limits the potential of service-based development in most provincial regions. Instead of a 'one size fits all' view on regional development, the chapter suggests that a regionally differentiated sectoral mix offers the best prospects for modernisation.

This notion is further examined in three contributions dealing with the transformation of specific economic sectors. Chapter 3, discussing the path-dependent processes of industrial development across CEE industrial regions, points out that while the global reintegration of CEE manufacturing has taken place under the dominant influence of FDI, the emerging structures are not spatially blind: they draw heavily upon strongly localised productive legacies, networks and production factors. The chapter also cautions that the post-socialist development model is nearing its limits, and may have insufficient upgrading potential without the revitalisation of endogenous, place-specific development potential.

Co-written by Zoltán Gál and Sándor Zsolt Kovács, Chapter 4 sheds light on development processes within the service sector, particularly business and finance services. The chapter confirms that the most valuable segments of the service-based economy show strong territorial concentration, locating in privileged metropolitan areas. However, Gál and Kovács also draw attention to emerging secondary cities which develop specialisation in certain fields, particularly services offshoring. Similar to manufacturing, the influence of FDI dominates development, and this external dependency has come with systemic vulnerabilities exposing CEE to exogenous shocks within the global economic system.

Chapter 5 by Péter Póla is concerned with a particularly difficult area of transition: the transformation of rural areas and agriculture. Not only have CEE's rural areas struggled with historical underdevelopment and peripherality, they have often been adversely impacted by both the socialist system and by post-socialist development. This chapter explores how this picture hides multiple layers of complexity, as development paths strongly diverge on the national level, and are further segmented by the conflicting development interests of local communities and external actors. The chapter draws attention to agricultural and non-agricultural land use as the two pillars of rural development, and places particular emphasis on the potential of the LEADER philosophy for involving and empowering local communities.

In Chapter 6, Balázs Páger investigates how entrepreneurial activity in CEE is increasingly influenced by a range of qualitative factors, reaffirming the significance of localisation and endogenous development. Following the end of the entrepreneurial boom of the 1990s, the most successful enterprises have been those which could successfully draw on their surrounding regional socio-economic environments. Competitiveness and entrepreneurship are not independent of their surroundings, and it is particularly higher education and the surrounding network of existing firms that make the difference in the post-crisis era.

Concluding the first part, Chapter 7 by Márton Czirfusz highlights the role of culture and the creative economy in the reproduction of uneven regional development. His contribution provides a critical look at how a trending topic in regional development shapes the economy and society of CEE cities as well as rural areas, and how cultural mega-projects and key events pose mounting challenges for effective governance across the macro-region. Well-informed by the theoretical and political debates surrounding his topic, Czirfusz draws special attention to the new social movements which oppose creativity policies and for-profit cultural redevelopment.

The second part of the book is concerned with social issues underlying regional development, as well as policy development at the heart of the macro-region's transformation. Chapter 8, the first of this thematic unit, is also of key importance in understanding the socio-economics of Central and Eastern Europe. Zoltán Hajdú, Réka Horeczki and Szilárd Rácz call attention to the historical character and rigidity of settlement systems, yet also note how political change after 1990 has had far-ranging consequences on not just public administration, but also the growth or decline of towns and cities. New growth poles have emerged in newly independent states, and national capitals have become the unambiguous winners of transition. Meanwhile, the position of secondary cities (regional centres) can be much more ambiguous, and in their case, historical settlement patterns continue to dominate. Only a few countries have a genuinely polycentric character, while others struggle with the weight of one or two dominant metropolitan areas. This feature of regional development has far-reaching implications for a host of social and economic issues, as already discussed. If we live in 'the urban century' (Nijkamp and Kourtit, 2013; Kourtit, Nijkamp and Scholten, 2015) or 'a metropolitan world' (Lux, 2015), there is a pressing need to rethink the chances of minor cities and areas which lie outside the hinterlands of the main metropolitan nodes.

The remaining four chapters in this part form two interrelated pairs. Chapters 9 and 10 investigate two areas of public policy: governance and managing regional disparities. In Chapter 9, Ilona Pálné Kovács takes an in-depth look into the transformation of administrative systems across CEE. This is one area where national models influenced by long-lasting historical legacies prevail, yet there are still some common lessons to be drawn. In spite of ongoing divergence, CEE administrative systems are linked not just by unsatisfactory performance, institutional weaknesses and rigidities, but also the growing problem of stalling or abandoned regionalisation (a process rooted in the failure of 'new regionalism') and a shift towards the efficiency-driven re-nationalisation and central control of governance. These changes, as Pálné Kovács highlights, have been exacerbated by the economic crisis. The expansion of state power, however, stands in contrast with the principles of subsidiarity and bottom-up social organisation. We have entered a new period where both local governments and medium-level governance structures face harsh financial and political challenges and uncertain prospects.

Yet centralisation is not restricted to administrative systems, nor is it a uniquely CEE phenomenon. Placing the macro-region's regional policy development in the context of ongoing

'EUropean' trends, Chapter 10 by László Faragó and Cecília Mezei scrutinises the centralist impulses which also seem to have infected Brussels decision-makers. They highlight how existing centre–periphery relationships are giving rise to a generation of regional policies which increasingly serve the EU's general political interests instead of the European (including Central and Eastern European) regions. As they argue, the discursive deficit of the 'new' EU member states has contributed to declining support for territorial cohesion, and an acceptance of significant development gaps. This chapter, an important pillar of our arguments for subsidiarity, also takes a bottom-up look at the efficiency of CEE regional policies. The authors propose that the effective mitigation of regional inequalities require a new commitment to territorial cohesion, and – echoing conclusions from the first part of the book – the exploitation of regional capital and endogenous development potential. Instead of focusing on regions where market-led change is already generating solid growth, regional policy should be focused on realising the potential of less developed regions, and it should be built on multilayered governance based on the principle of subsidiarity.

The second pair of chapters in this part deals with two heavily intertwined topics. James Wesley Scott's Chapter 11 offers a look at bordering processes as well as the rise and (to an extent) decline of cross-border cooperation (CBC) initiatives within the macro-region. He examines how the euphoria of open borders has given way to more sober and perhaps also more realistic routines in border areas, and how CBC is restricted by local realities. It seems that competing territorial logics at different levels, conflicting attitudes, and the limited means of (often underdeveloped) border areas all contribute to the steady, but slower than expected de-bordering of Central and Eastern Europe.

Not independent of the previous contribution, Chapter 12 by Nóra Baranyai scrutinises how (ethno-)regional movements fit into the puzzle of the lagging, sometimes barely visible, process of CEE regionalisation. The chapter traces the history, objectives and achievements of CEE initiatives to push for administrative reform, and/or achieve some form of autonomy for specific historical or ethnic–multicultural regions. In a sense, this chapter provides an analysis of a failure, since in contrast with their Western European counterparts, local and regional autonomies have failed to materialise within the centralised CEE countries, and initiatives to this end have met considerable resistance from central governments and nation-building efforts.

Over six chapters (and the conclusion), the third part of the book looks into questions related to the broadly understood concern of sustainability. These chapters form the most heterogeneous section of the book, but they are connected by a common interest: looking into issues which will have a major impact on the regional development of the coming decades.

Perhaps it seems strange at first to place the issue of migration and labour markets here, as they are usually discussed among the fundamental social structures within the scope of demography, or viewed through a resource-based perspective and evaluated on the basis of their potential contribution to economic growth. But there is a good reason that Chapter 13, co-written by Jan Sucháček and Mariola Pytliková, is found here: human potential, along with the ability to renew, attract and retain it, is perhaps the most important lynchpin of socio-economic development in knowledge-based societies. 'Who will build the future?' is the question of our age, and as reflected across several of the previous chapters, the quality and quantity of skilled, knowledgeable people forms a large part of the answer. The question of human capital, even as its role was attracting worldwide attention in academic discourse and public policy, was neglected during the two decades of post-socialist development, and has only come to the fore with the recent shortages of skilled labour and the persistent human capital losses brought about by westwards out-migration. Sucháček and Pytliková capture migration and labour market trends at a most important turning point, where CEE countries are at a crossroads between their former, no longer sustainable status as medium-low-wage, net emigration countries, and a position where improving wages provide incentives for strong human capital accumulation. The chapter also discusses how CEE is facing increased immigration, and how these processes have unfolded during the recent migration crisis. Last but not least, it looks into how the spatial patterns of CEE labour markets reflect and reinforce urban–rural polarisation, east–west gradients and the exclusive position of capital cities.

Chapter 14 by Zoltán Gál and Balázs Páger further elaborates on the challenges facing the emergence of knowledge-based societies in their study on higher education and innovation performance. Unsurprisingly, this is yet another aspect of development showing strong polarisation, particularly when it comes to the location preferences of high-tech employment. Yet, neither universities nor innovation are the monopoly of metropolitan centres: in their own way, they also make meaningful contribution to the development of the peripheries, where mid-range universities serve to supply local firms and society with vital knowledge. Accordingly, it is the challenge of the coming decades to narrow the gap, and create opportunities for knowledge-based development outside the main urban centres via investment into innovation and the networks that produce and disseminate it.

Another aspect of sustainability lies in the resilience of CEE regions, explored in Chapter 15 by Adam Drobniak, Adam Polko and Jan Sucháček. Their findings show a general improvement in economic, technological and environmental resilience across the macro-region, albeit with a significant development gap between CEE and the EU15. This, again, has much to do with institutional and path-dependent factors, which also show variance across different countries and regions; as well as the mismatch of local needs and existing administrative and political structures.

Ferenc Erdősi in Chapter 16 considers the situation of CEE through the major trends in transport space. He calls attention to how national and EU-level political support for motorway-oriented transport policies and a select number of main corridors have reshaped the transport space of a macro-region struggling with a weak post-socialist heritage. Indeed, just as the most polluting transport modes, motorway transport and air traffic, have gained enormously over recent decades – in a way that has scarcely benefited the peripheries – more sustainable railway and inland water transportation have been the losers of transition. Erdősi also examines how access to sea ports influences the transport policies of landlocked or largely continental countries, and how competition among Europe's less prominent eastern sea ports fits into the puzzle. Finally, he warns of the rising importance of interactions with (East) Asia, Africa and the Middle East, connections which have largely remained unexplored by CEE countries.

Chapter 17 by Andrea Suvák reflects on a more 'traditional' aspect of sustainability: the role of spatial planning in environmental policymaking. Suvák accentuates the influence of EU policy transfer in setting the agenda for post-socialist environmental policies and the surrounding institutional system, but also acknowledges the limits of policy uptake. Indeed, CEE environmental policies – here presented through a comparative analysis of policy texts in the Visegrad countries – reveal a different mix of motivations and concepts than their EU15 counterparts, and particularly the dominance of 'resource'-oriented narratives. Amidst rushed policymaking, the resulting policies and institutional systems are often incoherent and riddled with inherent conflicts.

Chapter 18 comes from Gyula Horváth, who had been the principal investigator of the research project serving as this book's foundation, helped develop the book proposal, and who passed away shortly after I could tell him the good news about our proposal's acceptance. In a way, this paper, an abridged and slightly edited version of a longer piece published in our institute's Discussion Papers series, has now become a coda to his life work, which had consistently revolved around the idea of decentralisation and the empowerment of regions and local communities. In the chapter, he charts the evolution of this idea, from its roots to the emergence, proliferation and institutional development of regional studies in Central and Eastern Europe. This question was always close to his heart, and as his student, colleague and friend, I hope to honour his legacy by the paper's inclusion.

Concluding the book but continuing the debate on CEE's future, Chapter 19 proposes *four emerging dilemmas which will impact the sustainability of development paths in the post-transition period*. Of these four, the first two highlight the contradictions of the macro-region's current development paradigm. First, as many chapters in this book have attested, scale issues will remain prominent, or even grow in their importance in the following decades. The worldwide rise of metropolitan areas poses hard challenges for regions with a less dense urban network, and it remains a question how the minor cities and small towns of these regions will integrate their hinterlands. Second, further attention should be dedicated to map the system of external dependencies characterising CEE regional development, and policies elaborated to reduce their negative effects while retaining their benefits. A renewed focus on endogenous development and capital accumulation – particularly of human capital – should serve to transform hierarchical centre–periphery relationships into mutually beneficial network linkages. The second two dilemmas concern the future of European integration. The third is that of European regional policy, where spatially aware development approaches with a stronger emphasis on empowering local communities and regions should return to focus. A new regional policy based on the principles of subsidiarity and territorial cohesion lies in the long-term interests of both the EU15 and CEE member states, contributing to an internally strong European Union. Fourth, the principle of subsidiarity should return to the heart of the CEE–EU relationship. The currently emerging divide between Brussels and national capitals is the product of top-down philosophies, and results in 'competing centralisms' which subvert the European integration process and do not serve the best interests of citizens and communities. Rather, the way forward lies in a renewed commitment to decentralisation on multiple territorial scales – perhaps the most important message presented in our book.

Acknowledgements

The publication of this research has been supported by project #104985 of the Hungarian National Research, Development and Innovation Office. While writing this paper, Gábor Lux was supported by the János Bolyai Research Scholarship of the Hungarian Academy of Sciences.

Notes

1 Experts argued for a clean break with the consciously discriminatory planning practices of state socialism (industry over services and agriculture; cities over rural areas; self-sufficiency over trade), but instead of planning's democratisation, they achieved its almost complete dismantling, and the rise of new market-driven inequalities.
2 Citing Albert's (1993) earlier work, Dunford (2005) refers to these alternatives as the 'neo-American' and 'Rhine' models of capitalism, and points out that the uncritical adoption of the former during early transition has produced lacklustre results across CEE.
3 Much more limited attention in academia has been dedicated to drawing lessons from East Asian developmental states, while their example has been studied with considerable eagerness as of late by Hungarian policymakers.
4 Grant #104985 of the Hungarian National Research, Development and Innovation Office, whose financial assistance is gratefully acknowledged. Further chapters were commissioned during the development of this volume.

References

Aalbers, M.B. and Rossi, U., 2007. A Coming Community: Young Geographers Coping With Multi-Tier Spaces of Academic Publishing across Europe. *Social & Cultural Geography*, 8(2), pp. 283–302.
Albert, M., 1993. *Capitalism Against Capitalism*. London: Whurr Publishers.
Amin, A., 2001. Moving On: Institutionalism in Economic Geography. *Environment and Planning A*, 33(7), pp. 1337–1341.
Bachtler, J., Downes, R. and Gorzelak, G., eds., 2000. *Transition, Cohesion and Regional Policy in Central and Eastern Europe*. Aldershot: Ashgate.
Bartoli, F., Rotondi, Z. and Tommasi, D., 2014. Mezzogiorno and Neue Bundesländer: What Lessons Can Germany Learn from Italy? S. Collignon and P. Esposito, eds. *Competitiveness in the European Economy*. Abingdon: Routledge, pp. 197–211.
Bohle, D. and Greskovits, B., 2004. Capital, Labor, and the Prospects of the European Social Model in the East. *CES Central & Eastern Working Paper*, 58.
Bohle, D. and Greskovits, B., 2006. Capitalism without Compromise: Strong Business and Weak Labor in Eastern Europe's Weak Transnational Industries. *Studies in Comparative International Development*, 41(1), pp. 3–25.
Boschma, R. and Frenken, K., 2006. Why Is Economic Geography Not an Evolutionary Science? Towards an Evolutionary Economic Geography. *Journal of Economic Geography*, 6(3), pp. 273–302.
Cimoli, M., Dosi, G. and Stiglitz, J.E., 2015. The Rationale for Industrial and Innovation Policy. *Intereconomics*, 50(3), pp. 126–132.
Czirfusz, M., 2011. Understanding Geographies of Post-Socialist Economies: Invoking the Arts. K. Kocsis, Á. Erőss and D. Karácsonyi, eds. *Regional Socio-Economic Processes in Central and Eastern Europe – 20 Years in Transition and 2 Years in Economic Crisis*. Budapest: HAS Geographical Research Institute, pp. 19–31.
Domański, B., 2004. West and East in 'New Europe': The Pitfalls of Paternalism and a Claimant Attitude. *European Urban and Regional Studies*, 11(4), pp. 377–381.
Drahokoupil, J. and Myant, M., 2015. Putting Comparative Capitalism Research in its Place: Varieties of Capitalism in Transition Economies. M. Ebenau, I. Bruff and C. May, eds. *New Directions in Critical Comparative Capitalisms Research*. London: Palgrave Macmillan, pp. 155–171.
Dunford, M., 2005. Old Europe, New Europe and the USA: Comparative Economic Performance, Inequality and Market-led Models of Development. *European Urban and Regional Studies*, 12(2), pp. 149–176.
ESPON ET2050, 2015. *Territorial Scenarios and Visions for Europe*. Luxembourg: ESPON and MCRIT.
Farkas, B., 2011. The Central and Eastern European Model of Capitalism. *Post-Communist Economies*, 1, pp. 15–34.
Gereffi, G., 1995. State Policies and Industrial Upgrading in East Asia. *Revue d'Économie Industrielle*, 71(1), pp. 79–90.
Gorzelak, G., 1996. *The Regional Dimension of Transformation in Central Europe*. London: Jessica Kingsley – Regional Studies Association.
Gorzelak, G., Bachtler, J. and Smetkowski, M., eds., 2010. *Regional Development in Central and Eastern Europe: Development Processes and Policy Changes*. Abingdon: Routledge.
Gorzelak, G. and Goh, C.C., eds., 2010. *Financial Crisis in Central and Eastern Europe: From Similarity to Diversity*. Warszawa: Wydawnictwo Naukowe SCHOLAR.
Gowan, P., 1995. Neo-Liberal Theory and Practice for Eastern Europe. *New Left Review*, 213, pp. 3–60.
Gowan, P., 1996. Eastern Europe, Western Power and Neo-Liberalism. *New Left Review*, 216, pp. 129–140.
Gutiérrez, J. and López-Nieva, P., 2001. Are International Journals of Human Geography Really International? *Progress in Human Geography*, 25(1), pp. 53–69.
Heenan, P. and Lamontage, M., eds., 1999. *The Central and Eastern European Handbook*. Chicago, IL: Fitzroy Dearborn Publishers.
Heimpold, G. and Titze, M., 2014. Economic Development in East Germany since German Unification: Results, Shortcomings and Implications for Economic Policy. S. Collignon and P. Esposito, eds. *Competitiveness in the European Economy*. Abingdon: Routledge, pp. 184–196.
Herrschel, T., 2007. *Global Geographies of Post-Socialist Transition: Geographies, Societies, Policies*. Abingdon: Routledge.
Horváth, Gy., 1998. *Európai regionális politika* [European Regional Policy]. Budapest–Pécs: Dialóg Campus Kiadó.

Horváth, Gy., 2012. *The German Mezzogiorno? Supplements to the Natural History of East German Regional Development*. Discussion Papers No. 86. Pécs: Hungarian Academy of Sciences Research Centre for Economic and Regional Sciences Institute for Regional Studies.

Horváth, Gy., 2015a. *Spaces and Places in Central and Eastern Europe. Historical Trends and Perspectives of Regional Development*. Abingdon: Routledge.

Horváth, Gy., ed., 2015b. *Kelet- és Közép-Európa régióinak portréi* [The portraits of Eastern and Central European regions]. Budapest: Kossuth Kiadó.

Hughes, J., Sasse, G. and Gordon, C., 2004. *Europeanization and Regionalization in the EU's Enlargement to Central and Eastern Europe*. London: Palgrave.

Kourtit, K., Nijkamp, P. and Scholten, H., 2015. The Future of the New Urban World. *International Planning Studies*, 20(1–2), pp. 4–20.

Lang, T., Henn, S., Sgibnev, W. and Ehrlich, K., eds., 2015. *Understanding Geographies of Polarization and Peripherization: Perspectives from Central and Eastern Europe and Beyond*. London: Palgrave.

Lengyel, B. and Bajmócy, Z., 2013. Regionális és helyi gazdaságfejlesztés az evolúciós gazdaságföldrajz szemszögéből [Regional and local economic development from the perspective of evolutionary economic geography]. *Tér és Társadalom*, 27(1), pp. 5–29.

Lovering, J., 1999: Theory Led by Policy: The Inadequacies of the 'New Regionalism' (Illustrated from the Case of Wales). *International Journal of Urban and Regional Research*, 23(2), pp. 379–395.

Lux, G., 2012. Saját tereikbe zárva? Közép- és Kelet-Europa a regionális Tudomány európai integrációjában [Locked into their own spaces? Central and Eastern Europe within the European integration of regional studies]. J. Rechnitzer and Sz. Rácz., eds. *Diálogus a regionális tudományról*. Győr: Széchenyi István Egyetem – Magyar Regionális Tudományi Társaság, pp. 151–170.

Lux, G., 2015. Minor Cities in a Metropolitan World: Challenges for Development and Governance in Three Hungarian Urban Agglomerations. *International Planning Studies*, 20(1–2), pp. 21–38.

MacKinnon, D., Cumbers, A., Pike, A., Birch, K. and McMaster, R., 2009. Evolution in Economic Geography: Institutions, Political Economy, and Adaptation. *Economic Geography*, 85(2), pp. 129–150.

Martin, R., 2000. Institutional Approaches in Economic Geography. E. Sheppard and T.J. Barnes, eds. *A Companion to Economic Geography*. Oxford: Blackwell, pp. 77–94.

Medve-Bálint, G., 2014. The Role of the EU in Shaping FDI Flows to East Central Europe. *Journal of Common Market Studies*, 52(1), pp. 35–51.

Nijkamp, P. and Kourtit, K., 2013. The 'New Urban Europe': Global Challenges and Local Responses in the Urban Century. *European Planning Studies*, 21(3), pp. 291–315.

Nölke, A. and Vliegenthart, A., 2009. Enlarging the Varieties of Capitalism: The Emergence of Dependent Market Economies in East Central Europe. *World Politics*, 61(4), pp. 670–702.

Paasi, A., 2013. Fennia: Positioning a 'Peripheral' but International Journal under Conditions of Academic Capitalism. *Fennia*, 191(1), pp. 1–13.

Peck, J. and Theodore, N., 2007. Variegated Capitalism. *Progress in Human Geography*, 31(6), pp. 731–772.

Pickles, J., 2008. The Spirit of Post-Socialism: 'What Is to Be Understood by It?' J. Pickles, ed. *State and Society in Post-Socialist Economies*. London: Palgrave, pp. 1–16.

Pogátsa, Z., 2014. Cultural Defeatism in Central Europe. *Visegrad Revue*, 27 January.

Pogátsa, Z., 2016. *Magyarország politikai gazdaságtana: Az északi modell esélyei* [The political economy of Hungary: The chances of the Northern model]. Budapest: Osiris.

Rodríguez-Pose, A., 2006. Commentary: Is There an 'Anglo-American' Domination in Human Geography? And, Is It Bad? *Environment and Planning A*, 38(4), pp. 603–610.

Rugraff, E., 2008. Are the FDI Policies of the Central European Countries Efficient? *Post-Communist Economies*, 20(3), pp. 303–316.

Soja, E.W., 2009. The City and Spatial Justice. *Justice Spatiale – Spatial Justice*, 1(1), pp. 1–5.

Sokol, M., 2001. Central and Eastern Europe a Decade after the Fall of State-Socialism: Regional Dimensions of Transition Processes. *Regional Studies*, 35(7), pp. 645–655.

Stenning, A., 2005. Out There and in Here: Studying Eastern Europe in the West. *Area*, 37(4), pp. 378–383.

Timár, J., 2004a. More than 'Anglo-American', It Is 'Western': Hegemony in Geography from a Hungarian Perspective. *Geoforum*, 35(5), pp. 533–538.

Timár, J., 2004b. What Convergence between What Geographies in Europe? A Hungarian Perspective. *European Urban and Regional Studies*, 11(4), pp. 371–375.

Turnock, D., ed., 2001. *East Central Europe and the Former Soviet Union: Environment and Society*. Abingdon: Routledge.

Part I
Economic transformation processes before and beyond the crisis

2
Reintegrating economic space
The metropolitan–provincial divide
Gábor Lux

Rescaling in the modern space economy

In the last decades, even the most developed European economies had to reconsider their development strategies due to increasing competition and the rescaling of the modern space economy. The pressures of 'unlimited globalisation' have been brought about by advances in transportation and info-communication technologies (ICT); massive worldwide deregulation; the appearance of several new actors in global economic integration; and the constantly increasing permeability of national borders. Controlled mainly by transnational corporations (TNCs), Foreign Direct Investment (FDI) flows have had an increasing role in shaping the development prospects of states and regions. Except for a few key players on the world stage, countries and their regions face adaptation pressure impossible to avoid without being threatened by marginalisation.

A process of rescaling has taken place, leading to increased concentration in global centres (Faragó, 2010). The new winners of worldwide agglomeration processes are the 'ideal' locations of space: *globalised metropolitan city-regions* which serve as frameworks for agglomeration economies (Gordon and McCann, 2000) and fulfil both hub and gateway roles in the distribution of transcontinental flows (Taylor, 1997; Derudder et al., 2003; Erdősi, 2003; Sassen, 2006; Gál, 2010). Their strengths, based on a spatially limited system of location advantages, enable them to collect the most advanced functions of the post-Fordist economy: knowledge-intensive business services (KIBS), the most advanced innovative technologies, command and control functions in both the commercial and the public sectors. The highest value-added economic branches show great concentration in these 'world cities' (Audretsch, 1998). In comparison, medium-sized metropolitan areas linked to the world city network tend to specialise only in a few activities, from finance (Frankfurt, Zurich) to fashion and culture (Milan). Their examples are often presented as idealised case studies or ready-made development recipes, without paying enough attention to their unique situation and capabilities. This problem has often led to the failure of new regionalist policies – a problem already discussed by Lovering (1999), and later by Moulaert et al. (2007).

Benefiting from state-led development policies (Gereffi, 1995), some – primarily East Asian – emerging economies have undergone significant upgrading from peripheral to global actors through attracting TNCs and supporting their own 'national champions'. Globally established companies possess special advantages when it comes to competitive strategies: they can optimise the factor intensity, the knowledge content and the added value of their activities on a worldwide scale. This unique 'bird's eye view' enables them to pay their taxes in tax havens; locate their labour-intensive production on low-cost sites while exploiting high-skilled labour, innovative activities and management close to the global centres; and to sell their products to advanced economies as well as the broadening global middle class. Economies of scale and their bargaining power grant them a position similar to that of the global centres with which they exist in symbiosis – while locality is increasingly on the defensive, even when reinforced by powerful economic networks such as clusters and industrial districts. In the world of global value chains, everyone stands alone against the pressure of the markets.

Non-metropolitan spaces and those outside the great global flows often experience threats of marginalisation and decline. 'Minor cities', second-tier urban centres without sufficient critical mass (Sucháček, 2010), find themselves in a precarious situation amidst losing ground to global champions and having to balance their development agendas between strong specialisation and a flexible economic structure (Lux, 2015).[1] 'For whoever has, to him more will be given; but whoever does not have, even what he has will be taken away from him' – so Mark (4:25) describes the essence of historical accumulation processes, and these words have never been more true than in our age. Even advanced economies in Western Europe and North America feel the ensuing development challenges. Unlimited competition results in a race towards a relatively low global average, and exerts a burden on welfare states (Kilicaslan and Taymaz, 2008; Milberg and Winkler, 2010). Wage stagnation, long-term job displacements and labour market insecurity, coupled with a structural shift towards post-Fordism and the crisis of traditional industrial regions, have together led to the erosion of previously secure medium-skilled jobs in both blue- and white-collar professions. The phenomenon of the 'vanishing' or 'disappearing middle' has been noted as a severe problem by numerous authors (Goos and Manning, 2007; Acemoglu and Autor, 2010; Tüzemen and Willis, 2013), prompting a search for effective development strategies representing a 'high road' of global competitiveness, characterised by high levels of social spending, employee skills, innovation and (consequently) productivity (Milberg and Houston, 1999).

In regional policy, the spatial interpretation of high-road development has encouraged an entire set of policies, a 'new consensus' on regional development relying on the collaboration of territorially embedded public and private networks aiming to foster learning and innovation (Humphrey and Schmitz, 2002). *Endogenous development* stresses the exploitation of locally rooted, hard-to-reproduce location advantages, primarily unique skills and knowledge, in achieving competitiveness in a selected industrial or tertiary niche. The central tenets of this development approach are a combination of the following factors:

- concentrating resources, exploiting agglomeration advantages, enabling less dense regions to realise benefits similar to those in metropolitan city-regions;
- increasing the regional embeddedness of production through an upgrading process;
- empowering local small and medium-sized enterprises (SMEs) and their networks; and
- preserving social cohesion and the welfare state.

This philosophy is expressed in a variety of instruments and in concentrated development units, like regional clusters and industrial districts, growth poles, regional innovation systems, learning regions, etc. These concepts are interrelated inasmuch as they attempt to encourage local resource accumulation and the generation of spillover or multiplier effects which, starting from a concentrated location, try to integrate a broader region into a production network, whether operated by local actors or external investors. Endogenous development has become a 'go-to' development approach of EU regional policy, with mixed success.

Like regional policy in general, the strategies of endogenous development are often applied haphazardly, without regard to local capabilities, historical antecedents or institutional development. In the last decade, even its success stories have been facing new challenges in the form of cost-based competition with post-socialist and particularly Far Eastern emerging economies. SME networks without effective niche strategies are increasingly disrupted by TNCs which have entered and captured the markets traditionally dominated by local enterprises. Furthermore, transnational private governance has introduced TNC-friendly legislation through the EU, representing Anglo-Saxon competitive philosophies in contrast to the continental model (Nölke, 2011). There has also been a cultural change characterised by weakening informal ties, less integrated firm networks and changing populations, particularly visible in Italian industrial districts (Parrilli, 2009). The result is the weakening of the environment which have allowed endogenous development models to succeed in non-metropolitan regions, the lower embeddedness of local companies, and the restructuring of local company networks into more hierarchical, centrally or even externally controlled formations.

This chapter aims to present the outcomes and the limitations of this worldwide rescaling and integration process in post-socialist Central and Eastern Europe. Reconfiguring the historical legacies of CEE regions, post-industrial development has produced territorially uneven results. While national capitals and their surroundings have emerged as advanced service economies integrated into European and worldwide networks of metropolitan growth areas (MEGAs), other regions have a more even balance between industrial and tertiary sources of competitiveness, or they experience hollowing-out processes which entail the dissolution of productive specialisations and long-term socio-economic decline. It is argued that in an era of globalisation and metropolisation, non-metropolitan regions in CEE face a significant risk of falling behind, which should be counteracted by comprehensive efforts to foster territorial reintegration and endogenous growth capabilities.

Territorially uneven structural changes under post-socialism

Socialist development policies prioritised industrialisation at all costs, while neglecting or outright suppressing consumption and business services. This led to overindustrialised national and regional economies. Not only were these structures oversized, they were also unsustainable, burdened with a host of insoluble problems (see Chapter 3). Accordingly, regional restructuring in Central and Eastern Europe in the post-socialist era coincided with a rapid transition to post-Fordism, the far-reaching tertiarisation of the overindustrialised economies, and a massive decline in industrial employment (de-industrialisation). Restructuring eliminated the dominance of industry on all levels of the space economy, and the tertiary sector has universally become the main source of production and employment, absorbing much of the industrial labour surplus.[2] Post-traditional ruralisation, i.e. labour returning from the cities to the countryside and from industry to agriculture (Kovács, 2003), was a feature of the first decade of transition in South-Eastern Europe, where the primary sector acted as a temporary buffer for the unemployed (Büschenfeld, 1999; Petrakos and Totev, 2000; Maniu, Kallai and Popa, 2001; Molnár, 2010). However, this was much more limited in the Visegrad countries where only Poland retained a large agricultural population. By the 2000s, this labour-absorbing role of rural areas was waning, although later it was again observed in Greece during the financial crisis.

However, the ubiquity of tertiarisation conceals important disparities: for example, those in the spread of service activities at both national and regional levels. Furthermore, these activities themselves show enormous differences with respect to their added value, innovation content, competitiveness and territorial integration. These differences are not only significant, they have

Gábor Lux

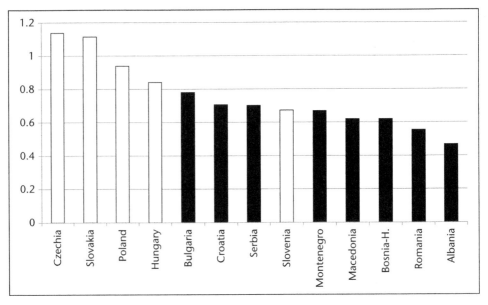

Figure 2.1 Changes in the share of industrial employment in CEE countries, 1990–2008 (1990 = 100%)
Source: Author's calculations based on data from national statistical yearbooks and Eurostat.
Legend: White columns represent Central European, black columns represent South-Eastern European countries.

also turned out to be rather persistent, and they can have a far-reaching influence on long-term regional development paths. In order to understand the socio-economic differentiation of CEE countries and regions, beyond looking at the basic structural indicators, we must assess the underlying quantitative and qualitative differences as well. As services can be found 'everywhere' and agriculture has a comparatively low share in employment and economic output,[3] industry has become the main sector representing regional differences.

Encompassing the main period of radical structural changes, Figure 2.1 shows the national differences in de-industrialisation between the first years of transition and the global financial crisis. It is apparent that the long-term decline in industrial employment was significantly lower in the Visegrad group of countries than elsewhere. In fact, Czechia and Slovakia even experienced minor reindustrialisation. These features point beyond the common characteristics of transition, calling attention to the differences in market processes, political and institutional contexts surrounding structural transformation.

- In the Visegrad countries, tertiarisation has been partially counterbalanced by reindustrialisation, driven by high FDI inflows into the manufacturing sector. The duration of the transformation recession here was shorter than in South-Eastern Europe or the post-Soviet countries.
- In South-Eastern Europe, particularly in the successor states of Yugoslavia, deeper de-industrialisation is explained by both the slower pace of political and economic transition, and the very outdated industrial structure of state socialism. This has led to a lower survival rate of companies, delays in the spread of FDI and severe socio-economic problems. Furthermore, the original degree of industrialisation was itself a statistical illusion, bolstered by the underdevelopment of the service sector.

The significance of national dissimilarities becomes even more prominent if we consider the qualitative differences behind the numbers, which prompts us to look for the internal differences of a universal phenomenon.

- First, tertiarisation is *the structural correction of overindustrialised economies*. The artificial industrialisation and the suppression of services under state socialism had produced an abnormal economic structure which was subsequently corrected by market-led restructuring in the post-socialist era. This interpretation of tertiarisation became mainstream in academia, policy and public discourse alike in the 1990s and 2000s.
- Second, tertiarisation represents a *modernisation process* corresponding to the global post-Fordist transformation. This interpretation highlights the increasing variety of service-based employment and business opportunities, new forms of consumption and an improving quality of life. However, the benefits of this modernisation are territorially uneven, particularly when it comes to KIBS which are concentrated in national capitals and a few metropolitan areas, trickling down to the peripheries only through a slow hierarchical spreading process. We can also observe the problems posed by inadequate critical mass: high value-added services tend to avoid minor cities with insufficient agglomeration impact, and thus may be limited to a few regions even in the long term (Chapter 4). Finally, as Audretsch (1998) showed on a global scale, multiple studies (e.g. Horváth, 2007, 2015; Gál, 2005) have concluded that the more innovative or 'valuable' a service is, the more likely it shows high concentration.
- The third variety of tertiarisation is less benign: it represents *peripherisation and hollowing-out* in less developed regions. Where declining industries were not replaced by high value-added services, structural change is merely a sign of economic decay and regional de-specialisation. Hollowing-out involves a loss of valuable economic functions, the 'emptying' of the space economy. In this case the new service economy is merely the dominant employer (partly due to the absence of alternatives), but not a genuine source of competitiveness.

Structural change is a layered process, and the above categories overlap in both space and time; there are important trade-offs and opportunity costs which may crop up during the restructuring process. Yet it is clear that while the structural correction effect has been universal in post-socialist CEE, the other two effects show a centre–periphery relationship both among countries and at subnational level. Concentrated mainly in metropolitan areas and a few well-integrated large cities, the benefits of modern service economies follow global trends, while the trends of hollowing-out mostly affect the peripheries.

Territorial structures and driving forces of the post-industrial economy

Both services and industry in the CEE macro-region are heavily influenced by the selective location decisions of Foreign Direct Investment. Although the Visegrad countries have been the most successful in attracting FDI over the years, and they have enjoyed an early advantage, some economies in South-Eastern Europe have also been catching up since the turn of the millennium, particularly in selected 'forerunner' or 'bridgehead' regions which serve as origins of dispersal processes.[4] The sectoral breakdown of gross value added (GVA) follows the split within the macro-region: FDI-dominated industry plays a stronger role in the economic performance of the Visegrad group and Slovenia, while the South-Eastern countries show a higher share of both agriculture and services (Table 2.1).

Table 2.1 Foreign Direct Investment (FDI) in post-socialist countries, 1995–2011, and the sectoral breakdown of gross value added (GVA), 2013

Country	1995	2005	2011	Agriculture	Industry and construction	Services
	USD/capita			%		
Albania	65	319	1462	n.a.	n.a.	n.a.
Bosnia-Herzegovina	n.a.	735	1791	8.2	26.2	65.6
Bulgaria	53	1785	6400	5.6	28.6	65.8
Croatia	106	3283	7026	4.1	26.4	69.5
Czechia	711	5936	11889	2.5	38.1	59.4
Hungary	1094	6137	8473	4.5	31.0	64.5
Macedonia	44	1025	2291	10.9	26.5	62.6
Montenegro	n.a.	n.a.	9178	8.8	17.9	73.3
Poland	203	2377	5158	3.4	33.6	63.0
Romania	36	1192	3281	5.0	34.2	60.8
Serbia	n.a.	n.a.	2321	9.9	31.2	58.9
Serbia and Montenegro	n.a.	592	n.a.	–	–	–
Slovakia	242	4394	9375	3.1	38.6	57.3
Slovenia	1316	3623	7442	2.0	33.0	65.0
Visegrad–4 + Slovenia	450	4493	8467	3.1	35.1	61.8
South-Eastern Europe	270	1276	4219	7.5	27.2	65.3

Source: Author's calculations and compilation based on investment data from UNCTAD, population and labour statistics from Eurostat and national statistical yearbooks.

Notes: Data for Serbia in the second column are from 2006–2007. Population data for Bosnia-Herzegovina are official; hence, strongly debatable.

At the subnational level, we can get a more accurate picture about the driving forces of regional development if we correlate the (percentage) data series of sectoral employment with nominal gross domestic product (GDP) per capita (expressed as a percentage of the EU average) in the CEE regions.[5] For the sake of comparison, data for national capitals and their surrounding regions were merged even if they are treated as two separate NUTS 2 units – i.e. Prague and Central Bohemia, Bratislava and Western Slovakia, or Bucharest and the South Region. These are henceforth referred to as *central regions*. The South-Eastern group of countries include here Bulgaria, Croatia and Romania. The dataset of 50 regions was subjected to correlation analysis, first for the group of all regions; then separately for non-central and central region groups. Employment figures were from 2013, and regional GDP from 2011.

Reproduced in Table 2.2, the results show evidence of both a macro-regional split and a divide between the central and the non-central regions.

- While correlation between primary sector employment and per capita GDP was universally negative, showing the underdevelopment of rural regions (Chapter 5), it was the least definite in the South-Eastern European countries. The relationship was reverse for services which seemed to have a positive effect on economic development.

Table 2.2 Correlation between sectoral employment and nominal GDP per capita

Region	Agriculture	Industry and construction	Services
All regions	−0.47	0.05	0.50
Visegrad–4 + Slovenia	−0.50	−0.02	0.56
South-Eastern Europe	−0.22	−0.24	0.37
Non-central regions	−0.47	0.31	0.35
Visegrad–4 + Slovenia	−0.46	0.35	0.25
South-Eastern Europe	−0.26	−0.16	0.37
Central regions	−0.87	0.35	0.54

Source: Author's calculations and construction based on data from Eurostat.

- The most interesting differences can be found in industry and construction. No notable correlation exists for the whole database, but the picture changes when we separate non-central and central regions. Non-central regions in the Visegrad group show medium-level positive correlation between industrial employment and nominal GDP, while the relationship is weaker and negative in South-Eastern Europe. *This difference highlights the relevance of different economic development paths in non-metropolitan CEE regions.* In the Visegrad group, industrial restructuring and FDI-based development have largely replaced the low-performing industrial branches (although there are important caveats, discussed in Chapter 3), and competitiveness is based on a mixture of industry and services – but more strongly on the former. In South-Eastern Europe, structural change has been less thorough, and low-road cost advantages still dominate: the development role of the secondary sector is more ambiguous. While the differences can be expected to diminish slowly, they will most likely remain in the coming decades.
- Finally, the group of the six central regions in our dataset is too small to draw valid conclusions. However, they seem to fit into the picture of the standard metropolitan growth path: service-based development dominates, while the positive value of industry might signify the joint relevance of construction, and some high value-added, knowledge-based industrial activities.

Based on our findings here and in further empirical work, we can speak of a *triple typology of regions*, pointing towards different, perhaps even divergent paths of regional development, and a varied economic landscape across the CEE macro-region (Figure 2.2). *Central regions*, including national capitals and their functionally linked agglomerations, are the most successful examples of economic transformation. They are integrated into global metropolitan networks and are specialised in KIBS, innovation, corporate and public command and control functions, besides having a limited number of high-innovation, high value-added manufacturing activities. However, it is a sobering fact that they still rank rather low in global city hierarchy (Csomós, 2011). The high territorial concentration of KIBS in central regions clearly shows the limits to competitive post-industrial development. Measured by employment in information and communication (NACE J) as well as in financial and insurance activities (NACE K) in 2014, these concentration values were very high in small countries with monocentric territorial structures. They reach over 60 per cent in Slovakia, Hungary, Bulgaria and Slovenia, and 76 per cent in Croatia. On the other hand, the distribution of high value-added services is more even in Poland, due to its emerging metropolitan network which can effectively counterweigh Warsaw and the Mazowieckie region (34 per cent). Romania's urban network has not yet shown a similar

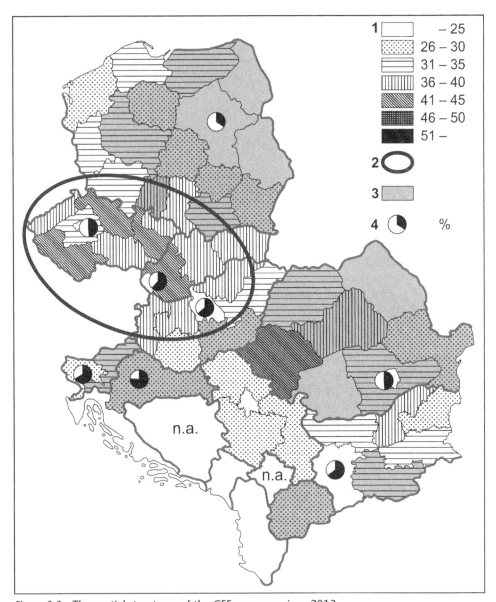

Figure 2.2 The spatial structures of the CEE macro-region, 2013

Source: Author's calculations based on data from Eurostat.

Legend: 1 – The share of industry and construction in total employment (per cent)
2 – Manufacturing core area
3 – Regions with high share of agricultural employment (>10 per cent)
4 – The main concentrations of advanced business services (percentage of national employment).

decentralisation process, but its large cities have the potential to emerge as competitive service hubs. Czechia has Brno as a potentially strong secondary pole (with 16 per cent vs. the 49 per cent of Prague and Central Bohemia). Since urban networks are among the most path-dependent territorial structures and there are no great prospects for CEE's large cities to grow (Chapter 8), it must be accepted that, while advanced tertiary activities can be expected to get slightly decentralised (cf. Gál and Sass, 2009, and Chapter 4), there are natural barriers to their territorial dispersion – and the regions losing out will have to look for other forms of development.

Combining industrial and tertiary sources of competitiveness, the second group of successful regions could be described as *intermediate* regions. Most of them are located in Central Europe's manufacturing core where beneficial productive legacies and new investments have together created a favourable environment for industrial development. This area – encompassing Czechia, Western Slovakia, Southwestern Poland and Northwestern Hungary – is increasingly integrated into global production networks, particularly into the production systems established in Germany, Austria and Northern Italy (Chapter 3). There are also other manufacturing hotspots, but they are less prominent and less connected. The competitiveness of intermediate regions is mainly driven by exogenous factors (FDI), but these investments are typically built on strong historical foundations, with path-dependent development which provides relatively favourable opportunities for endogenous growth. They are starting to show some growth in business services, as well as in manufacturing activities, while the relatively high incomes in these sectors also encourage beneficial spillover effects into local consumption services. However, the lack of critical mass and sufficient network density limits the development of tertiary activities in all but a handful of regional centres. It is likely that the industry–service mix will be the main source of regional development in the coming decades, and the main challenges will arise in network integration and in switching from low- to high-road forms of competitiveness.

The strong market selection in the course of transition radically divided non-metropolitan space. Regions unable to integrate into global networks – retaining their old industrial base or initiating endogenous accumulation processes on their own due to lack of local capital, knowledge and institutions – have experienced *hollowing-out*, leading to the emergence of a 'homogenous periphery'. Their homogeneity consists in their ability to offer only the same set of resources at any of their locations: cheap, mostly unskilled labour and basic infrastructure. This is insufficient for survival in the global race, even with low wages and low labour protection. The weakening and disappearance of industrial specialisation has been particularly severe in small and medium-sized towns which were largely avoided by FDI favouring large regional centres, and consequently they lost their role in integrating local economies, thus experiencing a 'disconnection' from their territorial context. Hollowed-out regions invariably have their competitive SMEs, FDI branch plants or the odd successful cluster: but these isolated cases are accompanied by an underdeveloped environment and may themselves be precarious, prone to closures or delocalisation.[6] Following EU-accession, and particularly during the financial crisis, the problems of the peripheries were aggravated by severe human capital losses, by out-migration to the EU15 (Chapter 13). This affected precisely the skilled, mobile, entrepreneurial workforce which might offer hope for future development.[7] Under the circumstances of general capital shortage, external development funding may be insufficient to create new evolutionary paths: linkage possibilities and social capital are in short supply, and re-specialisation faces strong challenges due to low network density.

Reintegrating socio-economic space

Apart from the European crisis, and also taking into account the lessons of global integration in the Western European economies, economic transition in the CEE macro-region poses some questions. Can we diminish the deep and persistent development differences which have emerged, particularly between a few metropolitan areas, successful regions and the rest? Are there ways for the peripheries to embark on successful development paths if they cannot benefit from the presence of strong urban centres and developed industrial networks? The current trends of regional development are heavily path-dependent and they seem to show signs of geographic determinism in some respects (Sucháček, 2010), where the advantages of the winners are insurmountable, while those falling behind have little chance to improve their lot. As the following chapters will show, FDI-driven development in the CEE regions has produced many success stories, but its benefits have been geographically uneven, and they even seem to have come with trade-offs and clear limits. After over a quarter of a century, and among the uncertainties of post-crisis Europe, the driving forces of post-socialism growth paths often seem to be exhausted. Where now for regional development?

Based on the available evidence, the author believes that it is time to rethink our development philosophies. Whereas the period of post-socialism has been characterised by attempts to attract and channel exogenous resources (from FDI to EU funds and imported know-how), now a deeper look into fostering better conditions for endogenous development should be taken. Although often invoked in discourse, very little of this paradigm has effectively entered into development practice. Concepts such as clusters, regional innovation systems or industrial districts are often used as a pretext to gain funding, but comprehensive philosophies of systematically investing into local human capital, entrepreneurship, socio-economic networks and embedded production are relatively hard to find. Human capital, in particular, has been a neglected issue of transition, while high-road development paths are unthinkable without strong human potential.

Endogenous development also has its specific importance due to the transforming sources of competitiveness, which are becoming increasingly localised, tied to a specific place or region in high-road development. While FDI can help develop its own production networks and will continue to play a crucial role in shaping regional development in CEE, endogenous development will always be the strongest and most sustainable in firms and socio-economic networks which emerge locally in an organic fashion. Domestic entrepreneurship, particularly medium-sized companies in supply networks or high value-added product niches, is indispensable in the long-term accumulation processes enriching regions. As even successful regions are starting to face the pressure of low-cost global competitors, they must formulate effective upgrading strategies to forestall the decline or loss of their current economic base, and they must do so in a sustainable way.

The long-term objective of endogenous development can be *the reintegration of socio-economic space* through building strong, resilient, locally embedded production networks. Its abstract illustration can be seen in Figure 2.3. Space was dominated by large, vertically integrated companies with strong central control under state socialism. They had few horizontal links to other local companies, although they developed their own local skill base and accumulated knowledge in their production networks. There were also smaller, isolated companies that did not engage in traditionally understood competition, although they served as conduits and foci of knowledge and capital accumulation, too. Market selection in the course of transition, followed by the uneven restructuring during and after the transformation recession, has increased territorial discrepancies. As described previously, this has resulted in regions undergoing global integration

Reintegrating economic space

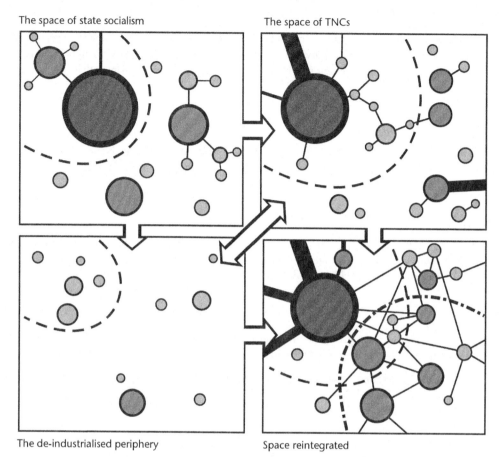

Figure 2.3 The transformation and reintegration of post-socialist space
Source: Lux, 2014, p. 43.

mainly through FDI inflows, with others experiencing hollowing-out through capital loss and the disintegration of their production networks. These two development paths are not the final stages of socio-economic evolution in CEE: transition from one to the other is possible, and new forms of integration can – and should – also emerge.

In this sense, endogenous development can be seen as a development paradigm enabling CEE economies to move towards *a reintegrated space economy*: the building of strong local networks which can provide sufficient added value for both TNCs and domestic enterprises. The key to these networks are the density and diverse directions of their connections, which can break one-sided dependent relationships and help these regions to get established as competitive players in both European and global contexts. Altogether, endogenous development and the reintegration of space should achieve three different, but closely connected goals:

- encourage re-specialisation in regions that have lost their previous focus;
- make it possible to transcend the limitations of FDI-based competitiveness;
- and, finally, open opportunities towards 'high-road' growth paths and the incremental improvement of socio-economic conditions.

There is no guarantee that endogenous development can prevent the problem of 'the disappearing middle' from emerging, or that it can offer full protection from global competitive pressure; and the metropolitan–provincial divide will continue to shape future growth opportunities. However, refocusing development can hopefully help regions learn to adapt – that is, to learn how to learn better.

Acknowledgements

This research has been supported by project #104985 of the Hungarian National Research, Development and Innovation Office. While writing this paper, Gábor Lux was supported by the János Bolyai Research Scholarship of the Hungarian Academy of Sciences.

Notes

1 Beyond a book outlining a research agenda (Sucháček, 2010), second-tier cities had enjoyed relatively little attention until an ESPON project (Parkinson et al., 2012), two highly relevant papers (Dijkstra, 2013; Dijkstra, Garcilazo and McCann, 2013), and more recently a special issue (Camagni, Capello and Caragliu, 2015).
2 This is the case even if, based on manufacturing's share in gross value added, Czechia, Slovenia and Hungary were still ranked as the three most industrialised economies of the enlarged EU in 2014 – followed by Germany in the fourth place.
3 While the significance of agri-business is nationally significant even in some advanced economies, its bulk is now found in the food industry-related and retail segments of the value chain (Buday-Sántha, 2001).
4 In Romania, the agglomeration of Bucharest concentrated 57.1 per cent and the West region (the 'Banat') 10.7 per cent of FDI in 2001. Between 2000 and 2008, Bucharest drew a full 63 per cent of subsequent inflows (Molnár, 2010). Other South-Eastern European states show similar investment patterns.
5 Kuttor and Hegyi-Kéri (2012) have arrived at similar results through different methods.
6 The gradual downsizing of the electronics industry in Hungary is a typical case. This industry had little upgrading potential and so it was heavily hit by the collapse of investors like Nokia, Elcoteq, etc.
7 In addition to the pull effect of radically higher wages, a significant push factor is also present due to non-performing housing loans which played a major role in encouraging emigration from Hungary and Poland.

References

Acemoglu, D. and Autor, D., 2010. Skills, Tasks and Technologies: Implications for Employment and Earnings. Working Paper.
Audretsch, D.B., 1998. Agglomeration and the Location of Innovative Activity. *Oxford Review of Economic Policy*, 14(2), pp. 18–29.
Buday-Sántha, A., 2001. *Agrárpolitika – vidékpolitika: A Magyar agrárgazdaság és az Európai Unió* [Agricultural and rural development policy: Hungarian agriculture and the European Union]. Budapest, Pécs: Dialóg Campus Kiadó.
Büschenfeld, H., 1999. Wirtschaftliche Transformationsprozesse in den Nachfolgestaaten Jugoslawiens. *Europa Regional*, 7(4), pp. 23–38.
Camagni, R., Capello, R. and Caragliu, A., 2015. The Rise of Second-rank Cities: What Role for Agglomeration Economies? *European Planning Studies*, 23(6), pp. 1069–1089.
Csomós, Gy., 2011. Analysis of Leading Cities in Central Europe: Control of Regional Economy. *Bulletin of Geography. Socio-Economic Series*, 16, pp. 21–33.
Derudder, B., Taylor, P.J., Witlox, F. and Catalano, G., 2003. Hierarchical Tendencies and Regional Patterns in the World City Network: A Global Urban Analysis of 234 Cities. *Regional Studies*, 37(9), pp. 875–886.
Dijkstra, L., 2013. Why Investing More in the Capital Can Lead to Less Growth. *Cambridge Journal of Regions, Economy and Society*, 6(2), pp. 251–268.

Dijkstra, L., Garcilazo, E. and McCann, P., 2013. The Economic Performance of European Cities and City Regions: Myths and Realities. *European Planning Studies*, 21(3), pp. 334–354.

Erdősi, F., 2003. Globalizáció és a világvárosok által uralt tér I [The space ruled by globalisation and metropolises I]. *Tér és Társadalom*, 17(3), pp. 1–27.

Faragó, L., 2010. Területi koncentráció és a jelentőségüket vesztő perifériák [Territorial concentration and the declining significance of the peripheries]. Gy. Barta, P. Beluszky, Zs. Földi and K. Kovács, eds. *A területi kutatások csomópontjai*. Pécs: Magyar Tudományos Akadémia Regionális Kutatások Központja, pp. 432–453.

Gál, Z., 2005. The Development and the Polarised Spatial Structure of the Hungarian Banking System in a Transforming Economy. In: Gy. Barta, É. G. Fekete, I. Kukorelli Szörényiné and J. Timár, eds. *Hungarian Spaces and Places: Patterns of Transition*. Pécs: Centre for Regional Studies, pp. 197–219.

Gál, Z., 2010. *Pénzügyi piacok a globális térben: A válság szabdalta pénzügyi tér* [Financial markets in global space: The crisis-segmented financial space]. Budapest: Akadémiai Kiadó.

Gál, Z. and Sass, M., 2009. Emerging New Locations of Business Services: Offshoring in Central and Eastern Europe. *Regions*, 274(1), pp. 18–22.

Gereffi, G., 1995. State Policies and Industrial Upgrading in East Asia. *Revue D'Économie Industrielle*, 71(1), pp. 79–90.

Goos, M. and Manning, A., 2007. Lousy and Lovely Jobs: The Rising Polarization of Work in Britain. *Review of Economics and Statistics*, 89(1), pp. 118–133.

Gordon, I.R. and McCann, P., 2000. Industrial Clusters: Complexes, Agglomeration and/or Social Networks. *Urban Studies*, 37(3), pp. 513–532.

Horváth, Gy., 2007. A régióépítés dilemmái Kelet-Közép-Európában [The dilemmas of creating regions in Central and Eastern Europe]. In: Z. Hajdú, I. Illés and Z. Raffay, szerk., *Délkelet-Európa: Államhatárok, határon átnyúló kapcsolatok, térstruktúrák*. Pécs: MTA Regionális Kutatások Központja, pp. 74–99.

Horváth, Gy., 2015. *Spaces and Places in Central and Eastern Europe*. London: Routledge.

Humphrey, J. and Schmitz, H., 2002. How Does Insertion in Global Value Chains Affect Upgrading in Industrial Clusters? *Regional Studies*, 36(9), pp. 1017–1027.

Kilicaslan, Y. and Taymaz, E., 2008. Labor Market Institutions and Industrial Performance: An Evolutionary Study. *Journal of Evolutionary Economics*, 18(3–4), pp. 477–492.

Kovács, T., 2003. *Vidékfejlesztési politika* [Rural development policy]. Budapest, Pécs: Dialóg Campus Kiadó.

Kuttor, D. and Hegyi-Kéri, Á., 2012. Sectoral and Regional Dimensions of Industrialization in East Central Europe. In: D. Pavelkova, J. Strouhal, M. Pasekova and J. Sucháček, eds., *Advances in Economics, Risk Management, Political and Law Science*. Zlin: WSEAS Press, pp. 290–299.

Lovering, J., 1999. Theory Led by Policy: The Inadequacies of the 'New Regionalism' (Illustrated from the Case of Wales). *International Journal of Urban and Regional Research*, 23(2), pp. 379–395.

Lux, G., 2014. Industrial Districts: Building Blocks of the Organised Economy. In: E. Somlyódyné Pfeil, ed. *Industrial Districts and Cities in Central Europe*. Győr: Universitas-Győr Nonprofit, pp. 27–45.

Lux, G., 2015. Minor Cities in a Metropolitan World: Challenges for Development and Governance in Three Hungarian Urban Agglomerations. *International Planning Studies*, 20(1–2), pp. 21–38.

Maniu, Mircea T., Kallai, E. and Popa, D., 2001. Explaining Growth. Country Report: Romania (1990–2000). *The Romanian Economy: From Communism through the Transition to a Market Economy*. Initiatives for Democracy in Eastern Europe (IDEE), Bucharest. www.eldis.org/static/DOC13116.htm (2007. III. 21)

Milberg, W. and Houston, E. 2002. The High Road and the Low Road to International Competitiveness. In Taylor, L. ed. *Globalization and Social Policy*. New York: The Free Press.

Milberg, W. and Winkler, D., 2010. Economic Insecurity in the New Wave of Globalization: Offshoring and the Labor Share under Varieties of Capitalism. *International Review of Applied Economics*, 24(3), pp. 285–308.

Molnár, E., 2010. Közép- és Délkelet-Európa határán: Gazdasági modernizáció és szerkezetváltás Romániában [On the borders of Central and South-Eastern Europe: Economic modernisation and restructuring in Romania]. In: G. Demeter and Zs. Radics, eds., *Kompországok – ahol a part szakad ... Szemelvények Köztes-Európa integrációs törekvéseiből (1990–2008)*. Debrecen: Didakt Kiadó, pp. 370–411.

Moulaert, F., Martinelli, F., González, S. and Swyngedouw, E., 2007. Introduction: Social Innovation and Governance in European Cities: Urban Development between Path Dependency and Radical Innovation. *European Urban and Regional Studies*, 17(3), pp. 195–209.

Nölke, A., 2011. *Transnational Economic Order and National Economic Institutions*. Köln: Max-Planck-Institut für Gesellschaftsforschung (MPIfG Working Paper, 11/3).

Parkinson, M., Meegan, R., Karecha, J., Evans, R., Jones, G., Sotarauta, M., et al. 2012. *Second-tier Cities and Territorial Development in Europe: Performance, Policies and Prospects.* Luxembourg: ESPON.

Parrilli, M.D., 2009. Collective Efficiency, Policy Inducement and Social Embeddedness: Drivers for the Development of Industrial Districts. *Entrepreneurship & Regional Development*, 21(1), pp. 1–24.

Petrakos, G. and Totev, S., 2000. Economic Structure and Change in the Balkan Region: Implications for Integration, Transition and Economic Cooperation. *International Journal of Urban and Regional Research*, 24(1), pp. 95–113.

Sassen, S., 2006. *Cities in a World Economy.* London: Sage.

Sucháček, J., 2010. On the Emergence of Minor Cities. In: J. Sucháček and J.J. Petersen, eds. *Developments in Minor Cities: Institutions Matter.* Ostrava: VŠB – Technical University of Ostrava, pp. 13–28.

Taylor, P.J., 1997. Hierarchical Tendencies amongst World Cities: A Global Research Proposal, *Cities*, 14(6), pp. 323–332.

Tüzemen, D. and Willis, J., 2013. The Vanishing Middle: Job Polarization and Workers' Response to the Decline in Middle-skill Jobs. *Economic Review*, 39(1), pp. 5–32.

3
Industrial competitiveness
Beyond path-dependence
Gábor Lux

Evolutionary change and regional industrial development

While the brave new world of post-Fordism – the global transformation of production, labour and consumption – is nothing less than a revolutionary force, it has involved much evolution: achievements in social sciences in the last decades have shown that the development patterns of global and local economies are as much rooted in embedded history, routines and traditions as in the disruptive influence of technological, economic and social change. Originating from Veblen and Schumpeter, the formative article of Nelson and Winter (1982) on evolutionary economics, Paul A. David's reflections on path-dependence (revisited in David 2007), and heterodox non-equilibrium economics (Boschma and Martin, 2010), the idea of an evolutionary investigation of socio-economic development has borne rich fruits in social sciences just as they were undergoing an 'institutional turn' (Martin, 2000). The spatial interpretation of social dynamics 'characterised by positive feedbacks and self-reinforcing dynamics' (David, 2007, p. 92) has led to the rise of evolutionary economic geography, which has integrated a number of strains in regional studies, coming up with its own explanations for the inner workings and long-term development of complex phenomena.

Industrial development has been one of the major study areas of evolutionary economic geography, particularly when the long-term success or failure of companies and industrial regions (milieus) are examined. Researchers have focused their interest on multiple areas; among others:

- The role of economic traditions, variety, exploratory behaviour and 'historical accidents' in forming the evolutionary paths of technologies, industries and regions, from path creation to selection, maturity and eventual decline (Martin and Sunley, 2010, 2011; Menzel and Fornahl, 2009).
- The consequences of bounded rationality in institutional dynamics for local industrial development (Boschma and Frenken, 2009; MacKinnon et al., 2009).
- The role of knowledge-based development (Asheim, 1996) and socio-economic networks (Markusen, 1996; Zeitlin, 2008; Potter and Watts, 2010) in the model of 'organised space', dominated by mutual dependences and a mixture of cooperation and rivalry instead of atomistic competition (Belussi and Sedita, 2009).

- The decline and eventual transformation of traditional industrial growth areas ('Old Industrial Regions' or OIRs), explained by structural deficiencies, institutional sclerosis or evolutionary cycles (Steiner, 2003), as well as overspecialisation and network-agglomeration-related models (Grabher, 1993; Grabher and Stark, 1997; Boschma and Lambooy, 1999).

A perspective stressing the importance of adaptation and adaptability is very important in a world facing many uncertainties. The fate of industrial regions under post-Fordism has not always been promising in the advanced market economies either. Formerly prosperous OIRs had often undergone decades of malaise attempting to approach their previous level of employment, only to discover that the new service economy brought low-wage jobs and part-time or 'flexible' employment for the new generation of employees. Low entry barriers in mass production have resulted in massive competition with newly industrialised countries (NICs), challenging traditional industrial heartlands in Europe and North America (Acemoglu and Autor, 2010). Freeman (2008) writes about 'the great doubling' of the world's globally integrated labour force from 1.46 to 2.93 billion people since the 1990s, and calls attention to the resulting change in the global capital/labour balance which decreased to 61 per cent of its previous level during the same time span.

However, the ability to adapt is but one side of development. It contrasts with the power of path-dependence whose trajectories carry the burden of history and institutions. This is both a curse and a blessing, as its self-reinforcing processes can create virtuous and vicious circles alike. The co-evolution of industry, governance and knowledge (the components of the well-known Triple Helix model) takes place in a strongly localised regional context, in which the mutual dependences and interactions create socio-economic networks that can greatly increase the competitiveness of economic actors; or, in the case of most OIRs, hinder them and prevent their renewal. These localities matter increasingly when we discuss competitiveness, as they affect the global value chains as well as the perspectives of locally embedded endogenous development.

Building on the results of Chapter 2, the present study is concerned with the regional development of manufacturing in Central and Eastern Europe in the post-socialist era. Relying on the analytical framework of evolutionary economic geography, it analyses how the structural crisis of the Fordist–post-Fordist shift unfolded in the context of systemic change, and how post-socialist economies integrated into the global division of labour under the dominance of Foreign Direct Investments. It is argued that these developments were less radical and much more path-dependent than it might seem, and that the underpinnings of the success stories can usually be found in the recovery and long-term evolution of pre-existing industrial milieus. We also argue that FDI-based development has had significant opportunity costs, too, and there are clear limits to its sustainable growth. Meanwhile, alternative paths towards competitive manufacturing – endogenous development, place-specific location advantages, focusing on empowering domestic entrepreneurship and building effective socio-economic networks – have been neglected, so their outcomes still leave much to be desired.

Structural crisis and transition

In Central and Eastern Europe, state socialism applied homogeneous development policies and dual isolation, from both market economies and one another. The lion's share of development funding went into industry, by way of redistributing resources from agriculture, infrastructure and civilian consumption. By the end of the period of forced industrialisation – the end of the 1950s in the more developed socialist states, and the following decades elsewhere – the new heavy industrial complexes had been established, and they overwhelmingly focused on producing investment goods (means of production) instead of consumption goods.[1] The resulting economic profile was formed as much by catching-up attempts as by political and military pressure, and was *ab ovo* burdened with severe problems and contradictions. This chapter does not seek to provide an in-depth critique of the dysfunctions plaguing planned economies (see Jánossy, 1971; Kornai, 1980, 1992; Winiecki 1986, and others); rather, it provides a summary of its regional operating mechanisms, its structural crisis from the 1970s and its similarities to the transformation of OIRs, particularly its socio-economic legacies which continue to influence modern development paths whether they are characterised by path-dependence or discontinuity.

One of the most crucial spatial consequences of socialist industrialisation lies in its *disruption of spatial networks*. In the 1950s and 1960s, visions of industrialisation and catching-up (not unique to socialist states, but a common goal of peripheral countries) were formulated on the basis of stressing the need for national self-sufficiency. Cross-border trade was limited to large package deals, never reaching the volume of the myriad small trades of market exchange: in practice, the bulk of trade within the COMECON was carried out through bilateral agreements between the individual CEE socialist states and the Soviet Union, trading manufactured goods for raw materials and energy. Attempts on part of the Soviets and the more developed states to increase national specialisation and establish a COMECON-wide division of labour were firmly rejected by the less developed countries, especially Romania (Valev, 1964; Turnock, 1986), which finally led to limited specialisation in differentiated industrial profiles.[2] This arrangement resulted in the emergence of parallel industrial structures, isolated from competitive pressure and innovation flows.

Directed from national capitals and following a vertical, top-down design of territorial integration, centralised development policies decoupled or greatly reduced local and regional linkages. The creation of large-scale manufacturing complexes achieved vast economies of scale, but deliberately constructed monofunctional productive milieus, suppressing local socio-economic variety and horizontal connections. At the local level, single enterprises dominated the economy of city regions to an unprecedented degree. Already by the 1960s, supplemental investments were needed in heavy industrial regions to combat massive hidden (informal) unemployment, particularly among women. The consolidation of heavy industrial plants required substantial follow-up investments which locked in national and regional development paths, and limited decision-making in economic planning. Heavy dependence on existing production profiles curtailed even the rarely formulated restructuring strategies. Moreover, the ever-present shortage of development resources at national level led to a development trap by considering the industrialised regions to be developed which thus did not receive further assistance to modernise and diversify their economies. Without complex policy interventions, 'modernisation' stalled and even concealed deeper layers of socio-economic underdevelopment which remained untreated until transition.

Poland and Hungary were exceptions to some extent, as they experimented with regional policy in the 1960s and 1970s, involving the industrialisation of peripheral regions (mostly smaller towns) via light and food industries (Bartke, 1971; Lijewski, 1985). These experiments, often reviving smaller-scale pre-war industries, were limited in scope and especially funding, and were cut short by the emerging crises of the 1970s, but they contributed to decentralising the industrial networks and diversifying the economy of the peripheries. Most importantly, they became seeds of innovative local development through path creation, and their legacies have continued to influence industrial trajectories and local entrepreneurship up to the present day.

By the late 1970s, the crisis of socialist industry became undeniable; and in the following decade it produced symptoms very similar to those of the Old Industrial Regions in market economies. The question can be raised whether these similarities are incidental (superficial) and so it is not reasonable to compare the evolution of separate economic systems as they are too different, or whether they indicate a deeper resemblance common to the transition from

Fordism to post-Fordism. It should be noted that it was not public intervention, rapid industrialisation or support for strategic sectors that had separated socialist and market economies, since these were also widespread under the Keynesian development model. Likewise, there were also substantial parallels when the crisis was unfolding:

- *In general*, we can speak of soft budget constraints (weak or missing 'natural selection' mechanisms), rent-seeking behaviour and 'non-market rationality' (defence-related, political and social concerns) in corporate decision-making and development policy. These are not system-specific, as they have been observed in the behaviour and restructuring of firms and regions in Fordist market economies, too.
- *Specifically*, we can easily and mostly successfully apply the relevant crisis theories explaining the restructuring problems of OIRs to the conditions of late socialism and transition – the details may be different, but the underlying mechanisms are easy to identify. Low economic diversity, overspecialisation and decline in adaptability; institutional, technological and political lock-in; as well as weak collective learning and institutional sclerosis are all well known in socialist and post-socialist OIRs.

However, we can also speak of symptomatic and systemic differences:

- The *dual economic isolation* in state socialism reduced selection pressures and isolated firms and regions from innovation, and most crucially from the paradigm shift of post-Fordist transformation, which led to a major transition shock.
- The previously discussed *regional dysfunctions* of state socialist development policy, the absence of local and regional economic diversification, and the dominance of large companies fostered a culture of dependence which not only extended to employment, but involved companies in providing services, consumption goods and infrastructure as well.
- The *new conditions of transition* also prevented effective restructuring strategies. The sudden collapse of the socialist system brought about a confluence of several different, untreated crises, each requiring enormous funding, in a period when public planning and industrial policy had lost much of their political legitimacy and public support.

Altogether, the similarities are stronger than the differences, which are more of degree than kind. However, these differences still had important consequences for industrial transformation: post-socialist crises were broader, affected more regions and they had deeper socio-economic consequences than in the west, and, most importantly, were not followed by the kind of complex, in-depth restructuring policies that took place in Western and Southern Europe. As Rugraff (2008) notes, post-socialist countries had dismantled their own instruments which could have served as a basis for a strategic industrial policy, and placed their fate in the hands of TNCs and international organisations. Transition was quickly followed by the elimination of trade barriers, the privatisation of high-performing state-owned companies (potential national champions) and the banking system which could have been harnessed to finance domestic industrial development. Industrial restructuring was *dominantly market-driven instead of being managed*, with only marginal, mainly preventive policy interventions in selected crisis areas to forestall a socio-economic catastrophe. Comparative data from Illés (1994) show that industrial production was the main victim of the transformation recession in all post-socialist states, particularly in the South-Eastern countries. But export volumes also declined sharply between 1989 and 1993, by 20 per cent in Hungary, 54 per cent in Bulgaria and 70 per cent in Romania.

More successful manufacturing regions demonstrated *strongly path-dependent evolution,* the industrial legacies of state socialism having been 'reused' and reconfigured through privatisation and large-scale Foreign Direct Investment inflows. Meanwhile, less successful manufacturing regions (typically OIRs and mining areas) as well as peripheral regions faced *destructive de-industrialisation and hollowing out* (cf. Chapter 2) as previously dominant industries were shuttered or became marginalised without being replaced by competitive alternatives. The first years of the transformation recession were characterised by a selection process that can only partially be described as Schumpeterian creative destruction (disintegration followed by reorganisation, new economic functions filling emptied niches, and the decline of low value-added industries such as textiles, leatherworking and woodworking). It was simultaneously an extinction process showing negative hysteresis (Martin, 2012) and the disappearance of state-owned and newly privatised companies.[3] With the disintegration of productive networks and the westward re-orientation of export activity (pursued most aggressively and certainly over-eagerly in Hungary), inter- and intra-industrial linkages loosened, bringing about further bankruptcies. The survival strategies of the remaining companies were often rooted in 'defensive restructuring' (Pavlínek and Smith, 1998) and 'peripheral reintegration' (Lux, 2009), retreating from high-complexity own-brand production, shedding knowledge-intensive R&D functions, jettisoning subsidiaries, and respecialising with a focus on low value-added subcontracting work and basic assembly. These had prepared the ground for investments by TNCs that could now find an abundance of 'loose' production factors on the cheap, but had destroyed domestic corporate innovation and led to de-specialisation and the demise of previously important industries such as Hungary's autobus manufacturing (from over 13,000 units sold annually to 8,000 in 1990, 1,500 in 1995, and 1,000 in 2005) and information and communications technology sector (the loss of 2,000 R&D staff in the flagship company Videoton alone) or Czechia's heavy machinery manufacturers. This loss was not considered in the first decades of transition; today, looking at these enormous opportunity costs, we can only shake our heads.

Global industrial reintegration through Foreign Direct Investment

Under post-socialism and its market-driven restructuring processes, the decline of pre-1990 structures coincided with the new wave of European and global integration. In the absence of effective and well-financed state policies, and as the stock of attractive companies offered up for privatisation dwindled, this change was dominated by the location preferences and space-shaping role of Foreign Direct Investment. As Barta, Czirfusz and Kukely (2008) have shown, overall CEE was a winner of the global relocation processes between 1995 and 2005. At the same time, Kiss (2007b) demonstrates that delocalisation before the crisis was limited, even if it might have had substantial effects on specific local economies (typically small towns which had borne the costs of light industry downsizing).

On the one hand, particularly at the visible level, FDI represents positive discontinuity in regional development (cf. Gorzelak 1998's regional typology): it has been coupled with substantial technological upgrading, the introduction of modern management methods, access to world markets and a massive increase in productivity. Having 'ready-made' channels of access and standardised recipes for growth, FDI-based companies have found it much easier to integrate into global product markets than the struggling, undercapitalised ones in domestic ownership. The local actors of global value chains benefit from this competitive but highly specialised and weakly embedded growth path (Humphrey and Schmitz, 2002). On the other hand, looking into the deeper layers of regional economies beyond the firm level, we can see that investment decisions and new manufacturing plants do not emerge from nothingness as Athena sprang from Zeus's head. Investors build on productive legacies, reusing the production factors, the socio-economic networks and the institutions of their predecessors, demonstrating path-dependent behaviour. The 'retreat' and dissolution of domestic companies give way first to a decoupling of production networks, then to new configurations. However, the resulting paths are not independent of regional histories. Where the untrained eye sees random events, the historian often discovers abundant precedents.

The results of FDI-based restructuring have often been described as 'dual economies', characterised by deep imbalances between the capitalisation, knowledge base, market position, export activity and further vital characteristics of foreign and domestic corporations (Barta, 2005; Havlik, 2005; Kiss, 2007a). Therefore, the space-shaping role of manufacturing companies with foreign ownership greatly outstrips that of the domestic ones, and their decisions to locate in or to avoid specific regions have far-reaching consequences on employment, competitiveness, the development of industrial milieus and the emergence of productive networks. FDI has consistently been the major driving force behind reindustrialisation, with the exception of Slovenia that based its development model on endogenous growth, the dominance of local ownership, and a domestically owned banking system. There is a clear and strong relationship between FDI inflows and economic performance. Nölke and Vliegenthart (2009) revealed that among the Visegrad Four, the share of FDI in the 2007 GDP production was the highest in Hungary (51.8 per cent), followed by Czechia (48 per cent), Slovakia (31.5 per cent) and Poland (24.9 per cent), outstripping several developed economies such as e.g. Austria (22.7 per cent) and Germany (16.4 per cent). As the EU's industrial structural report (Competing in Global Value Chains, 2013) shows, the industries of Slovakia, Hungary and Czechia have integrated most deeply into global value chains within the European Community, and this also holds true for individual industries (electronics and optical industry, machinery). Due to its robust domestic market and strong domestic entrepreneurship, Poland was close to the median in these rankings. As a hint of warning, the report calls attention to the very high import content of gross manufacturing exports which in 2009 was the highest in Hungary with 52 per cent, followed by Czechia (44 per cent), Slovakia (40 per cent), Poland (33 per cent) and Romania (29 per cent).[4]

There are not only firm-level but also geographic differences in FDI-based 'post-socialist' development paths. Chapter 2 has shown how the industrial–tertiary divide has shaped Central and South-Eastern Europe's economic space. The majority of FDI – indeed, a higher than average level of manufacturing employment – is concentrated in Central Europe's 'manufacturing core', arranged around its manufacturing centres and major transport corridors (Smith and Ferenčíková, 1998; Turnock 2001; Domański, 2003; Kiss, 2007b). This spatial formation is another evidence of path-dependent development: in a way it can be traced back to early capitalism, the 'upwards triangle' drawn among Łódz, Erfurt and Budapest. However, there are also other factors at play: the dominant industries of the 'core' (machine industry and electronics) and the peripheries (traditional light and food industries) form relatively clear spatial divisions, both contributing to competitiveness in their own way (Figure 3.1). The industries of the 'core' are the local manifestations of globally integrated value chains. German automotive manufacturers play a particular role in creating a 'complete space', an integration zone covering the complete range of production activities in broader Central Europe (Frigant and Layan, 2009). The location patterns of the automotive industry and its suppliers can mostly be found in the 'core' area where the pre-1990 machine production hubs had been located (Worrall, Donnelly and Morris, 2003; Pavlínek, Domanski and Guzik, 2009; Molnár, 2012; Krzywdzinski, 2014; Wójtowicz and Rachwał, 2014).

There is also a less apparent but nevertheless notable reintegration process specific to Eastern and South-Eastern peripheries which is usually tied to the food and light industries (particularly textiles). Light industry has been a traditional field of global economic integration. While it has undergone a major decline in the Visegrad countries (not the least thanks to its displacement by

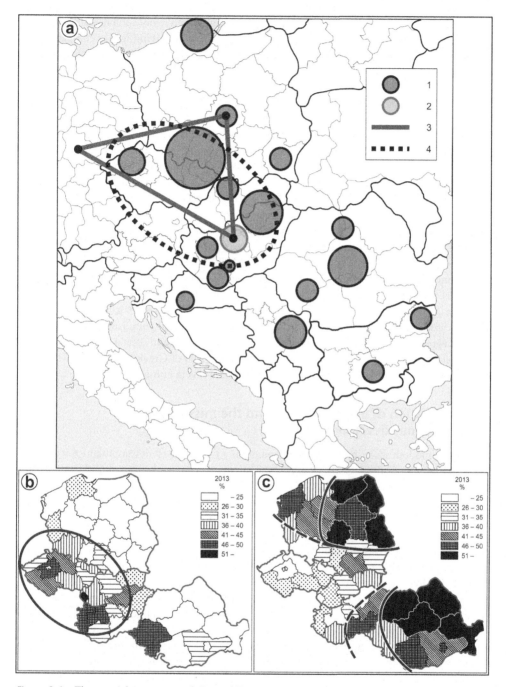

Figure 3.1 The spatial structures of Central European industry, 2013 (%)

Source: Author's construction; (b–c) based on data from Eurostat.
Legend: (a) Main spatial structures: 1 – Old Industrial Regions, 2 – central region with some OIR characteristics, 3 – traditional industrialisation core, 4 – new manufacturing core
 (b) The share of machine and electronics industries in industrial employment
 (c) The share of light and food industries in industrial employment.

more competitive branches), it has continued to be prominent in regions still showing significant cost advantages, particularly in South-Eastern Europe. Here, surviving textile companies have become suppliers to the western apparel industry or acted as intermediaries towards the post-Soviet macro-region (Kalantaridis, Slava and Sochka, 2003, Smith et al., 2005). However, the majority of these companies tend to occupy the lower segments of global value chains. While Yoruk (2004) predicted the gradual upgrading of textile industry companies from assembly towards own brand production and design, Evgeniev (2008) and Anić, Rajh and Teodorović (2008) have revealed that unlike the Turkish clothing industry, Bulgarian and Croatian textile producers have shown little evidence of successfully changing over to more valuable activities.

Industrial development in South-Eastern Europe shows patterns different from those in the Visegrad countries (and, increasingly, Romania). Reaching a volume of €10 billion in a decade, the role of investments coming from Greece and Turkey differed from those in their Western European, North American and Far Eastern counterparts. These investments have targeted low and medium-tech branches and labour-intensive activities which have shown close technological and cultural proximity to the ones in the host countries. Many Greek capital exporters (Kaditi mentions 3,600 active firms) are SMEs, who find it easier to integrate into their environment and build stronger networks than several western firms. However, they find it hard to protect their intellectual property, which leads to valuable knowledge spillovers (Totev 2005; Monastiriotis and Alegria, 2011; Kaditi, 2013). Greek capital exports gathered momentum from the turn of the millennium and peaked during the financial abundance period of 2005–2008. Regrettably, the Greek crisis had not only led to a decline in capital exports, but turned into capital repatriation by 2013. If we also consider the role of Greek banks in the South-East European banking system, we can anticipate even more severe consequences.

The localisation of manufacturing and the missing pillar of endogenous development

Taken in a modern sense, industrial development is a form of local development. According to the 'global–local paradox', global industrial competitiveness is deeply rooted in the competitive advantages of localities (Lengyel, 2010), and this localisation is intrinsically tied to evolutionary development paths. The mainstream development model of localised resource concentration – reflected in the concepts of clusters, regional innovation systems and industrial districts – is typically based on the idea of endogenous development. However, exogenous (FDI- or global value-chain-based) actors also tend to favour regions where they enjoy good location advantages, including skilled labour, the benefits of business networks and good institutions. Furthermore, instead of individual factors like cost-advantages or geographic proximity, industrial competitiveness is increasingly built on bundled 'packages' of multiple location advantages found in the local/regional context of companies, educational institutions, local governments and society. Identifying and exploiting local capabilities have become an important focus of regional development in order to gain territorially embedded competitive advantages which transcend low wages. This can mean building on pre-existing economic potential (such as the modernisation or conversion of existing industries) or mobilising previously unexploited resources (e.g. strengthening the knowledge transfer role of universities).

CEE has had to face major impediments in the localisation of competitiveness: whereas state socialism greatly devalued locality and disrupted socio-economic networks, post-socialism would often bring de-specialisation and hollowing-out, both acting against strong localisation. The 'homogenised peripheries' of post-socialism (Chapter 2) are caught in a development trap of low development level and weak growth potential. Although positive examples are starting to emerge, many FDI plants have shown weak territorial integration and local linkages, due either to the large technological gap or the limited integration potential of their manufacturing activity (i.e. low-level assembly). Meanwhile, domestic manufacturing companies lack the critical mass in size, resources and network density to make a difference. Although the author's empirical research has identified an emerging group of mid-sized companies that play an increasingly important role in local economic development and whose senior managers are rising actors in regional development coalitions (Lux, 2015), these developments cannot be compared meaningfully to the uninterrupted evolutionary paths of Western Europe's industrial milieus.

Rebuilding a full landscape of regionally differentiated competitive localities is a hard task, and while it has been much discussed in literature and enshrined in development strategies, the results have been very modest. The most successful examples of assembling competitive packages of location advantages have usually been based on path-dependent economic traditions, some from the pre-war period and some from the socialist legacy. Contrary to the mainstream ideas of the 1990s, which were very pessimistic about their development prospects (see e.g. Gorzelak, 1998), the positive examples include multiple OIRs which had been able to mobilise their latent technological know-how, human potential, firm networks and institutions to form new technological complexes. The most important ones among them are Upper Silesia and Moravian Silesia on the Polish–Czech border (Sucháček et al., 2012; Drobniak, Kolka and Skowronski, 2012; Gwosdz, 2014), but to some extent also Łódz in Poland (Páger, 2013), Central Transdanubia and Miskolc in Hungary, and Košice in Slovakia (Pástor, Šipikal and Rehák, 2013; Sokol, 2013). Unfortunately, the slow renewal of the latter two cities has had little effect on the depressed peripheral regions surrounding them. The development of these OIRs has been based on a mixture of old industries (machinery, coal mining and metallurgy in Silesia) and new activities (most markedly infocommunication technologies), as well as on the strong technological foundation provided by local technical universities and the fostering of local or regional business networks. There are other examples of renewal as well: traditions of the food industry have been rejuvenated in the rural regions of Eastern Hungary and Poland where domestic entrepreneurship has been more prominent, and industrial estates have followed the example of Italian industrial districts.

Smart specialisation (S3) policies, a concept rooted in the literature surrounding the transformation, have been embraced by the EU as a new policy paradigm in regional development shortly after their initial introduction (Foray, David and Hall, 2011; Foray, 2015; McCann and Ortega-Argilés, 2013). Although possessing limited theoretical novelty, the smart specialisation concept has positive implications for the development of CEE industry. Some of its progressive features which make it well-suited for the development of non-core regions include:

- the willingness to break with the universal 'cure-all' interpretation of a few branches (e.g. ICT, bio- and nanotech, tourism), and focus on local strength and resources;
- abandoning the sectoral neutrality of traditional competitiveness policies in favour of sector preference;
- an emphasis on the 'seeking' nature of regional development and the entrepreneurial discovery process, resulting in new combinations;
- the ability to absorb innovation into traditional, mature industries.

Smart specialisation, however, also comes with three risks, which must be resolved when deploying them as policies:

- Unless handled carefully, there is a *structural risk* of S3 reproducing previous overspecialisation or lock-in patterns, and increasing vulnerability to exogenous shocks.

- Hollowed-out regions may not in fact be able to come up with new, valuable specialisation patterns, and the benefits of S3 strategies risk *being captured* by external actors from core regions (c.f. Camagni and Capello, 2013).
- There is a particular *implementation risk* in CEE, associated with the prevailing centralised logic of development policy. S3 strategies can be – and as early experiences suggest, *have been* – subverted by top-down development agendas, effectively disrupting their exploratory nature, and subordinating the will of local actors to national policy goals formulated in central regions.

Weighing the risks and benefits, S3 seems to be the appropriate tool for a new generation of development strategies across the macro-region. However, their effectiveness depends as much on the reform of regional development institutions as on the capabilities of local actors.

New path creation, or recovery, from destructive de-industrialisation, is a much rarer phenomenon than building on existing legacies. Several failed efforts to build clusters or attempts to establish innovative industries on the periphery with EU funds point to the problems of inadequate critical mass, low network density and the pull effect of more successful regions. Even these peripheries have some highly competitive economic actors, but these examples often remain remote, individual success stories without strong spillover effects.

We undertook a complex survey on new regional development phenomena in our project, including questions about industrial location factors. Covering our study area evenly at national level, questionnaires were sent out to 540 actors involved in regional development.[5] The questions discussed in this chapter were answered by 90 to 101 respondents (n). Although the number of valid responses precludes country-level breakdowns, we can still get a picture of the state of competitiveness and development needs in the CEE macro-region. The results are given in Table 3.1.

Respondents were first asked for an evaluation of their region's location factors ('*Please evaluate your region within the Central and Eastern European context according to the following industrial location factors*') on a scale of 1 (very weak) to 5 (very strong). Then they were asked to select up to five of the location factors they thought most important to be improved and rank them according to their importance ('*In your opinion, in which factors does your region need most improvement to increase its competitiveness?*'). The first question produced few surprises; the average rating was 3.4, respondents tended to have a favourable impression of their region's geographic location and market access, while they were least satisfied with the availability of tax benefits and support schemes, high-quality business services, and the strength of partnerships and business support institutions. In the case of the second question, innovation-related needs were mentioned most often, followed by partnership and business support institutions, as well as tax benefits and support schemes. It is notable that the 'general' factors – mostly country-level location advantages (macroeconomic conditions, road accessibility, tax benefits and support schemes) – were ranked first, and 'locally embedded' location factors, which are localised and feature prominently in endogenous growth paths (innovation-related factors, skilled workforce, flexibility and adaptability, as well as partnership and business support), came in as second and third. While both groups of location advantages are important, the first one represents the current model of industrial competitveness in CEE more closely, whereas the second is becoming increasingly relevant in this decade, forming the foundation for locally differentiated upgrading paths. Regions and localities able to successfully develop their endogenous development capabilities can expect to emerge as the new winners of the post-transition era.

Table 3.1 Evaluation of industrial location factors in Central and Eastern Europe

Evaluation of current location factors			Improvement needs of location factors					
	Scale 1–5	n	Rank 1 n = 99	Rank 2 n = 97	Rank 3 n = 90	Rank 4 n = 78	Rank 5 n = 70	Average %
Macroeconomic conditions	3.2	100	14	4	7	5	6	8
Tax benefits and other support	3.0	99	16	10	3	11	3	10
Availability of high-quality business services	3.2	100	8	8	6	7	6	8
Road accessibility, logistic potential	3.5	99	15	11	8	4	1	9
Air accessibility, airport logistics	3.3	100	5	10	5	2	5	6
Geographic location, market access	4.0	101	1	0	4	1	4	2
Reasonable labour costs	3.5	99	0	7	3	5	5	5
Skilled workforce, industrial know-how	3.6	98	9	10	12	4	4	9
Flexibility, adaptability, reliability	3.3	99	10	8	11	6	6	9
Innovation background, R&D, higher education	3.4	100	12	20	12	14	2	14
Organisational background, partnership and business support institutions	3.2	99	8	7	12	13	11	12
Proximity of industrial districts, clusters, industrial estates	3.4	98	1	2	4	4	10	5
Other urban advantages	3.6	90	0	0	3	2	7	3

Source: Author's calculations and construction based on online survey.

The limited upgrading potential of the post-socialist development model

The global integration of manufacturing is a dynamic, qualitative process. The location of competitive activities is only second in importance to their upgrading potential and eventual sustainability. The factor intensity of production (Guerrieri, 1998; Soós, 2002) and the position of CEE firms and regions in the global division of labour are worth mentioning here. In the early years of transformation, the dominant form of integration was labour-intensive production, coupled with the weakening of specialisation profiles and decreased territorial embeddedness. However, by the 2000s, new investments and the upgrading of manufacturing activity had led to a more stable integration model based on scale-intensive, specialised suppliers and, to a limited extent, knowledge-intensive roles and R&D functions (Soós, 2002; Lefilleur, 2008; Jürgens and Krzywdzinski, 2009; Pavlínek, 2012). These manufacturing companies are specialised mainly in medium-tech and medium value-added segments, and employ medium-skilled labour.

Upgrading and the changing competitive environment exert pressure on the remainder of the macro-region's industry, as low value-added forms of production are getting 'priced out of the market' (Pavlínek and Ženka, 2010). These development patterns are evident in CEE's deepening integration into continental economic networks, partially surpassing a simple core–periphery model. There is evidence of upgrading processes in competitive industrial branches, resulting in increasing factor intensity. Simultaneously, growing supply networks have also increased the territorial embeddedness of FDI plants, although outside Poland's robust domestic supply sector most of these suppliers are themselves based on foreign capital.

The European economic crisis did not alter the FDI-based development trajectory fundamentally. Although export-driven industries were short-term losers of the initial shock, leading to swift corporate downscaling and massive redundancies, this was followed by a rebound of exports. Meanwhile, the contraction of domestic markets – except in Poland where they served as a potent shock absorber – had a lasting, but less significant, negative impact (Barta and Lőcsei, 2011; Lengyel, 2014). The post-crisis world saw a new wave of investments by manufacturing TNCs, leading to continuing internationalisation. As Lengyel demonstrates, foreign ownership in Hungarian manufacturing increased from 62 per cent of gross value added (GVA) in 2008 to 73 per cent in 2011. Partly in a bid to follow the successful German example, countries in CEE have undertaken industry-friendly structural reforms, from adopting Germany's dual vocational training model (Hungary and Romania) to attempts at encouraging knowledge transfer through innovation vouchers and the development of technological knowledge centres (Moravian Silesia in Czechia, Upper Silesia in Poland). The main trend in the space economy is slow-paced but steady reindustrialisation, even more than in the accession period (Figure 3.2).

While the FDI-based development model can be viewed as an overall success story, there are two important qualifiers that deserve to be mentioned. First, the territorial unevenness of

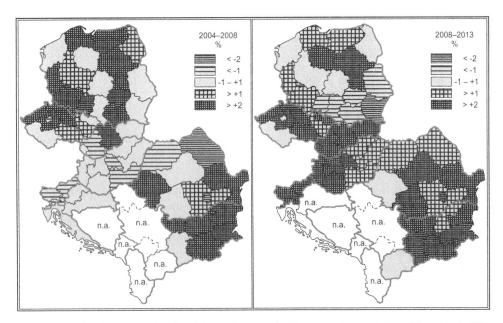

Figure 3.2 The dynamics of industrial employment in the accession and the crisis periods (%)

Source: Author's calculations based on data from Eurostat.

integration has been a factor in increasing regional differences: only a limited number of regions have benefited from high capital inflows and modernisation, while those missing out have often experienced destructive de-industrialisation, a loss of productive capacities without new industries or competitive services to replace them. There have been few comparable success stories outside the scope of exogenous, path-dependent development. At subnational level this poses a major problem and a significant impediment where regional development gaps should be reduced, as the investment behaviour of TNCs follows the formula of 'concentrated dispersal': global networks establish their offshoots in a very concentrated manner, favouring a handful of privileged sites. Substantial spatial dispersion is only found in the lower segments of global value chains, characterised by low-waged activities with diminishing returns (Potter and Watts, 2010) and very modest knowledge content. Iammarino and McCann (2010) describe the relationship between multinational firms and localised innovative development as the complexity of technical know-how acting as a 'filter' in selecting potential production sites. At the same time, the processes of knowledge creation are especially strongly localised, 'sticky' activities. The prospects of the peripheries to attract large-scale capital investments and valuable manufacturing activities are very slim indeed. The location of such activities is limited even in supply networks. R&D and innovation on the peripheries are at a low level, and although they are present in the form of 'on-site' process innovation, they are less formally tied to the R&D centres that are mainly located in the TNCs' home countries (Šipikal and Buček, 2013; Pavlínek, 2016; Pavlínek and Žižalová, 2014).

Second, the development process is strongly dependent on external capital, which is starting to pose problems in the development phase when cost-based competitive advantages are no longer sufficient and companies as well as regions need to explore 'high-road' strategies to maintain or improve their positions. Thus the success factors underpinning CEE's current industrial competitiveness can hinder its further upgrading, and in some way they are antithetical to endogenous development paths.[6] Wage pressure can lead to increasing risks of delocalisation and a new wave of company closures, which was already experienced during the post-2010 decline of the low value-added electronics industry.

Third, upgrading through endogenous development faces strong barriers. In addition to the weak local networks discussed in the previous section, even the basic conditions of this development path stand on weak foundations. Data from the World Bank's *Knowledge Assessment Methodology* (2012) show that the Knowledge Economy Indices of the EU15 countries had an average of 8.7 on a global scale of 0–10; while the V-4 and Slovenia had an average of 7.8, and South-Eastern Europe only 5.8. The components of the index – i.e. economic incentives and institutional regulation, the innovation system, education and human resources, infocommunication – showed the same deficiencies. Although no subnational breakdown is available, a plethora of evidence supports the assumption that the situation is significantly worse outside the metropolitan areas. There are weaknesses in the support institutions of endogenous development as well: the strong traditions of centralisation and weak local power limit both effective development planning and the emergence of strong development coalitions or urban regimes (Pálné Kovács, 2011, and Chapter 9). We can find only a few exceptional cases such as Upper Silesia in Poland, Győr in Hungary or the Banat in Romania.

Towards sustainable development paths in the post-transition period

Beyond the European crisis, and taking into account the lessons of global integration, industrial competitiveness should be rethought in the CEE context to be able to foster high-road industrial development, particularly for the benefit of non-metropolitan regions. FDI will continue to play a dominant role in shaping industrial production, but it cannot be relied on as the be-all and end-all source of competitiveness. The *virtuous path-dependence* of manufacturing has been a boon to a number of successful regions, but the benefits are very unevenly distributed. Even where it contributes to current regional development, regions cannot rely on exploiting the existing paths without risking a new structural crisis (which e.g. might affect the automotive industry in the 2020s). Specialisation – an agenda best served by smart specialisation strategies – must be counteracted by diversification, an increase in related and unrelated variety. There is a need to discover and exploit latent but unrecognised or untapped potentials, as well as to plant the seeds of new industries that may lead to new path creation. Domestic entrepreneurship, particularly medium-sized companies in supply networks or high value-added product niches, should also be supported. This might need to be supplemented by a limited number of large national champions that can integrate their own SME networks. Altogether, a wave of development policies based on innovative and sustainable activities is needed, with special focus on endogenous growth potential. This leads us to two potential approaches to reindustrialisation:

- *Direct reindustrialisation* aims to improve the local or regional business environment. Through the logic of resource concentration, it attempts to build industrial districts and regional clusters in order to channel and concentrate localisation advantages, and to encourage endogenous capital accumulation or external investments. The final aim is to *respecialise* the region, creating a production system able to generate sufficient spillovers for multiple enterprises and remain competitive in the global environment.
- *Indirect reindustrialisation* builds on the innovative development of the local factor supply, particularly skills, knowledge and learning ability, endeavouring to increase the general adaptation capability of society, institutions and economic actors. This philosophy is based on the concept of economic *diversification*, the continuous exploration of alternative growth paths, and on results in improved economic resilience.

These alternative philosophies are complementary and ideally should be pursued together, but they represent a hard choice for non-metropolitan cities and peripheral regions which are too small and have too few resources to maintain either diverse or specialised economic profiles like the ones found in metropolitan regions. It is possible but difficult to achieve a balance, especially because of domestic capital shortage. Effective institutional solutions are needed to manage development cooperation, either in the form of a relatively informal *development coalition* focusing on specific, narrowly defined development tasks, or in a formalised *neo-corporativist* model of interest articulation, following the Austrian or German model, in order to foster long-term restructuring processes.

Acknowledgements

This research has been supported by project #104985 of the Hungarian National Research, Development and Innovation Office. While writing this paper, Gábor Lux was supported by the János Bolyai Research Scholarship of the Hungarian Academy of Sciences.

Notes

1 By 1960, compared to 1938, the share of industries producing investment goods had increased from 51 to 59 per cent in Czechslovakia, from 47 to 57 per cent in Poland, from 38 to 66 per cent in Hungary, from 30 to 63 per cent in Romania, and from 24 to 47 per cent in Bulgaria (Markos, 1951; Enyedi, 1978).

2 These included electronics and chemicals in the German Democratic Republic; heavy machinery, household machinery and arms manufacturing in Czechoslovakia; coal mining and chemical industry in Poland; non-metallic machinery, public transport vehicles, pharmaceuticals and communications technologies in Hungary; petrol-based chemicals in Romania; and light industries, later electronics in Bulgaria.
3 In Czechia and Hungary, strict bankruptcy laws were responsible for a particularly harsh selection environment. Data by Barta (2002) show that the survival rate of privatised enterprises in Hungary was a mere 20–25 per cent, much less than could be explained by 'economic necessity'.
4 On the flipside, Romania's rating is not a sign of high value-added domestic manufacturing, but relates to low product complexity and significant raw material exports.
5 Of the 540 questionnaires, 154 were returned, with a response rate of 18 per cent; consisting of 98 full and 86 partial answers: 33 per cent of the responses came from universities, 31 per cent from development organisations, 16 per cent from statistical agencies and the rest from city governments and other actors. Some 17 per cent of the responses came from capital regions, 83 per cent from outside them; 30 per cent of the respondents were from Poland, 19 per cent from Czechia, 15 per cent from Hungary, 10 per cent from Bulgaria, and below 10 per cent from other countries.
6 As Humphrey and Schmitz (2002) have shown, this has happened to numerous Western European assembly sites, too.

References

Acemoglu, D. and Autor, D., 2010. Skills, Tasks and Technologies: Implications for Employment and Earnings. Working Paper.

Anić, I-D., Rajh, E. and Teodorović, I., 2008. Full Manufacturing versus Subcontracting Business Models in the Croatian Textile and Clothing Industry. *Economic Review/Ekonomski Pregled*, 59(7–8), pp. 325–343.

Asheim, B., 1996. Industrial Districts as 'Learning Regions': A Condition for Prosperity. *European Planning Studies*, 4(4), pp. 379–400.

Barta, Gy., 2002. *A magyar ipar területi folyamatai 1945–2000* [The regional development processes of Hungarian industry, 1945–2000]. Budapest, Pécs: Dialóg Campus Kiadó.

Barta, Gy., 2005. The Role of Foreign Direct Investment in the Spatial Restructuring of Hungarian Industry. In: Gy. Barta, É.G. Fekete, I. Kukorelli Szörényiné and J. Timár, eds. *Hungarian Spaces and Places: Patterns of Transition*. Pécs: Centre for Regional Studies, pp. 143–160.

Barta, Gy. and Lőcsei, H., 2011. The Effects of the Recent Economic Crisis on the Spatial Structure of Hungarian Industry. *Regional Statistics*, 51(1), pp. 99–109.

Barta, Gy., Czirfusz, M. and Kukely, Gy., 2008. Újraiparosodás a nagyvilágban és Magyarországon [Reindustrialisation in the world and in Hungary]. *Tér és Társadalom*, 22(4), pp. 1–20.

Bartke, I., 1971. *Az iparilag elmaradott területek ipari fejlesztésének főbb közgazdasági kérdései Magyarországon* [Selected economic questions of industrial development in Hungary's industrially backwards regions]. Budapest: Akadémiai Kiadó.

Belussi, F. and Sedita, S.R., 2009. Life Cycle vs. Multiple Path Dependency in Industrial Districts. *European Planning Studies*, 17(4), pp. 505–528.

Boschma, R. and Lambooy, J., 1999. The Prospects of an Adjustment Policy Based on Collective Learning in Old Industrial Regions. *Geojournal*, 49(4), pp. 391–399.

Boschma, R. and Frenken, K., 2009. Some Notes on Institutions in Evolutionary Economic Geography. *Economic Geography*, 85(2), pp. 151–158.

Boschma, R. and Martin, R., 2010. The Aims and Scope of Evolutionary Economic Geography. In: R. Boschma and R. Martin, eds. *The Handbook of Evolutionary Economic Geography*. Cheltenham: Edward Elgar, pp. 3–39.

Camagni, R. and Capello, R., 2013. Regional Innovation Patterns and the EU Regional Policy Reform: Toward Smart Innovation Policies. *Growth and Change*, 44(2), pp. 355–389.

Competing in Global Value Chains: EU Industrial Structure Report 2013 (2013). European Commission Directorate-General for Enterprise and Industry, Luxembourg.

David, P.A., 2007. Path Dependence: A Foundational Concept of Historical Social Science. *Cliometrica*, 1(2), pp. 91–114.

Domański, B., 2003. Industrial Change and Foreign Direct Investment in the Postsocialist Economy. The Case of Poland. *European Urban and Regional Studies*, 10(2), pp. 99–118.

Drobniak, A., Kolka, M. and Skowroński, M., 2012. Transition and Urban Economic Resilience in Poland's Post-industrial Cities: The Case of Katowice. *Regions*, (286), pp. 13–15.

Enyedi, Gy., 1978. *Kelet-Közép-Európa gazdaságföldrajza* [The economic geography of East Central Europe]. Budapest: Közgazdasági és Jogi Könyvkiadó.

Evgeniev, E., 2008. *Industrial and Firm Upgrading in the European Periphery. The Textile and Clothing Industry in Turkey and Bulgaria*. Sofia: Professor Marin Drinov Academic Publishing House.

Foray, D., 2015. *Smart Specialisation: Opportunities and Challenges for Regional Innovation Policy*. Abingdon: Routledge.

Foray, D., David, P.A. and Hall, B.H., 2011, Smart Specialization: From Academic Idea to Political Instrument, the Surprising Career of a Concept and the Difficulties Involved in Its Implementation. *MTEI Working Papers* 1. École Polytechnique Fédérale de Lausanne, Lausanne.

Freeman, R.B., 2008. The New Global Labor Market. *Focus*, 26(1), pp. 1–6.

Frigant, V. and Layan, J-B., 2009. Modular Production and the New Division of Labour within Europe. The Perspective of French Automotive Parts Suppliers. *European Urban and Regional Studies*, 16(1), pp. 11–25.

Gorzelak, G., 1998. Regional Development and Planning in East Central Europe. In: M. Keune, ed. *Regional Development and Employment Policy: Lessons from Central and Eastern Europe*. Budapest: International Labour Organization, pp. 62–76. www.ilo.org/public/english/region/eurpro/budapest/publ/_book/regdev_toc.htm (2007. III. 21.)

Grabher, G., 1993. The Weakness of Strong Ties. The Lock-in of Regional Development in the Ruhr Area. In: G. Grabher, ed. *The Embedded Firm. On the Socioeconomics of Industrial Networks*. London: Routledge, pp. 255–277.

Grabher, G. and Stark, D., 1997. Organizing Diversity: Evolutionary Theory, Network Analysis, and Post-Socialism. In: G. Grabher and D. Stark, eds. *Restructuring Networks in Post-socialism. Legacies, Linkages and Localities*. Oxford: Oxford University Press, pp. 1–32.

Guerrieri, P., 1998. Trade Patterns, Foreign Direct Investment, and Industrial Restructuring of Central and Eastern Europe. In: J. Zysman and A. Schwartz, eds. *Enlarging Europe: The Industrial Foundations of a New Political Reality*. Berkeley: University of California, pp. 130–156.

Gwosdz, K., 2014. *Pomiędzy starą a nową ścieżką rozwojową. Mechanizmy ewolucji struktury gospodarczej i przestrzennej regionu tradycyjnego przemysłu na przykładzie konurbacji katowickiej po 1989 roku*. Kraków: Uniwersytet Jagielloński w Krakowie, Instytut Geografii i Gospodarki Przestrzennej.

Havlik, P., 2005. Central and East European Industry in an Enlarged European Union: Restructuring, Specialisation and Catching-up. *Économie Internationale*, 102(4), pp. 107–132.

Humphrey, J. and Schmitz, H., 2002. How Does Insertion in Global Value Chains Affect Upgrading in Industrial Clusters? *Regional Studies*, 36(9), pp. 1017–1027.

Iammarino, S. and McCann, P., 2010. The Relationship between Multinational Firms and Innovative Clusters. In: R. Boschma and R. Martin, eds. *The Handbook of Evolutionary Economic Geography*. Cheltenham: Edward Elgar, pp. 182–204.

Illés, I., 1994. *A kelet-közép-európai országok 1990–1993. évi gazdasági fejlődésének összehasonlítása* [Comparing the economic development of East Central European countries in 1990–1993]. Pécs: MTA Regionális Kutatások Központja. Mimeo.

Jánossy, F., 1971. *The End of the Economic Miracle: Appearance and Reality in Economic Development*. White Plains, NY: International Arts and Sciences Press.

Jürgens, U. and Krzywdzinski, M., 2009. Changing East–West Division of Labour in the European Automotive Industry. *European Urban and Regional Studies*, 16(1), pp. 27–42.

Kaditi, E.A., 2013. Foreign Investments and Institutional Convergence in South-Eastern Europe. *International Economic Journal*, 27(1), pp. 109–126.

Kalantaridis, C., Slava, S. and Sochka, K., 2003. Globalization Processes in the Clothing Industry of Transcarpathia, Western Ukraine. *Regional Studies*, 37(2), pp. 173–186.

Kiss, E., 2007a. Foreign Direct Investment in Hungary. Industry and its Spatial Effects. *Eastern European Economics*, 45(1), pp. 6–28.

Kiss, E., 2007b. The Impacts of Relocation on the Spatial Pattern of Hungarian Industry. *Geographica Polonica*, 80(1), pp. 43–61.

Kornai, J., 1980. *Economics of Shortage*. Amsterdam: North-Holland.

Kornai, J., 1992. *The Socialist System. The Political Economy of Communism*. Oxford: Oxford University Press.

Krzywdzinski, M., 2014. How the EU's Eastern Enlargement Changed the German Productive Model. The Case of the Automotive Industry. *Révue de la Régulation*, 15(1), pp. 1–20.

Lefilleur, J., 2008. Geographic Reorganisation of the European Automobile Sector. What Role for the Central and East European Countries in an Enlarged European Union? An Empirical Approach. *Eastern European Economics,* 46(5), pp. 69–91.

Lengyel, I., 2010. *Regionális gazdaságfejlesztés. Versenyképesség, klaszterek és alulról szerveződő stratégiák* [Regional economic development: Competitveness, clusters and bottom-up strategies]. Budapest: Akadémiai Kiadó.

Lengyel, I., 2014. Reorganizing of Hungarian Manufacturing Sector: Impacts of EU Accession and Global Crises. *European Journal of Business Research,* 14(2), pp. 93–100.

Lijewski, T., 1985. The Spread of Industry as a Consequence of the Location of New Factories in Poland, 1945–1982. *Geographia Polonica,* 51, pp. 199–206.

Lux, G., 2009. Divergent Patterns of Adaptation among Central European Old Industrial Regions. *European Spatial Research and Policy,* 16(1), pp. 145–157.

Lux, G., 2015. Minor Cities in a Metropolitan World: Challenges for Development and Governance in Three Hungarian Urban Agglomerations. *International Planning Studies,* 20(1–2), pp. 21–38.

McCann, P. and Ortega-Argilés, R., 2013. Transforming European Regional Policy: A Result-driven Agenda and Smart Specialization. *Oxford Review of Economic Policy,* 29(2), pp. 405–431.

MacKinnon, D., Cumbers, A., Pike, A., Birch, K. and McMaster, R., 2009. Evolution in Economic Geography: Institutions, Political Economy, and Adaptation. *Economic Geography,* 85(2), pp. 129–150.

Markos, Gy., 1951. *A népi demokratikus országok gazdasági földrajza* [The economic geography of the People's Democratic Republics]. Budapest: Közoktatásügyi Kiadóvállalat.

Markusen, A., 1996. Sticky Places in Slippery Space: A Typology of Industrial Districts. *Economic Geography,* 27(3), pp. 293–313.

Martin, R., 2000. Institutional Approaches in Economic Geography. In: E. Sheppard and T.J. Barnes, eds. *A Companion to Economic Geography.* Oxford: Blackwell, pp. 77–94.

Martin, R., 2012. Regional Economic Resilience, Hysteresis and Recessionary Shocks. *Journal of Economic Geography,* 12(1), pp. 1–32.

Martin, R. and Sunley, P., 2010. The Place of Path Dependence in an Evolutionary Perspective on the Economic Landscape. In: R. Boschma and R. Martin, eds. *The Handbook of Evolutionary Economic Geography.* Cheltenham: Edward Elgar, pp. 62–92.

Martin, R. and Sunley, P., 2011. Conceptualizing Cluster Evolution: Beyond the Life Cycle Model? *Regional Studies,* 45(10), pp. 1299–1318.

Menzel, M.P. and Fornahl, D. 2009. Cluster Life Cycles – Dimensions and Rationales of Cluster Evolution. *Industrial and Corporate Change,* 19(1), pp. 205–238.

Molnár, E., 2012. Kelet-Közép-Európa az autóipar nemzetközi munkamegosztásában [East Central Europe in the international division of labour of the automotive industry]. *Tér és Társadalom,* 26(1). pp. 123–137.

Monastiriotis, V. and Alegria, R., 2011. Origin of FDI and Intra-industry Domestic Spillovers: The Case of Greek and European FDI in Bulgaria. *Review of Development Economics,* 15(2), pp. 326–339.

Nelson, R.R. and Winter, S.G., 1982. *An Evolutionary Theory of Economic Growth.* Cambridge, MA: Harvard University Press.

Nölke, A. and Vliegenthart, A., 2009. Enlarging the Varieties of Capitalism: The Emergence of Dependent Market Economies in East Central Europe. *World Politics,* 61(4). pp. 670–702.

Páger, B., 2013. Ismét az ígéret földje? Łódz gazdasági átalakulása a rendszerváltást követően [Does 'the Promised Land' return? The economic transformation of Łódz in the post-socialist period]. In: J. Rechnitzer, E. Somlyódyné Pfeil and G. Kovács, eds. *A hely szelleme – A területi fejlesztések lokális dimenziói.* Győr: Széchenyi István Egyetem Regionális- és Gazdaságtudományi Doktori Iskola, pp. 447–458.

Pálné Kovács, I., 2011. *Local Governance in Hungary: The Balance of the Last 20 Years.* Pécs: Centre for Regional Studies of Hungarian Academy of Sciences. (Discussion Papers, 83).

Pástor, R., Šipikal, M. and Rehák, Š., 2013. Knowledge Creation and Knowledge Acquisition in the Software Industry in Slovakia: The Case Study of Košice Region. *Regional Science Policy &Practice,* 5(4), pp. 401–415.

Pavlínek, P., 2012. The Internationalization of Corporate R&D and the Automotive Industry R&D of East-Central Europe. *Economic Geography,* 88(3), pp. 279–310.

Pavlínek, P., 2016. Whose Success? The State–Foreign Capital Nexus and the Development of the Automotive Industry in Slovakia. *European Urban and Regional Studies,* 23(4), pp. 571–593.

Pavlínek, P. and Smith, A., 1998. Internationalization and Embeddedness in East-Central European Transition: The Contrasting Geographies of Inward Investment in the Czech and Slovak Republics. *Regional Studies,* 32(7), pp. 619–638.

Pavlínek, P. and Ženka, J. 2010. The 2008–2009 Automotive Industry Crisis and Regional Unemployment in Central Europe. *Cambridge Journal of Regions, Economy and Society*, 3(3), pp. 349–365.

Pavlínek, P. and Žižalová, P., 2014. Linkages and Spillovers in Global Production Networks: Firm-level Analysis of the Czech Automotive Industry. *Journal of Economic Geography*, 16(2), pp. 331–363.

Pavlínek, P., Domański, B. and Guzik, R., 2009. Industrial Upgrading through Foreign Direct Investment in Central European Automotive Manufacturing. *European Urban and Regional Studies*, 16(1), pp. 43–63.

Potter, A. and Watts, D., 2010. Evolutionary Agglomeration Theory: Increasing Returns, Diminishing Returns, and the Industry Life Cycle. *Journal of Economic Geography*, 11(3), pp. 417–455.

Rugraff, E., 2008. Are the FDI Policies of the Central European Countries Efficient? *Post-Communist Economies*, 20(3), pp. 303–316.

Šipikal, M. and Buček, M., 2013. The Role of FDIs in Regional Innovation: Evidence from the Automotive Industry in Western Slovakia. *Regional Science Policy & Practice*, 5(4), pp. 475–491.

Smith, A. and Ferenčíková, S., 1998. Inward Investment, Regional Transformations and Uneven Development in Eastern and Central Europe: Enterprise Case-studies from Slovakia. *European Urban and Regional Studies*, 5(2), pp. 155–173.

Smith, A., Pickles, J., Begg, R., Roukova, P. and Bucek, M., 2005. Outward Processing, EU Enlargement and Regional Relocation in the European Textiles and Clothing Industry: Reflections on the European Commission's Communication 'On the Future of the Textiles and Clothing Sector in the Enlarged European Union'. *European Urban and Regional Studies*, 12(1), pp. 83–91.

Sokol, M., 2013. Silicon Valley in Eastern Slovakia? Neoliberalism, Post-socialism and the Knowledge Economy. *Europe-Asia Studies*, 65(7), pp. 1324–1343.

Soós, K.A., 2002. Az átmeneti gazdaságok EU-exportja nemzetközi összehasonlításban, 1993–2000 [The EU export of the transition economies in international comparison, 1993–2000]. *Közgazdasági Szemle*, (49)12, pp. 1063–1080.

Steiner, M., 2003. Modernizing Traditional Industries in Declining Regions – Concepts of Transformation in Old and New Market Economies. In: S. Michael, ed. *From Old Industries to New Regions. Policies for Structural Transformations in Accession Countries*. Graz: Leykam Buchverlagsgesselschaft, pp. 9–24.

Sucháček, J., Krpcová, M., Stachoňová, M., Holešinská, L. and Adamovský, J., 2012. Transition and Resilience in Czech Post-industrial Towns: The Case of Ostrava and Karviná. *Regions*, (286), pp. 17–19.

Totev, S., 2005. Foreign Direct Investment in Bulgaria: Advantages and Disadvantages to Investment. *South-East Europe Review*, (4), pp. 91–104.

Turnock, D., 1986. *The Romanian Economy in the Twentieth Century*. London: Croom Helm.

Turnock, D., 2001. Location Trends for Foreign Direct Investment in East Central Europe. *Environment and Planning C: Government and Policy*, 19(6), pp. 849–880.

Valev, E.B., 1964. Problemi economichescova razvitiya pridunayskih raionov Ruminii, Bolgarii i SSSR. *Vestnik Moskovskova Universiteta*. Seriya V/2, pp. 56–69.

Winiecki, J., 1986. Az ipar túlméretezettsége a kelet-európai szocialista országokban: tények, okok, következmények [Oversized industries in the East European socialist countries: facts, causes, consequences]. *Közgazdasági Szemle*, 33(5), pp. 579–592.

Wójtowicz, M. and Rachwał, T., 2014. Globalization and New Centers of Automotive Manufacturing: The Case of Brazil, Mexico, and Central Europe. *Prace Komisji Geografii Przemysłu*, 25, pp. 81–107.

World Bank, 2012. *Knowledge Assessment Methodology*. http://web.worldbank.org

Worrall, D., Donnelly, T. and Morris, D., 2003. Industrial Restructuring: The Role of FDI, Joint Ventures, Acquisitions and Technology Transfer in Central Europe's Automotive Industry. Reinventing Regions in a Global Economy. RSA Conference, Pisa. www.regional-studies-assoc.ac.uk/events/pisa03/worrall.pdf (2007. III. 28.)

Yoruk, D.E., 2004. Patterns of Industrial Upgrading in the Clothing Industry in Poland and Romania. F. McGowan, Slavo, Radosevic and von N. Tunzelman, eds. *The Emerging Industrial Structure of the Wider Europe*. London: Routledge, pp. 95–110.

Zeitlin, J., 2008. Industrial Districts and Regional Clusters. In: G. Jones and J. Zeitlin, eds. *The Oxford Handbook of Business History*. Oxford: Oxford University Press, pp. 219–243.

4
The role of business and finance services in Central and Eastern Europe

Zoltán Gál and Sándor Zsolt Kovács

Introduction

The development of the service sector is considered to be a demand-driven phenomenon, which was to a large extent responsible for productivity growth and rising incomes in the period between 1970 and 2000. This sector became the most important economic sector in terms of share of employment and gross domestic product (GDP) in developed market economies. Business and financial services have gone through the fastest development in the most advanced countries (UN, 2001). During the 1990s the accelerated growth of the tertiary sector was faster than average growth in real terms, and it nowadays accounts for about 60 per cent of employment (Begg, 1993; Stare, 2007). The service sector also continues to capture an increasing share of global Foreign Direct Investment, usually above half. The structural shift towards these services started later in Central and Eastern Europe.

This study focuses on the economic role of business and finance services in the Central and Eastern European economies. The chapter examines the structural and systemic change in this sector during and after economic transition. In this period, these economies became open market economies; consequently, FDI played a key role in their economic transformation. Therefore, we also focus on the special role of FDI in these countries.

The role and development of the tertiary sector in the focus area

European economies are in a process of structural change and two major trends have characterised the period of the past two to three decades. The first is de-industrialisation and the shrinking of agricultural activities, while the other, in parallel, is the significant strengthening of the tertiary sector, with the strong service sector becoming the incubator of business innovation in this region in the last decades (Kox and Rubalcaba, 2007). According to Soubbotina and Sheram (2000, p. 52) these changes were generated by the development of social needs and values. Another benefit of the growing service sector is that by using fewer natural resources than agriculture or industry, it puts less pressure on the local, regional and global environment, hence the importance of the tertiary sector in sustainable economic development. Along with these processes, the role of these activities in spatial development has become incontestable, with the tertiary sector becoming a determining factor in complicated and complex production and distribution systems and a part of manufacturing activities (Raffay, 2004). It is an organiser of spatial processes, because as a general feature not all services are spread over all territorial units: services are concentrated in centres and central settlements thanks to economies of scale and the optimal locating of service activities (Beluszky and Győri, 2004; Kovács, 2014; and Chapter 2). This trend holds in the focus area as well. In the years of systemic change, the performance of capital cities was the best in these countries: well-educated workforces and higher incomes were accompanied by the growth of the service sector (Maurel, 2006).

The examined Central and Eastern European countries are characterised by more significant agricultural and industrial sectors in the national economies – as inherited from the socialist period – than Western European countries in the early years of the 1990s (Havlik, 2005). Furthermore, their lower urbanisation level and lack of strong metropolitan areas outside the national capitals have affected the development opportunities of tertiary activities (Chapter 8). Structural change of the economy followed the political changes. However, the growth of the service sector only partially compensated for the decrease of agricultural and industrial value added and employment. There was a visible effective increase of services between 1995 and 2003. As a result, the economic structure of CEE countries became similar to those of Western European countries; however, substantial differences are present in service activities, particularly in the field of financial and other business activities (marketing, consulting and accounting). The development of commerce, tourism and real estate services was similar to that of modern economies (Landesmann, 2000). Demand for knowledge-intensive business services (KIBS) was very limited in the socialist era, but increased rapidly after 1989, and the GVA-share of KIBS now measures more than 30 per cent in these countries (Balaž, 2004).

The transition to the market economy helped the growth of the services sector through the increased amount of Foreign Direct Investment. In the 1990s the largest part of this FDI was committed to the service sector (Table 4.1). In the examined countries – with the exception of Poland – the share of the service sector is the largest in the sectoral distribution of FDI (Vidović, 2002).

The sectoral distribution of GDP shows the strengthening of the service activities in the national economies after 1990. Based on data from 2013, the share of the tertiary sector in the national GDP was higher than 50 per cent; and in Romania higher than 60 per cent. At the same time, a significant decrease of the agricultural GDP-share was experienced.

Table 4.1 The distribution of Foreign Direct Investment (FDI) by economic activity, 1990–2000 (%)

	Agriculture, fishing and forestry	Industry and construction	Services and other non-classified activities
Bulgaria	0.20	49.50	50.30
Czechia	0.20	44.90	55.80
Croatia	0.00	41.25	58.75
Hungary	1.10	47.80	51.10
Poland	0.10	50.30	49.60
Romania	3.60	50.70	45.70
Slovakia	0.10	49.80	50.10
Slovenia	0.00	41.50	58.50
Serbia*	0.74	23.40	75.86

Source: Authors' calculations based on Vidović, 2002; Jovančević, 2007; and Stefanović et al., 2009.

Note: * Data from 2004.

The role of business and finance services

The sectoral distribution of gross value added (GVA) changed in a way similar to GDP in this period (Havlik, 2005); however, the dynamic of GVA growth is more interesting. According to the regional statistics and publications of Eurostat (and the statistical office of Serbia) we can say that GVA growth in the service sector was higher than in Western European regions in the last decades, with the exception of some statistical regions: Continental and Adriatic Croatia, the Czech North-East, regions of Slovakia, and Vojvodina. However, only the core regions reach the European average of service activities' GVA and some traditional agricultural (Vojvodina, Romanian North-West and Center) and industrial (Romanian West, Central Transdanubia, Central Moravia) regions have less than 75 per cent of the European average of service sector GVA.

The high sectoral GDP and GVA levels were followed by the restructuring of employment structure. In almost all regions the service sector is the largest employer (Figure 4.1), but as we have seen in Chapter 2, important qualifiers apply. The lowest share of services in employment

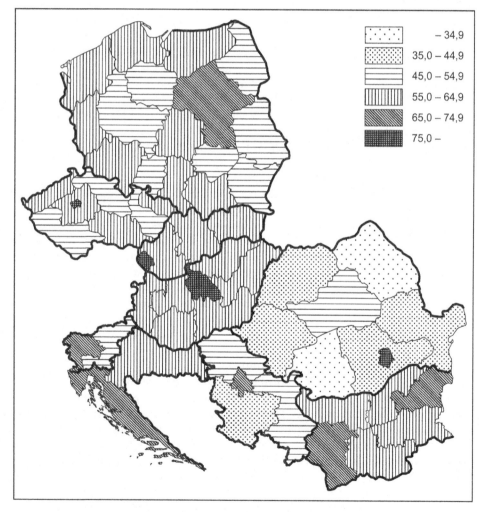

Figure 4.1 Share of service activities in employment, 2013 (%)

Source: Authors' construction based on data from national statistical yearbooks and Eurostat.

is observed in some Serbian and Romanian regions, while this indicator is more than 65 per cent in the capital cities' regions and more than 75 per cent in some well-developed NUTS-2 (Nomenclature of Territorial Units for Statistics) regions (Horváth, 2015). Based on a Globalisation and World Cities (GaWC) Research Network survey, Beaverstock, Taylor and Smith (1999) show in their study the important role of Warsaw, Prague and Budapest in regional service networks (Csomós, 2011). The high employment rate is the result of several factors: (1) the majority of enterprises operate in the service sector; (2) these provide higher incomes due to higher human resource investment (Soubbotina and Sheram, 2000); (3) these enterprises are more attractive for the participants of the labour market. The dispersion of income possibilities is different between service activities: in the examined 65 regions the highest incomes are in financial and consulting services and the lowest are in the public services and administration.

Business and financial services offshoring in Central and Eastern Europe

The relocation of service offshoring-related activities, such as outputs, value added, employment, Foreign Direct Investments and exports in services, has grown rapidly, particularly after 2000, and especially in the Visegrad countries – though 'latecomers' such as Romania and Bulgaria have also begun to act as hosts for this type of investment (Gál and Sass, 2009).

The relocation of services also conceptualises the types and impacts of Foreign Direct Investments in business services within the Global Production Network (GPN) (Fernandez-Stark et al., 2011). A bulk of research examines offshoring as a part of the worldwide structural shift toward service-based FDI (Bryson, 2007; Grote and Täube, 2006; Bevan and Estrin, 2004; Hardy, 2007). Due to problems with collecting data on business service investment, statistics are supplemented with qualitative research in recent studies (Hardy, 2006; Capik, 2008; Fifekova and Hardy, 2010; Sass and Fifekova, 2011).

Global service enterprises appeared in the Central and Eastern European countries and established their business services centres and shared service centres in the last 10–15 years. These centres are highly export-oriented (their export intensity approaches 100 per cent). Accordingly, trade data provides the relatively most relevant proxy for calculating the extent of offshoring and the outsourcing of these services. Investors came from various countries: 41 per cent of total business centres were established by US companies, 45 per cent by Western European, 3 per cent by Indian and 11 per cent by other investors. Service fields vary as well, from IT services (33 per cent) to supply chain management (3 per cent). Some 40 per cent of these business service centres provide their services globally, 52 per cent to selected regions and 8 per cent only to one county (ABSL, 2015). CEE countries are very popular among global service enterprises because of recent graduates with foreign language knowledge and wages and salaries that are lower than in Western Europe (Marciniak, 2014).

In 2003, only 5 per cent of service-related global FDI projects were realised in the Visegrad countries, while in 2006, 22 per cent of related FDI projects went to regions in these countries. However, the number of current projects in Western Europe continues to exceed CEE projects – which numbered 1,600 and 220 respectively by 2009 (Sass, 2008; Gál and Sass, 2009).

The scale and sectoral characteristics of services offshoring in NMS-6 (New Member States: Bulgaria, Croatia, Czechia, Hungary, Poland and Romania) can be examined in measuring the actual significance of NMS in offshoring services through using trade data. Following the international methodology (UNCTAD, 2005; Amiti and Wei, 2005; Ghibutiu and Dumitriu, 2008), Gál (2014) has measured the approximate size of trade in offshorable services. Information and computer technology (ICT) services and other business services (OBS) are the most inclusive categories that can be regarded as potentially offshorable services. Exports in services in the

The role of business and finance services

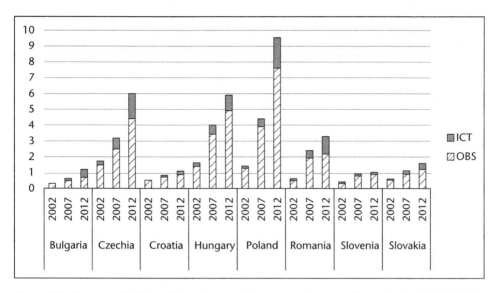

Figure 4.2 Exports of offshorable services and the sectoral composition of such, 2002, 2007 and 2012 (€bn)

Source: Authors' calculations based on Eurostat BoP data and Gál, 2014.

Notes: ICT – Information and Communication Technology Services
OBS – Other Business Services.

NMS-6 have been expanding from a very low base, amounting to €63 billion by 2007 and €82 billion by 2012, almost four times higher than in 1996. The share of the NMS-6 in the global service exports is still modest (2.4 per cent in 2012). In absolute terms, Poland, Czechia and Hungary are the leaders in this field.

The share of offshorable services (other business and ICT services) within total service exports steadily grew from 16 per cent to 33 per cent between 1997 and 2012. The total value of offshorable services in the NMS-6 was equal to €27.2 billion in 2012. Within this aggregate, the overwhelming dominance of business services (78 per cent on average) is striking. In absolute terms, Poland is the largest trader, followed by Hungary and Czechia (Figure 4.2).

The analysed services trade data support the preliminary assumption that offshoring generated expanding exports in particular service categories, and further, that a large proportion of business export services in the NMS has been associated with offshoring.

The following paragraphs examine the implications of offshoring for the home locations. Besides the general home market effect, the process of selecting and opening new locations is similarly important as offshoring has a strong impact on the cities selected. Location strategy-making is a multifaceted process, with different indicators coming into play as the focus is narrowed from macro-regions to countries, cities, districts and finally the individual property level. As far as the geographical distribution of the investments in business services is concerned, there is a strong spatial concentration in all the analysed countries. Capital cities are the main hosts to business services companies, and some secondary centres also emerged (Kraków, Brno). This is understandable: demand-led horizontal investments locate in the capital city where there is the highest demand for their services. For supply-driven vertical investments, the main attracting factor in capital cities is the large supply of suitable workers (Hardy et al., 2011; Gál, 2014).

The metropolitan transformation accompanied by both rapid de-industrialisation and the expansion of services has resulted in the concentration of high-level business and financial services in capital cities (Lux, 2010; Chapter 2). Capital cities are the main hosts to business service companies, but there is a new trend whereby global enterprises establish their European service centres in the secondary towns. As such, we are seeing a number of centres in Kraków, Brno, Poznań, Košice, Debrecen, Cluj-Napoca and so on (Gál, 2014). EU accession, competitive infrastructure costs and strong education systems are favourable preconditions that supported the first group of capital cities, such as Prague, Budapest and Warsaw, in the first wave of the offshoring boom. They have recently been followed by Bucharest and Sofia due to the saturation of the first wave of capitals. At the beginning, costs (labour cost, real estate prices and facilities) were the most important driver in selecting these locations for mostly routine offshoring activities. Despite growing wages, cost differentials with Western Europe are still significant, making these cities attractive still for higher value-added nearshore service activities. The first wave of cities in the offshoring boom, including the first-tier capital cities such as Warsaw, Prague and Budapest, are becoming saturated in terms of skilled labour and office space supply, to the point that raises the opportunity for the second-tier cities to extend offers in alternative locations. The ability to tap unexploited graduate pools has become an increasingly important operational issue, particularly when we consider the migration of skilled labour into EU core regions (Chapter 13). Vertical investments have more propensity to locate in the countryside, especially if suitable workers are not found in capital cities. Numerous cities, particularly in Poland, Czechia and, to a lesser extent, Hungary and Romania, are emerging as new destinations for outsourcing (see Figure 4.3).

While Poland has developed the most extensive network of offshoring locations in secondary cities (Wrocław, Kraków, Łódź, Poznań, Katowice and Gdańsk), spatial concentration is more pronounced in Hungary and Slovakia where the capital cities are the main locations of this type of investment. More vendors maintain operations in both the capital city and secondary cities in Hungary. Budapest is leveraged for higher value work, while lower level processing is accomplished in secondary locations that offer much lower costs and significantly lower attrition sites. Cities with the right combination of location factors will be the winners in the future waves of services investment into CEE. The challenge for individual cities will be to build their attractiveness and competitiveness by investing in their ICT infrastructure, education systems and business environment.

After the overall discussion of the Central and Eastern European service sector, the next section will examine one of the most important service fields, the general processes, geographical aspects and country-specific features of financial services.

The role of FDI in the financial services sector

Global financial capital has played an important role in all transition economies. Foreign Direct Investment in the banking sector and business services is closely connected to the transition process in CEE, and has received considerable attention from both a theoretical and an empirical perspective (Bonin et al., 1998; Wachtel, 1997a, 1997b; Claessens et al., 2001; Buch and Golder, 2001; Buch et al., 2003; Berger et al., 2003; Gál, 2004; Várhegyi, 2002; Banai et al., 2010; Csaba, 2011). Much less attention has been devoted to the post-transition period and the impact of the crisis, which has become the most serious challenge to transition models in national banking sectors (Gál, 2013).

The FDI development path in the Central and Eastern European Countries (CEECs) followed the pattern of the dependent market economy (DME) and the (semi)-peripheral type of capitalism (e.g. Raviv, 2008; Nölke and Vliegenthart, 2009; Myant and Drahokoupil, 2011; or Sokol,

The role of business and finance services

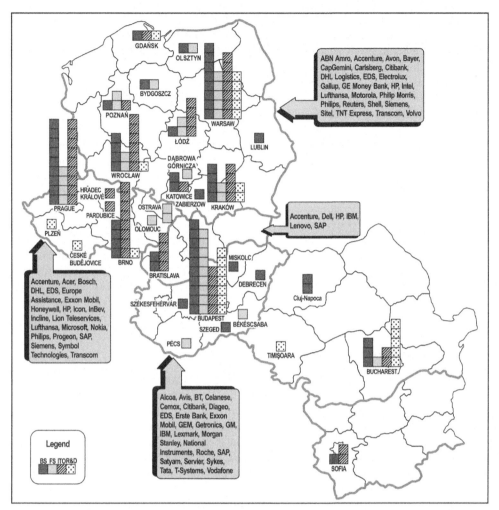

Figure 4.3 Geographical and sectoral breakdown of the major services offshoring sites, 2010

Source: Drawn by the authors based on data of PAIiIZ, Czehinvest, ITD-Hungary and Gál, 2014.

Legend: BS – Business services; FS – Financial services; ITO – Information Technology Outsourcing; R&D – Research and development, Knowledge process outsourcing; One box is equal to one offshoring site.

2001, 2013), which Raviv calls 'inherently predatory', noting that 'Western financial institutions in CEE were able to extract rent incomes far in excess of their profits in the West' (Raviv, 2008, p. 299). There was a shift of ownership of the banking sector from public to private and at the same time from domestic to foreign owners in the framework of privatisation. We underline that this had special effects on the analysed countries' economies during the 2007–2009 crisis.

The inflow of inward FDI has increased in CEE in the past 20 years to become the most common type of capital flow. FDI inflow into CEE economies has been a vital factor in the first stage of privatisation, and FDI became the predominant type of inward capital investment in the first stage of transition. This process served to facilitate the restructuring and transformation of not only centrally planned economies, but also the privatisation process, i.e. the increase of the share of private at the expense of state ownership. The banking and insurance sectors were the

53

primary targets of strategic foreign investors, resulting in significant inflows of FDI in these sectors, connected mainly to the privatisation of state-owned banks. Similar to global processes, foreign bank entry has been geographically/regionally concentrated, and the main investor banks have come from traditional/strong economic and trading partner countries (from mainly eurozone countries) of the host countries.

FDI inflows have resulted in dramatic changes of ownership structures. In 1994, in the wake of the early transition crises, an overwhelming majority of financial intermediaries in the post-communist countries were still publicly owned. In contrast, in 2007, more than a decade later, private foreign ownership already accounted for about 80 per cent of financial intermediaries' assets in the CEE region. These data are especially striking when we compare them with the average of the then 25 member states of the EU, which is less than one-quarter (the share of foreign-owned banking assets in total), and 15.5 per cent in the euro area. Even the average of non-Organisation for Economic Co-operation and Development (OECD) countries is only 50 per cent (Claessens and van Horen, 2012). This share of foreign banks is large even compared to the level of economic development of the analysed countries.

FDI inflows have been substantial in the financial services sector of the Visegrad countries and Slovenia, resulting in a dominant share of foreign capital and a large share of the sector in the stock of FDI already in the pre-crisis era. On the other hand, in Slovenia the role of foreign investors is comparatively much lower, resulting in a predominantly domestically owned financial services sector.

Transformation of financial services

The socialist regimes established single-tier banking systems with strong central banks after the Second World War (Myant and Drahokoupil, 2011). These systems changed: Central and Eastern European countries introduced the modern, two-tier bank system in the early years of transition, separating central banking from commercial banking functions (Gál, 2005). A unique aspect of banking systems in transitional economies is that financial markets do not emerge as a result of organic development. The creation of the CEE two-tier banking system was an artificial measure supervised by a central authority, and as such was a top-down process. Therefore, in the first years of its operation this system already operated in a highly centralised fashion with a considerable degree of territorial concentration, like in international practices (Gál, 2005). Furthermore, several follow-up trends could be observed in this period:

- the initiation of the two-tier banking system separated the functions of central banks and commercial banks;
- elimination of sectoral and operational restrictions;
- enabling the establishment of private banks;
- enabling the operation of foreign-owned and joint-owned banks;
- liberalisation of bank foundations;
- establishment of supervisory agencies.

In the new market environment, the number of banks increased rapidly and foreign ownership grew (Myant and Drahokoupil, 2011). The new, modern banking systems of these countries have no uniform structural features, because the analysed countries have different macro-economic performances, levels of liberalisation and privatisation methods. The stability and period of consolidation of these new banking systems are significantly dependent on specific processes: (1) a series of banking crises and bankruptcies, which have undermined public trust; (2) the degree of market liberalisation after governmental bank consolidation; and (3) the chosen strategy of privatisation. Active foreign privatisation shortened the period of transition crises in some analysed countries (Gál, 2004). The development of the banking sector depended significantly on the time horizon of economic and institutional transition (Balcerowicz, 2001). The development of modern banking systems took place over centuries in Western Europe and decades in Japan, but this process was completed in only a few years in CEE and the former Soviet Union (Boot and van Wijnbergen, 1995). However, the idea that new transition banking systems would simply become stable and efficient in a shorter period than the Anglo-Saxon banking networks proved to be naive. This idea also stated that during a very short period, less than one and a half decades, these countries would go through more stages of development of the banking system than any other region in world history (Kornai, 2005).

These transformations of banking systems took place at different times and with different structures. The Hungarian, Czech and Polish reforms took place initially, in the early 1990s, with the Baltic States following in the mid-1990s, and Bulgaria and Romania in the second half of the decade. The Balkan States are a little bit different, because some of these countries became independent states and market economies at the same time. The main elements of Slovenian reform were: (1) state guarantees for deposits in the central bank in Belgrade; (2) the increased role of the state in the early phase of reform; and (3) nationalisation of the biggest banks. As a result of consolidation, the number of banks decreased by 36 per cent between 1995 and 2012. As Croatia already had a two-tier banking system before gaining independence, adaptation to market conditions and international standards played the prime role in the initial period of transition and restructuring (Kovács, 2015). During the general recession of the early 1990s, market competition strengthened because of the mass appearance of foreign banks, and the role of state-owned banks decreased significantly, as seen in Figure 4.4 (Csáki, 1997). The transition phase was succeeded by the so-called 'post-transition' phase at the end of the 1990s in the pioneering countries, entailing further reforms and market openings. In this period, the structural problems of old EU member states – such as slowing expansion, concentration of services and the rationalisation-driven shrinking of the branch network – were transmitted to the transition countries. In addition, a retreat of banking intermediary activities was observed, caused by the new market players and institutional investors, which caused a more competitive financial market (Szabó, 2006).

The high importance of foreign ownership in the banking sector is indicated by the share of foreign ownership as a percentage of total banking sector assets. In comparison with the low level of penetration of foreign investors in the early stage of transition, the predominant share of foreign ownership became highly conspicuous by the millennium in Central and South-Eastern

Table 4.2 Branch density indicators, 2012

	Bank branches per 100 km²	Bank branches per 100,000 inhabitants
Bulgaria	3.39	51.75
Czechia	2.71	20.32
Croatia	2.16	28.52
Hungary	3.49	32.76
Poland	4.95	40.17
Romania	2.30	25.72
Slovakia	2.56	23.21
Slovenia	3.11	30.72
Serbia	2.44	26.37

Source: Authors' calculations based on data from ECB and national banks.

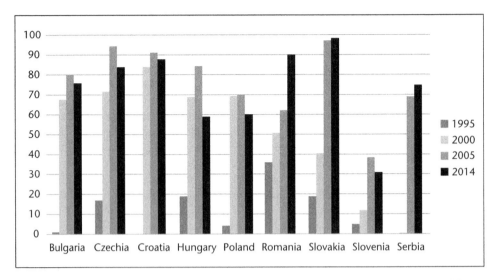

Figure 4.4 Foreign ownership as a percentage of total banking sector assets in CEE, 1995–2014

Source: Authors' construction based on data from Raiffeisen International.

European countries by the mid-2000s. The overwhelming dominance of foreign-owned banks is apparent in especially Czechia, Hungary, Croatia and Slovakia (and later in Romania). Moreover, their presence and dominance is reinforced by the fact that the market share of the top five banks in 2011 was above 60 per cent in Czechia, Slovakia and Slovenia, above 50 per cent in Hungary and above 40 per cent in Poland, according to Raiffeisen research. As Smith and Swain (2010) note, these countries became economically vulnerable due to the significant increase of foreign ownership in this period.

A significant part of post-crisis external adjustment was completed with painful fiscal austerity measures, and was accompanied by bank ownership restructuring. This was followed by the increasing role of state intervention through tightening regulations and increased taxation on banking (for example in Hungary and Slovakia). The Hungarian government launched a major re-nationalisation programme in the banking sector in 2010 in line with plans to strengthen local financial structures. Share of foreign ownership dropped to 49 per cent by 2015 in Hungary and decreased to lower levels in Poland and Slovenia (Figure 4.4).

A new financial intermediary structure emerged in the macro-region as a consequence of the processes described above, characterised by the following trends in the early 2000s:

- small bank market (banking assets in the NMS accounted for merely 2 per cent of the EU) (Gál, 1998; von Hagen and Dinger, 2005);
- low level of banking intermediation (Árvai, 2002), corporate and household lending;
- high foreign ownership (Herrmann and Jochem, 2003);
- significant potential of development and intensifying market competition (Yildirim and Philippatos, 2003).

The low level of banking intermediation can be detected in the deficiencies of the branch networks of the commercial banks and cooperative financial institutions. In the literature, branch density is measured with two indicators: the first is the number of branches per 100,000 inhabitants (social approach) and the second is the number of branches per 100 square kilometres (geographical approach). European integration and the appearance of foreign bank groups in this area had a positive effect on financial service coverage in the region, but deficiencies remain until today. The social approach indicator shows that the examined countries – with the exception of Bulgaria and Poland – lag behind the stronger Western European bank networks (the indicator is 45 in Germany and 57 in France). Using the geographical approach, results are even weaker: the Western European average is higher than 10, but the best value in the focus area is merely 4.95 (Poland) (Table 4.2). Because of the low level of financial intermediation, a gap emerged in the availability of the mainstream bank products and services (e.g. SME loans) between the core areas and the peripheral regions (Bečicová and Blažek, 2015).

Corporate lending has developed slowly: the trend is growth, but slower than the eurozone's tendency. The depth of the banking sector is still low. Loans for the private corporate sector have grown from €77.1 billion (1999) to €645.3 billion (2014), but in per cent of GDP terms, it has only moved from 19.4 to 26.4. Comparatively, this indicator increased by more than 60 per cent in the eurozone in the last few years. Loans in foreign currency (mainly EUR and CHF) increased rapidly in Poland, Hungary, Romania and Croatia: this trend changed in 2009–2010, but has grown again in Poland since 2012. The value of these loans was equal to 15–25 per cent of the GDP in Czechia, Hungary, Poland and Romania, 33 per cent in Serbia, 39 per cent in Bulgaria and 63 per cent in Croatia in 2014. On the other hand, the CEE countries are characterised by higher deposit bases, as bank deposits exceeded 40 per cent of GDP in 2007 and even after the crisis (Myant and Drahokoupil, 2011).

Geographical coverage is another important question (alongside the size of the bank branch network). Concerning the territorial dimension, it can be declared that, similarly to the public administration and governance structure of a specific country, the structure of the banking network may also be either centralised or decentralised, and generally the two systems resemble each other in this respect.

The geographical aspects of financial systems of these countries changed in various ways after the regime change. Some different cases are the following: in Poland and Croatia, a much more decentralised bank model emerged, with strong regional centres and non-capital bank headquarters; the Czech bank sector became mono-centric, like in most of the countries in this study, where financial and business activities were concentrated in the largest urban centre (Bristow et al., 1998) This concentration process was realised during the last 10–15 years, when banks relocated from 13 middle-size cities into the capital city of Czechia (Blažek, 1997, Bečicová and Blažek, 2015) In these models the regional directorates have merely informal roles. The middle ground lies in Slovakia and Romania. Their banking systems contain a few smaller monetary institutions which did not develop their managing bases in the capital city, yet these can hardly be called significant: they are centred generally on one specific sector or branch. The Hungarian situation is not simple. The commercial banks' headquarters are in the capital city, Budapest, with the exception of some small banks. Local cooperative banks have headquarters in some rural settlements, but the umbrella organisation (apex bank) of these cooperative savings banks is in Budapest. Warsaw, Prague and Budapest have stock exchanges, so these cities have multiple roles in their financial markets.

Pre- and post-crisis FDI trends in financial sector in Visegrad-4 countries and in Slovenia

This section aims to demonstrate the strong dependence on foreign banks and their resources (external liabilities). We would like to show the dominance of foreign capital in the banking systems of the analysed countries. It is important to have a look at the FDI data for financial

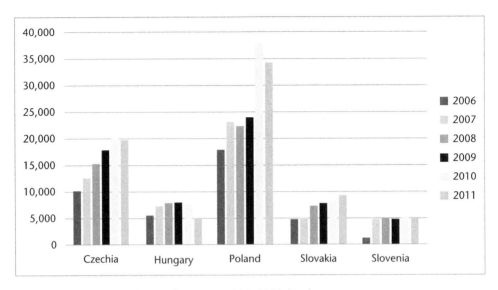

Figure 4.5 FDI stock in financial services, 2006–2011 (€m)

Source: Authors' calculation based on data from national banks.

activities in the pre-crisis period. In terms of the stock of FDI, Poland stood out in 2007 with more than €20 billion of foreign investment, while in per capita terms, the smallest country, Slovenia, had the highest amount of €2,175 (Figure 4.5).

FDI movement during the crisis period was substantial in the analysed countries. The FDI stock in financial services was mainly characterised by growth, especially in Slovakia and in Czechia (which had a slight decline in the last years). In Hungary and Slovenia, this indicator stagnated or even slightly decreased over the examined period.

As a result, in the post-crisis period, changes are prevalent compared to the pre-crisis period in the five countries. Stock has increased substantially in Czechia and Poland, slightly in Slovakia and Slovenia, and decreased in Hungary. In per capita terms, Slovenia is still the leader, followed closely by Czechia, compared to the pre-crisis period.

A better indicator of the importance of foreign banking capital in the banking sectors of the analysed countries is the share of foreign-owned economic actors in gross value added. Here data are available only for Czechia and Hungary, but even these indicate foreign dominance in the analysed sector in the analysed period (Sass et al., 2012).

Accumulated imbalances and shock transmissions in the banking sector

This section identifies the extent to which the CEEC's banking systems' integration into the EU and the eurozone contributed to regional imbalances. During the 2000s, the growing deficit in current accounts was increasingly financed by inflows of bank capital as the shift from FDI towards credit financing occurred. This process in CEE coincided with the global process of financialisation (Epstein, 2005; Lapavitsas, 2009; Sokol, 2013). Financialised capitalism led to excessive borrowing in CEE, increasing indebtedness and generating real estate bubbles that in the short term encouraged higher growth rates (Smith and Swain, 2010) but in the long run the financialisation-driven crisis deepened the gap between the European core and the periphery

(Sokol, 2013). Already by the year of EU-accession in 2004, some CEE countries had witnessed a massive debt accumulation. The increased imbalances in the financial sector through credit-fuelled consumption and mortgage-related indebtedness was further exacerbated by the exchange rate risk (Gál, 2013). We compare the pre- and post-crisis periods, identified as up to 2007 as 'pre-crisis' and 2010 and afterwards as 'post-crisis'. The credit boom in mortgage and consumer credits fuelled excessive growth in liabilities throughout the CEE macro-region during the pre-crisis period. The share in total volumes of liabilities grew from 8 per cent in 2005 to 25 per cent in 2008. The process of deleveraging was already visible in 2009, resulting in a drop in foreign funding.

In general, in 'normal' pre-crisis times, peripheral EU countries (Portugal, Ireland, Italy, Greece and Spain (PIIGS) and CEE) found it easy to finance their balance of payments deficits because banks in the core area were eager to provide funds. This financing scheme was not able to boost convergence within Europe, as capital flows were not directed to finance productive investment.

The crisis and its consequences in the less developed parts of the eurozone has in turn modified the incentives for CEE countries which are not (yet) part of the Economic and Monetary Union (EMU) (e.g. Poland, Hungary, Czechia) to access the eurozone. In a few CEE countries, catching up in the first half of the 2000s was generally accompanied by macroeconomic stability, but most countries of the region became increasingly vulnerable due to the unsustainable trajectories of huge credit, housing and consumption booms, high current-account deficits and rapidly rising external debt (a large proportion of which was denominated in foreign currencies). Foreign currency indebtedness channelled through interlinkages occurred between the West European parent banks and their local subsidiaries. This process has had implications for internal and external imbalances within the EU banking system. The highly integrated banking system prevalent in the CEE countries is more prone to transmit adverse shocks across borders and serves as a set of propagation channels for potential regional shocks, which might be transmitted throughout CEECs.

The household sector's foreign currency borrowing is a recent development in CEE, and, actually, it can be regarded as a carry trade: households borrow in a lower-yield foreign currency and invest in a high-yield domestic currency (e.g. in the form of a home mortgage), while expecting the domestic currency to continue to appreciate. Despite its immediate benefits – lower interest rates, longer maturities – for the non-banking sector, these loans carry a significant exchange rate risk. These challenges became apparent only after the sudden deterioration of the economic environment. Indeed, foreign currency indebtedness is a systemic problem with a crippling effect on the whole economic sector and economic policy (Yeşin, 2013; Schepp and Szabó, 2015).

During the credit boom that characterised the CEE region, credit growth significantly exceeded the growth rate of deposits – i.e. the growth rate of core liabilities – therefore banks' borrowing from external sources (non-core liabilities) increased. As a result, the share of deposits within banks' total liabilities may fall significantly. This causes a problem primarily because deposits are more stable compared to non-core liabilities (interbank and foreign funds); they are a more predictable source of funding and less exposed to the adverse shocks of changes during crisis times. A credit boom that was extensively financed by the foreign parent banks of local subsidiaries fuelled consumption in many CEECs (Holló, 2012).

The high exposure of the liability structure may not only increase banks' 'funding' liquidity risk, but may also raise contagion risks via the interbank market. During the first year of the financial crisis the default risk of banking subsidiaries was not only influenced by the extent to which their credit losses and 'funding' liquidity risks increased due to various shock transmissions across borders, but also by the fund-constraining reaction of parent banks to the shocks affecting the system. Besides the Baltic States', Hungary's and Slovenia's external funding exposures were the highest, reaching about one-third of total liabilities in 2009. On the other hand, the banking systems of Czechia and Slovakia largely relied on their internal savings (local deposits): the external liabilities were only 8.5 and 3 per cent in these countries. The significant asset and liability side imbalances of the Hungarian banking system started to emerge in 2005; the current problems facing Hungarian banks stem from the large imbalances on the assets and liabilities side (excessive credit growth and significant increase in the share of secondary liabilities within total liabilities), which characterised the period between 2005 and 2008.

The direction of shock transmissions and potential contagion affected the macroeconomically most fragile countries. In the first stage of the financial crisis, which started in the autumn of 2008, due to the increased reliance of CEE countries' (in its most severe form in the Hungarian) banking sector(s) on foreign funding to finance rapid credit expansion and the risks associated with liquidity. Crises that occurred at the parent banks were transmitted immediately to their foreign subsidiaries. In the second stage in Hungary (from mid-2009) foreign currency indebtedness captured potential spillover effects directly in the host countries, with the Hungarian banking sector left exposed. This local debt crisis affecting all sectors (indebtedness of the state, corporate and private households) generated propagation channels for potential regional shocks, which might be transmitted within the whole CEE region. The Vienna Initiative was created in January 2009 to prevent the further spread of the financial crisis in CEE. The initiative helped prevent a systemic banking crisis in the region and limited the negative fallout from nation-based uncoordinated policy responses. It further helped avoid a massive and sudden deleveraging by cross-border bank groups.

However, the crisis altered the future growth prospects of these CEE countries. Monetary and fiscal policies have been on a tightening course for several years, and there has been little room for powerful countercyclical policy responses. External capital inflow suddenly and significantly dropped and the recovery of internal markets has been sluggish. However, many countries executed very fast and significant external adjustment in order to diminish their external imbalances, which resulted in a situation where all indebted economic players were deleveraging. This process was dramatically rapid in Hungary, which had the second-highest net external liabilities in 2011 (Hungary's external debt was the seventh highest while Czechia had balanced net debt despite the large share of FDI finance) (Figure 4.6).

As the new wave of the eurozone crisis unfolded towards the end of 2011, signs of rapid deleveraging in an emerging Europe immediately followed. Furthermore, some serious gaps in regulatory coordination between home and host countries' authorities once again became apparent. The main objective of the Vienna Initiative 2.0 in 2011 was to better coordinate the host and home countries' supervision in order to manage the deleveraging process to minimise systemic risks in emerging Europe. In spite of the success of the Vienna initiatives, funding availability and costs remain a constraint for CEE banking.

Funding reductions of western banks vis-à-vis those in Central, Eastern and South Eastern Europe (CEE/SEE) were further moderated after the initiative, in 2012, but did not stop or reverse the ensuing trend of funding withdrawals. CEE/SEE – excluding Russia and Turkey – lost funding equivalent to 4.6 per cent of GDP over this period. Bulgaria, Croatia, Hungary, Latvia, Lithuania, Slovenia and Serbia were most affected, with losses in excess of 5 per cent of GDP. In 2012, the global banking sector reduced its external position vis-à-vis all sectors to Hungary by far above the regional average, by about 18 billion USD, or 14.2 per cent of the Hungarian GDP. For comparison, the reduction in Spain was 14.3 per cent of GDP. Accelerated deleveraging in the banking system resulted in a dramatic decline in bank lending in many CEE countries, with the highest declines in the Baltic States and Hungary, countries which previously had the highest proportion of foreign funding in banking. As a legacy of the foreign-funded

The role of business and finance services

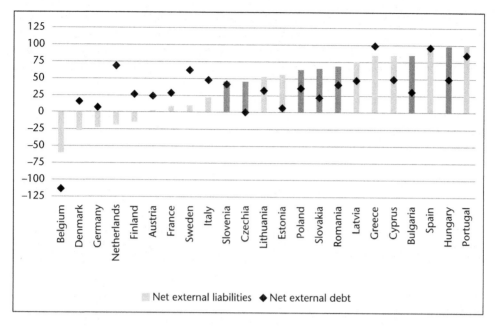

Figure 4.6 Dependencies on foreign bank resources (net external liabilities and debt), 2011
Source: Authors' calculation based on data from Eurostat.

credit boom, and despite the funding withdrawals since, the stock of outstanding financing from foreign banks remains large. For CEE/SEE, excluding Russia and Turkey, this amounts to 26 per cent of GDP or 29 per cent of private credit.

In summary, the finance-led growth model further strengthened the ownership, as well as the creditor-debtor control, over CEECs. In these financially integrated debtor countries, the crisis of 2008 further increased the systemic vulnerabilities that were earlier generated by the transition model (Gál and Schmidt, 2017).

Conclusions

The tertiary sector and financial services play a key role in the economies of the Central and Eastern European countries. These economies became open market economies in the 1990s and the transition processes provided the opportunity for the tertiarisation and development of advanced producer services activities. During the transition, more and more foreign capital (mainly in the form of FDI) flowed into this region, accompanied by a dramatic change in ownership structure. This led not only to an unprecedented transfer of property rights from local society to foreign investors, but also increased external dependencies and caused inherent imbalances in the financial sector through indebtedness and risk.

The financial market of the CEE countries is characterised by the dominance of the banking sector, while the other financial activities are underdeveloped in the global context. Nevertheless, the size of the banking market is still only 2 per cent (in total bank assets) of the EU total. Foreign ownership reached more than 50 per cent in all examined countries (except Slovenia and Serbia) by 2001. By 2008, more than a decade later, private foreign ownership accounted for about 80 per cent of financial intermediaries' assets in the CEE region. The international financial market integration of CEE was accompanied by foreign borrowing and financial inflows in order to finance household consumption and private sector activity, and covered the deficits on many countries' current accounts. Nevertheless, it became an important driver for short-term economic development in the years up to 2008. However, during the crisis foreign banks, with their risky cross-border financial exposures (foreign currency loans), played a significant role in the transmission of contagion to CEE. Funding reductions of western banks vis-à-vis that in the most affected countries of Central, Eastern, and South Eastern Europe (CEE/SEE) exceeded 5–6 per cent of GDP. Stability and predictability can help restore bank lending to a more normal level but the low growth environment, together with a middle income trap, is a huge challenge. Financialised growth escalated in the years up to 2008 in those countries that lacked domestic deposit bases. This, then, was a transient phase that ended with the world financial crisis, leaving a number of countries, and among them the CEECs, with uncertain futures.

While predominantly market-seeking, financial sector FDI can be characterised as horizontal, foreign investment that has flown into other business and ICT sectors is highly export-oriented and a vertical type of investment. Offshoring has been a stimulus to develop CEE as an important destination for resources seeking services investment. In spite of these highly export-oriented environments, shared services centres are less embedded into their local economies, while the embeddedness of offshoring locations in CEE (capital cities and some secondary cities) into the global value chains have intensified.

Acknowledgement

The publication of this research has been supported by project #104985 of the Hungarian National Research, Development and Innovation Office.

References

Amiti, M. and Wei, S.-J., 2005. Fear of Outsourcing: Is it Justified? *Economic Policy*, 20(4), pp. 308–348.
Árvai, Zs., 2002. A banki közvetítés mélysége [The depth of bank intermediation]. *Közgazdasági Szemle*, 49(7–8), pp. 621–640.
Association of Business Service Leaders (ABSL), 2015. *Business Services in Central & Eastern Europe 2015*. Warszawa: ABSL, 68. www.absl.cz/docs/CEE_report_final.pdf [Accessed 10 October 2015].
Balaž, V., 2004. Knowledge-intensive Business Services in Transition Economies. *Service Industries Journal*, 24(4), pp. 83–100.
Balcerowicz, L., 2001. Post-Communist Transition: Some Lessons. The Wincott Memorial Lectures Series No. 31. London: Institute of Economic Affairs.
Banai, Á., Király, J. and Várhegyi, É., 2010. *A rendszerváltás 20 évének egy egyedi fejezete: külföldi bankok dominanciája a kelet-közép-európai régióban, különös tekintettel Magyarországra* [Foreign banks in the CEE region, the special case of Hungary]. Budapest: Magyar Nemzeti Bank. (MNB Tanulmányok, 89)
Beaverstock, J.V., Taylor, P.J. and Smith, R.G., 1999. A Roster of World Cities. *Cities*, 16(6), pp. 445–458.
Bečicová, I. and Blažek, J., 2015. Is There a Credit-gap in a Periphery? The Perception of This Problem by Small Entrepreneurs. *Journal of Rural Studies*, 42(1), pp. 11–20.
Begg, I., 1993. The Service Sector in Regional Development. *Regional Studies*, 27(8), pp. 817–825.
Beluszky, P. and Győri, R., 2004. A magyar városhálózat funkcionális versenyképessége. [The functional competitiveness of Hungarian city network] In: Gy. Horváth, ed. *Régiók és települések versenyképessége*. Pécs: MTA Regionális Kutatások Központja, pp. 236–293.
Berger, A., Qinglei, D., Ongena, S., and Smith, D., 2003. To What Extent Will the Banking Industry Be Globalized? A Study of Bank Nationality and Reach in 20 European Nations. *Journal of Banking & Finance*, 27(3), pp. 383–415.
Bevan, A. and Estrin, S., 2004. The Determinants of Foreign Direct Investment into European Transition Economies. *Journal of Comparative Economics*, 32(4), pp. 775–787.

Blažek, J., 1997. The Development of the Regional Structure of the Banking Sector in the Czech Republic and its Implications for Future Regional Development. *Acta Universitatis Carolinae: Geographica*, 32 (Special Issue), pp. 265–283.

Bonin, J., Mizsei, K., Székely, I. and Wachtel, P., 1998. *Banking in Transition Economies, Developing Market Oriented Banking Sectors in Eastern Europe*. Cheltenham: Edward Elgar, p. 208.

Boot, A.W.A. and van Wijnberge, S., 1995. Financial Sector Design, Regulation and Deposit Insurance in Eastern Europe. In: J. Rostowski, ed. *Banking Reform in Central Europe and the Former Soviet Union*. Budapest: Central European University Press, pp. 42–57.

Bristow, G., Gripaios, P. and Munday, M., 1998. Financial and Business Services and Uneven Economic Development: Some Welsh Evidence. *Tijdschrift voor Economische en Sociale Geografie*, 90(2), pp. 156–167.

Bryson J., 2007. The Second Global Shift: The Offshoring or Global Sourcing of Corporate Services and the Rise of Distanciated Emotional Labour. *Geographiska Annaler, Series B. Human Geography*, 89(1), pp. 31–43.

Buch, C. and Golder, S., 2001. Foreign versus Domestic Banks in Germany and the US: A Tale of Two Markets? *Journal of Multinational Financial Management*, 11(4–5), pp. 341–361.

Buch, C., Heinrich, R. and Schertler, A., 2003. External and Internal Financial Structures in Europe: A Corporate Finance Perspective. EIFC – Technology and Finance Working Papers No. 19.

Capik, P., 2008. Offshoring and Outsourcing: New Trends in the Service Sector Foreign Direct Investment in Poland. Economic and Society Trust Working Paper.

Claessens, S. and van Horen, N., 2012. Foreign Banks: Trends, Impact and Financial Stability. IMF Working Papers 12/10. Washington, DC: International Monetary Fund.

Claessens, S., Demirguc-Kunt, A. and Huizinga, H., 2001. How Does Foreign Entry Affect Domestic Banking Markets? *Journal of Banking & Finance*, 25(5), pp. 891–911.

Csaba, L., 2011. Financial Institutions in Transition – The Long View. *Post-Communist Economies*, 23(1), pp. 1–14.

Csáki, Gy., 1997. Magyar bankrendszer: konszolidáció után, a privatizáció lezárása előtt? [The Hungarian bank system: After consolidation and before the closing of privatisation?] *Társadalmi Szemle*, 52(1), pp. 3–14.

Csomós, Gy., 2011. A közép-európai régió nagyvárosainak gazdaságirányító szerepe [The role of cities in Central Europe as leaders of the economy]. *Tér és Társadalom*, 25(3), pp. 129–140.

Epstein, G.A., 2005. Introduction: Financialization and the World Economy. In: G.A. Epstein, ed. *Financialization and the World Economy*. Cheltenham: Edward Elgar, pp. 3–16.

Fernandez-Stark, K., Bamber, P. and Gereffi, G., 2011. *The Offshore Services Global Value Chain Economic Upgrading and Workforce Development*. Durham, NC: Duke University Center on Globalization, Governance and Competitiveness.

Fifekova, M. and Hardy, J., 2010. *Business Service. Foreign Direct Investment in Central and Eastern Europe – Trends, Motives and Impacts*. Project Report, p. 28. www.czechinvest.org/data/files/fdi-project-report-1981-en.pdf [Accessed 25 May 2013].

Gál, Z., 1998. A pénzintézeti szektor területfejlesztési kérdései Magyarországon [The questions of financial services sector in Hungary from the point of view of regional policy]. *Tér és Társadalom*, 12(4), pp. 43–67.

Gál, Z., 2004. Spatial Structure and the Expanding European Integration of the Hungarian Banking System. In: A.D. Kovács, ed. *New Aspects of Regional Transformation and the Urban–Rural Relationship: XIII. Polish–Hungarian Geographical Seminar Tokaj*. Pécs: MTA Regionális Kutatások Központja, pp. 74–104.

Gál, Z., 2005. The Development and the Polarised Spatian Structure of the Hungarian Banking System in a Transforming Economy. In: Gy Barta, É.G. Fekete, I. Kukorelli-Szörényiné and J. Timár, eds. *Hungarian Spaces and Places: Patterns of Transition*. Pécs: Centre for Regional Studies of the Hungarian Academy of Sciences, pp. 197–219.

Gál, Z., 2013. Role of Financial Sector FDI in Regional Imbalances in Central and Eastern Europe. In: A. Gostyńska, P. Tokarski, P. Toporowski and D. Wnukowski, eds. *Eurozone Enlargement: Challenges for the V4 Countries*. Warsaw: Polish Institute of International Affairs, pp. 27–35.

Gál, Z., 2014. Relocation of Business Services into Central and Eastern Europe (Evidence from Trade and Location Statistics). *Romanian Review of Regional Studies*, 10(1), pp. 67–78.

Gál, Z. and Schmidt, A., 2017. Geoeconomics in Central and Eastern Europe: Implications on FDI; In: Mark Munoz (Ed.) *Advances in Geoeconomics*. Abingdon: Routledge, pp. 76–93.

Gál, Z. and Sass, M., 2009. Emerging New Locations of Business Services: Offshoring in Central and Eastern Europe. *Regions*, 274(1), pp. 18–22.

Ghibutiu, A. and Dumitriu, I., 2008. The Effects of Offshoring on Trade in Services. Evidence from Romania. The European Trade Study Group (ETSG) Working Paper. www.etsg.org/ETSG2007/papers/dumitriu.pdf [Accessed 14 November 2015].

Grote, M. and Täube, F., 2006. Offshoring the Financial Services Industry: Implications for the Evolution of Indian IT Clusters. *Environment and Planning A*, 38(7), pp. 1287–1305.

Hardy, J., 2006. Bending Workplace Institutions in Transforming Economies: Foreign Investment in Poland. *Review of International Political Economy*, 13(1), pp. 129–151.

Hardy, J., 2007. 'Learning' or 'Coercive' Firms? Foreign Investment, Restructuring Transforming Economies and the Case of ABB Poland. *International Journal of Management Concepts and Philosophy*, 2(3), pp. 277–297.

Hardy, J., Sass, M., and Fifekova, M., 2011. Impacts of Horizontal and Vertical Foreign Investment in Business Services: The Experience of Hungary, Slovakia and the Czech Republic. *European Urban and Regional Studies*, 18(4), pp. 427–443.

Havlik, P., 2005. *Structural Change, Productivity and Employment in the New EU Member States*. Vienna: Vienna Institute for International Economic Studies (wiiw Research Reports, 313).

Herrmann, S. and Jochem, A., 2003. The International Integration of Money Markets in the Central and East European Accession Countries: Deviations from Covered Interest Parity, Capital Controls and Inefficiencies in the Financial Sector. Discussion Paper No 07/03. Berlin: Deutsche Bundesbank.

Holló, D., 2012. *A System-wide Financial Stress Indicator for the Hungarian Financial System*. Budapest: Magyar Nemzeti Bank (MNB-tanulmányok, OP 105).

Horváth, Gy., ed., 2015. *Kelet- és Közép-Európa régióinak portréi* [Portraits of the regions of Eastern and Central Europe]. Budapest: Kossuth Kiadó.

Jovančević, R., 2007. The Impact of Foreign Investment Flows on Croatian Economy – A Comparative Analysis. *Ekonomski Pregled*, 58(12), pp. 826–850.

Kornai, J., 2005. Közép-Kelet-Európa nagy átalakulása. [The great transition of Central and Eastern Europe]. *Közgazdasági Szemle*, 52(12), pp. 907–936.

Kovács, S.Zs., 2014. Elérhetőség és kirekesztés Magyarországon a pénzügyi szolgáltatások aspektusából [Availability and exclusion from the aspect of Hungarian financial services]. *Területfejlesztés és Innováció*, 8(3), pp. 28–35.

Kovács, S.Zs., 2015. The Spatial Structure of Financial Services in the South Pannonian Region. In: Á. Bodor and Z. Grünhut, eds. *Cohesion and Development Policy in Europe*. Pécs: Institute for Regional Studies, Centre for Economic and Regional Studies, Hungarian Academy of Sciences, pp. 109–118.

Kox, H. and Rubalcaba, L., 2007. *Analysing the Contribution of Business Services to European Economic Growth*. Bruges: European Economic Studies Department, College of Europe (Bruges European Economic Research Papers, 9).

Landesmann, M., 2000. Structural Change in the Transition Economies, 1989–1999. *Economic Survey of Europe*, 2(3), pp. 95–117.

Lapavitsas, C., 2009. Financialised Capitalism: Crisis and Financial Expropriation. *Historical Materialism*, 17(2), pp. 114–148.

Lux, G., 2010. Location Differences of Services and Industry: A Central European Dichotomy. *Prace Komisji Geografii Przemysłu Polskiego Towarzystwa Geograficznego*, 16, pp. 29–37.

Marciniak, R., 2014. Global Shared Service Trends in the Central and Eastern European Markets. *Entrepreneurial Business and Economics Review*, 2(3), pp. 63–78.

Maurel, M.C., 2006. Területi egyenlőtlenségek Európában. A bővítés mint a kohézió próbája. [Territorial inequalities in Europe. Cohesion to the test of enlargement] *Tér és Társadalom*, 20(4), pp. 169–182.

Myant, M. and Drahokoupil, J., 2011. *Transition Economies: Political Economy in Russia, Eastern Europe and Central Asia*. Hoboken, NJ: Wiley.

Nölke, A. and Vliegenthart, A., 2009. Enlarging the Varieties of Capitalism: The Emergence of Dependent Market Economies in East Central Europe. *World Politics*, 61(4), pp. 670–702.

Raffay, Z., 2004. Az üzleti szolgáltatások szerepe a regionális versenyképesség erősítésében [Role of business services in the strengthening of the regional competitiveness]. In: I. Pálné Kovács, ed. *Versenyképesség és igazgatás*. Pécs: MTA Regionális Kutatások Központja, pp. 69–90.

Raiffeisen International, various dates. *CEE Banking Sector Report*. October 2004; June 2009; May 2013; June 2015. Vienna: Raiffeisen Bank International.

Raviv, O., 2008. Chasing the Dragon. East: Exploring the Frontiers of Western European Finance. *Contemporary Politics*, 14(3), pp. 297–314.

Sass, M., 2008. A szolgáltatások relokációja – európai folyamatok [The relocation of services – European processes]. *Európai Tükör*, 13(7–8), pp. 85–100.

Sass, M. and Fifekova, M., 2011. Offshoring and Outsourcing Business Services to Central and Eastern Europe: Some Empirical and Conceptual Considerations. *European Planning Studies*, 19(9), pp. 1593–1609.

Sass, M., Antalóczy, K. and Éltető, A., 2012. Emerging Multinationals and the Role of Virtual Indirect Investors: The Case of Hungary. *Eastern European Economics*, 50(2), pp. 41–58.

Schepp, Z. and Szabó, Z., 2015. Lakossági svájcifrank-hitelek árazása – narratíván innen és túl? [Pricing Swiss franc-denominated mortgage loans: Beyond the narratives]. *Közgazdasági Szemle*, 62(11), pp. 1140–1157.

Smith, A. and Swain, A., 2010. The Global Economic Crisis, Eastern Europe, and the Former Soviet Union: Models of Development and the Contradictions of Internationalization. *Eurasian Geography and Economics*, 51(1), pp. 1–34.

Sokol, M., 2001. Central and Eastern Europe a Decade after the Fall of State-socialism: Regional Dimensions of Transition Processes. *Regional Studies*, 35(7), pp. 645–655.

Sokol, M., 2013. Towards a 'Newer' Economic Geography? Injecting Finance and Financialisation into Economic Geographies. *Cambridge Journal of Regions, Economy and Society*, 6(3), pp. 501–515.

Soubbotina, T. and Sheram, K.A., 2000. *Beyond Economic Growth: Meeting the Challenges of Global Development*. Washington, DC: World Bank (WBI Learning Resources Series).

Stare, M., 2007. Service Development in Transition Economies: Achievement and Missing Links. In: J.R. Bryson and P.W. Daniels, eds. *The Handbook of Service Industries*. Cheltenham: Edward Elgar, pp. 168–187.

Stefanović, S., Radović-Marković, M. and Ćirilović, T., 2009. Foreign Direct Investment as a Factor of Economic Growth and Development of the Serbia. In: *Business Opportunities in Serbia: The Case of Italian Business Sector and the Role of Management Education*. Belgrade: Belgrade Banking Academy and Institute of Economic Sciences, pp. 251–257.

Szabó, Cs., 2006. A közép-kelet-európai bankszektor elemzése [Analysing of Central and Eastern European bank sector]. *Közgazdász Fórum*, 9(2), pp. 7–21.

UNCTAD, 2005. *World Investment Report 2005: Transnational Corporations and the Internationalization of R&D*. Geneva: United Nations.

United Nations (UN), 2001. *Services in Transition Economies*. Trade and Investment Guides 6. Geneva: United Nations, Committee for Trade, Industry and Enterprise Development.

Várhegyi, É., 2002. Hungary's Banking Sector: Achievements and Challenges. In: A. Riess, ed. *The Financial Integration of an Enlarged EU: Banking and Capital Markets*. EIB Papers, 7(1), pp. 75–91.

Vidović, H., 2002. The Services Sectors in Central and Eastern Europe. wiiw Research Report No. 289. Vienna: Vienna Institute for International Economic Studies, p. 89.

von Hagen, J. and Dinger, V., 2005. *Banking Sector (Under?) Development in Central and Eastern Europe*. Bonn: Center for European Integration Studies, University of Bonn (ZEI Working Paper, B06).

Wachtel, P., 1997a. A külföldi bankok szerepe a közép-európai átmeneti gazdaságokban I. [The role of foreign banks in the Central European transition economies I]. *Közgazdasági Szemle*, 44(1), pp. 13–30.

Wachtel, P., 1997b. A külföldi bankok szerepe a közép-európai átmeneti gazdaságokban II. [The role of foreign banks in the Central European transition economies II]. *Közgazdasági Szemle*, 44(2), pp. 124–141.

Yeşin, P., 2013. Foreign Currency Loans and Systemic Risk in Europe. *Federal Reserve Bank of St Louis Review*, 95(3), pp. 219–235.

Yildirim, S. and Philippatos, G.C., 2003. Competition and Contestability in Central and Eastern European Banking Markets. *Managerial Finance*, 33(3), pp. 195–209.

5

The transformation of rural areas in Central and Eastern Europe

Péter Póla

The rural areas of Central and Eastern Europe

Before the regime change of 1989–1990, the state of the rural areas of the communist countries belonging to the Soviet sphere of interest was fundamentally defined by agriculture. While from a distance this region appeared homogeneous, significant differences existed in agricultural performance due not only to divergent characteristics, but – especially in the case of Hungary – unique agricultural models as well. For this reason, transformation and the move to a market economy were launched from various foundations, and a number of similarities have been to this day accompanied by differences. Since EU accession in 2004 (and in 2007, 2014) similar development policies have been implemented with political–ideological differences, especially in the operations of the institutional system. Thanks to divergent levels of efficiency agrarian and rural policy actions and their effects on the state of rural areas cannot be understood by viewing the region as a whole. The goal of this study is primarily to examine which unique characteristics of CEE's rural areas apply to most states, and what problems and paths these states have in common, alongside identifying the presence of notable country-specific problems and opportunities. One of the key issues, and one of the causes of differences, is the degree to which rural areas can successfully modernise their agrarian economies while maintaining the most important economic bases of these areas. Another key issue is the degree to which countries have been able to moderate the agricultural character of rural areas with economic differentiation.

A pronounced macro-regional distribution – related to the level of development, among other things – can be measured across rural areas in CEE. Continuous developed areas can be observed in the southern parts of Czechia and in Slovenia, yet these only exist as isolated pockets in other countries, including agglomerations around capital cities (with their varying radii effects), the Budapest–Győr–Bratislava axis, Ciechanowsko-plocki in Poland, and Istria and Dubrovnik (due to tourism). However, underdeveloped rural areas are still dominant in the macro-region, mostly prevalent in the individual regions' central and eastern halves (Farkas and Szabó, 2014). Using the EU category system, Figure 5.1 illustrates the dominance of rural areas. Similarly, the population density figures for CEE also indicate the dominance of rural areas in the macro-region (Figure 5.2).

Ignoring for the moment atypical rural areas – such as those dynamically developing rural areas with significant labour force opportunities in the proximity of large cities, or settlements

The transformation of rural areas

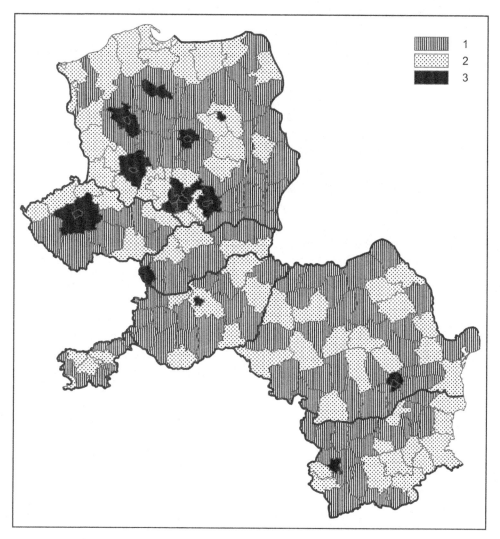

Figure 5.1 Urban–rural typology of NUTS-3 regions

Source: Compiled by the author based on Eurostat data.
Legend: 1 – Predominantly rural; 2 – Intermediate; 3 – Predominantly urban.

with tourist attractions – CEE's *typical* rural areas are characterised by the loss of their economic base through the modernisation of agriculture, and primarily due to the crises experienced since the regime changes (Buday-Sántha, 2010), which they have not been able to reconstruct since. The difficult situation of rural areas is at the same time not the result of conscious destruction, and in a sense the process is only unfavourable from the point of view of the countryside, while it is more characteristically a progressive process for society as a whole (Beluszky, 2010) – even in the CEE macro-region.

The agricultural economy has traditionally been of definitive significance in the countries of CEE, and a large portion of the macro-region experiences a more significant agricultural weight in the entirety of the economy than in the western half of Europe. The proportion of agriculture

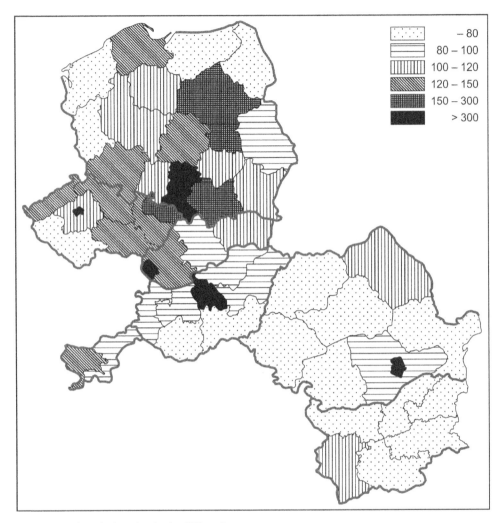

Figure 5.2 Population density in CEE regions
Source: Compiled by the author based on Horváth, 2015.

in national income production and in the employment picture exceeds the averages for developed countries – though we might add that behind the more significant role of the branch we can observe differences across countries, among which natural geographic features play a role, and variance in these terms is illustrated by the proportion of cultivated land, where significant divergences exist (Table 5.1).

The contribution of agriculture to GDP shows a declining trend in the macro-region, but years after EU accession it is still nearly twice the EU average (Figure 5.3). The proportion of agricultural workers after regime changes was over 15 per cent in the early 1990s, and 11 per cent at the time of the eastern expansion (which is close to three times the average for the EU15). Romania is particularly significant (close to 30 per cent): the 1990s here showed slight growth, which is explained by the flow of workers to agriculture as a result of the industrial

Table 5.1 The proportion of cultivated land, 2013

	Proportion of cultivated land (%)	Proportion of ploughed land within cultivated land (%)
Hungary	65.1	50.3
Romania	62.0	39.1
Poland	52.0	40.6
Bulgaria	47.8	39.1
Czechia	46.0	34.1
Slovakia	40.3	27.9
Slovenia	24.3	8.8

Source: Author's construction based on Eurostat.

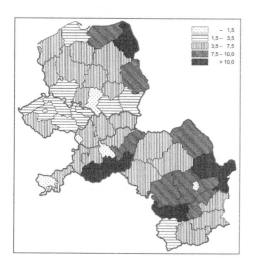

The share of agriculture in GDP

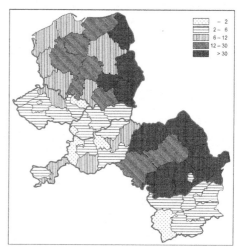

The share of agriculture in employment

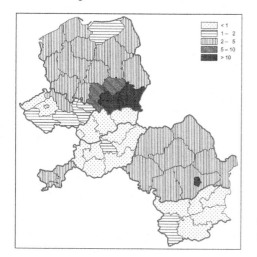

The proportion of agricultural GDP and employment

Figure 5.3 The proportion of agriculture in GDP and employment (%) in the CEE macro-region

Source: Compiled by the author based on data from Horváth, 2015.

crisis in the years after the revolution. The second most important agricultural employer is Poland. Although the indicators for Czechia, Slovakia and Hungary are closer to the EU average, the 3.5–5 per cent agricultural labour force in CEE countries is a very sensitive area in *economic*, *social* and *political* aspects. Small-scale entities that do not meet efficiency criteria, however, have remained characteristic and are often of a 'sanctuary' nature – especially in Romania and Poland (Darrot, 2008). At the turn of the millennium the growth of agriculture in Bulgaria and Romania is explained by the acute crisis in the labour market (Bourdeau-Lepage and Bazin, 2011). The significance of the agricultural sector is also supported by the fact that almost all the macro-region's important agricultural products can be said to be significant contributors to world production (Csáki, 1998).

Rural areas are not equal to agriculture, but given that agriculture is the most adequate branch for rural areas (Buday-Sántha, 2011), the situation of agriculture fundamentally defines the level of development, and developmental opportunities of rural areas. The modernisation of agriculture, despite its progressiveness, in itself put dominantly agrarian areas in a difficult situation. This is especially the case when the modernisation of agriculture was unsuccessful, or where the developing structure makes the emergence of efficient management difficult. The heritage of the CEE macro-region and the political and economic difficulties, as well as the economic and agricultural policy missteps of transition all contributed to a situation whereby the state of agriculture and, by extension, rural areas have not improved meaningfully: the significant erosion of some peripheral areas under globalisation's effects has been observed (Káposzta, 2014); rural communities have lost their service and community-building institutions, rural infrastructure has eroded, and the ability to maintain resources locally has weakened.

The role of agriculture in the labour market and its contribution to the GDP (Figure 5.3) diverge significantly in several countries in the CEE macro-region, which signifies the low productivity of management and a level of efficiency that lags behind the global average.

A significant role in this lag is played by the heritage of political and agricultural history, which led to the development of the current structure. The EU's agricultural policy and its rural and regional policies are also factors: given the rigidity of developed structures, they have not been able to fundamentally moderate development level differences, and have instead conserved them. As Pospěch (2014) stated, before 1989 the communist regime and then from 2004 EU programmes and policies, standardised agriculture. Lieskovský et al. (2015) claimed that one of the serious problems in the Common Agricultural Policy (CAP) is that its adaptation has been significantly beneficial exclusively to large-scale concerns and, as such, the numerous small-scale concerns in the macro-region could not have expected it to improve their position meaningfully.

An integrated countryside capable of developing is a basic condition for a unified and sustainable society and economy. In the CEE macro-region, it is necessary to create such foundations. This has been successful in some states in the region, while in others it has characteristically not been the case. The recipe for success is complicated and perhaps not replicable, but essentially two paths exist: one is integration into the economy of large cities with substantial development capacity, while the other is the creative utilisation of endogenous resources. Neither, but primarily the latter, can ignore the fact that although the countryside is not synonymous with agriculture, without the modernisation of the branch and agro-business-oriented development, a liveable and developing countryside is impossible. As such, there are fundamentally two tasks. One is to understand that without developing agriculture as a branch, the countryside cannot be successful. While there are differences between CEE states in terms of the ownership structure of the branch and competitiveness indicators, nowhere can the successful development of the sector ignore that the efficiency race cannot be avoided, that the branch requires infrastructural and technological development, and that it is reasonable to make competitiveness conditions secondary to structural and support policy issues. The second task is to fortify the multifunctional character of rural areas, with agriculture holding a primary position, but not allowing it exclusive status and thus diversifying the economy of rural areas. One of the most important conditions for such is the fortification of local communities and the enhancement of activity, for which one of the foundations is a rise in the level of training and education. Below we will use the two tasks as a starting point for the examination of the roots and direction of the diversification of agriculture and rural areas, and local development.

Agriculture and the heritage of rural areas

The presentation of hundreds of years of developmental history of CEE's agricultural economy is beyond the scope of this chapter, and as such we will start with a consideration of the heritage of the structure developed after the Second World War, which proceeded after it entered the Soviet Union's political and economic sphere of influence. At the same time it is expected that the previous period's agricultural history, economic and social structures had had an effect on how the agricultural sectors of CEE states developed within the frameworks of communist regimes, as structures that were locked into place at the time influenced the process of transformation that took place after the regime change.

Significant structural differences existed in the macro-region before the state socialist period. Bulgaria essentially lacked large estates, which were present in Romania and in the other parts of the Balkans. But in these areas large estates did not possess real operational forms (lease and tilling of the landowners' lands by peasants), which, given the lack of ample high-quality cultivable land and the large scale of the agrarian population, was a source of significant poverty. At the same time, in the Habsburg-held areas of the macro-region, large estates were significant in both ownership and operations aspects, and more than half the land in these parts of the macro-region was owned in this form (Illés, 2002).

Following the communist takeover of power the agricultural policy of state-parties was tightly bound to industrial policy considerations. Leaders hoped to execute large-scale investments through the subjection of agriculture, drawing necessary capital and labour from the agricultural sector. The path to such led through collectivisation in agriculture.

In several countries in the macro-region, this process proceeded quickly. Collectivisation was unsuccessful in Poland, where private property remained definitive in agriculture, although the effective operation of such was obstructed by numerous limitations. Similar significant limitations also characterised efficient operations in the member states of the former Yugoslavia, including *Slovenia*. Elsewhere collectivisation and the state and cooperative system became dominant. Collectivised agriculture not only lost its harmony as a production factor, but a significant value of fixed assets fell out of production. Hungary diverged from the dominant model and was the most successful country in doing so. Attempts to collectivise were undertaken twice between 1949 and 1956, but this succeeded fully only during the Kádár restoration period following the events of 1956. One of the features of the period was the divergence of the Hungarian agricultural model from the Soviet model that dominated in the macro-region, with Hungary following a unique path. One of the consequences of this was the sector's dynamic development, with results that were noteworthy even in the global context. Hungarian success was definitively driven by the unique ownership and entrepreneurial structure (the so-called multi-sector agricultural branch). Cooperatives, which gradually broke from the Soviet model, represented the largest proportion in agricultural production and integrated significant and successful household-scale operations. Family farms that were integrated as such (not formally, but essentially as small private enterprises) by the 1980s were providing a considerable proportion of agricultural products in

Table 5.2 The proportion of cooperative and state ownership of agricultural land before 1989

	Cooperative (%)	State (%)
Czechia	61	38
Bulgaria	58	29
Hungary	80	14
Poland	4	19
Romania	59	29
Slovakia	69	26

Source: Author's construction based on data from Illés, 2002.

the country. A portion of the fixed assets mentioned above was returned into production as such. The integration of small- and large-scale plants as well as the processing industry led to the emergence of a unique but highly functional construction in Hungarian agriculture which served the development of the sector well. The Hungarian model contained private (or so-called private), cooperative and state ownership elements concurrently. This was dominant in Hungary and Lithuania (Maciulyte, 2004), although it integrated less successfully in the latter.

It is important to emphasise that in the state socialist planned economy period significant differences existed in the macro-region (Table 5.2), although from afar it might have appeared homogeneous. The explanation for the structural differences lies not only in the divergent models defined by the political sphere, but also in the fact that a variety in the product structure influenced by natural and geographic features among other things was also experienced.[1] In Bulgaria, where certain territories have outstanding soil quality, pluralistic production became possible, but early collectivisation and characteristically unsuccessful agricultural industry complexes defined the agricultural structure. Overall Poland has middling soil features. Private ownership remained dominant, but could not function efficiently given the strict framework it was forced to operate in. A similar structure existed in the member states of the former Yugoslavia, Slovenia among them.

One of the most significant economic problems collectively burdening all countries was the lack of a land market, which also had an adverse effect on rational land use. Another issue was that collectivisation in effect dissolved small peasant economies. The macro-region appears unified in terms of a large-plant structure preferring large-area agriculture becoming dominant. Alongside its numerous advantages, this approach had problems which still have an effect today.

The period, however, had three unambiguous positive characteristics which laid the foundation for the development of the macro-region's agriculture. First was the Soviet Union's enormous market, capable of absorbing almost all products – especially agricultural and food industry products. The second was cheap energy from the Soviet Union, which not only laid the foundation for industrialisation, but also was the basis of the development of agricultural technology. It was no accident that the communist regime and the Soviet Union's economic and political collapse buried with itself the macro-region's agriculture. The third feature was that even though only a small portion of cooperatives managed their operations efficiently (and not in all countries), there was still a meaningful effect on welfare in rural areas.

Transformation after the regime change

The macro-region's agricultural economy from the first half of the 1990s was characterised by a general and significant drop in production. The 1989–1990 political regime changes in the countries of CEE were accompanied by the transformation of the economic system. The shift from socialist planned economies based on state ownership to a free market was problematic everywhere: the positive experiences of the regime changes were overshadowed by economic crises in all the countries involved. The crisis had disproportionately affected agriculture. Differences between individual development paths were already apparent in the years of transformation, and these became more visible in the agriculture policies of the countries of the macro-region.

Regarding state and cooperative lands, privatisation was the definitive policy in all affected countries. In fact, in most countries this meant full restitution, i.e. the return of lands to earlier owners. In Hungary this restitution was only partial, given that many one-time actual owners had no land to their name, and were given restitution vouchers (essentially financial compensation) instead. Land privatisation had varying effects in countries of different levels of development with different agricultural structures. In those countries where modern, industrialised agriculture based on strong cooperatives had developed – Hungary and the former Czechoslovakia are included in this group – the earlier agrarian population had left the branch and for the most part had left rural areas altogether. In these countries, after privatisation, new landowners were not in a situation where they could cultivate their own lands for lack of motivation and skills/knowledge. It seemed rational that the earlier, still functioning large concerns would continue to use the land, leasing it back from the new owners. As such, the inverse of the historical model emerged in the macro-region (Illés, 2002): whereas in the past the concentrated ownership structure was accompanied by a number of smallholders cultivating a small portion of land leased from estate owners, today typically a fragmented ownership structure is accompanied by a small number of large-scale leasers (large producers), which, from the point of view of management of large producers, is hardly an ideal state. Earlier large producers have transformed into various economic entities. Today the ban on these economic entities acquiring land ownership is at the centre of acute political debate. The solution to this problem would significantly simplify the management of large producers and make their activities more predictable. It would also increase investment incentives in areas where soil improvement programmes, inland water syphoning and irrigation are becoming increasingly necessary.

In those countries where cooperative lands were returned to the original owners, and in Poland, where despite efforts at collectivisation the proportion of cooperative land had remained low, the majority of small plots do not meet the criteria for competitiveness thanks to a lack of capital. In Romania there are a number of microenterprises that do not even produce for the market. In countries where land purchases are not banned (Croatia, Slovenia) the picture is heterogeneous, which was generated primarily by capitalist relations. Capital-rich producers have further fortified and expanded, while those that lacked resources have fallen behind and in many cases have collapsed.

As a result of heritage and privatisation processes, by the turn of the millennium significant differences emerged in holding sizes across the countries of the macro-region (Table 5.3). The concentration of holdings is exceptional in Czechia and significant in Slovakia, while ownership structures are highly fragmented elsewhere. Microholdings (plots less than 5ha) are present at a much higher proportion than in the western half of Europe. Microholdings make up close to nine-tenths of cultivated land in Hungary, Poland, Bulgaria and mostly Romania.

In the years following the regime change (between 1990 and 1995) the GDP produced by the branch decreased by close to 6 per cent per year in the macro-region, and on the whole fell to only 60 per cent of the 1989 level (Csáki, 1998). The decrease in production was characteristic of all the areas of the macro-region in this study. In most CEE countries the recession of agricultural production was more significant than drops in GDP. The low point came around 1993, although Poland and Romania[2] bottomed out slightly earlier (Burgerné Gimes, 2001).

The following are causes of the production decrease (Burgerné Gimes, 2001):

- the shrinking of the market: the decline of the Soviet market was more significant than the increase in exports to the West by several orders of magnitude, while the economic crisis led to a drop in demand in domestic markets;
- a drop in state support;
- decreasing income resulted in the use of less fertilisers and pesticides, which led to weaker crops;
- poor profitability and privatisation had cumulatively resulted in an increase of uncultivated land;
- chaos caused by privatisation.

Burgerné Gimes (2001) identified a number of factors that made future growth possible, with the following being key:

- existing expertise and knowledge, a significant number of well-trained agricultural intellectuals;
- the characteristics of land privatisation resulted in the expansion of the leasing system and forced an increase in the area of land available for lease and a price drop;
- from the mid-1990s support increased and the credit situation of the branch improved to a degree.

The transformation of the branch was unfolding very slowly, and there is no consensus when judging the results. The joint effect of ownership uncertainty and low profitability in many cases has resulted in a *visible* increase in the area of uncultivated lands in the first years of the transition. The use of the descriptor *visible* indicates that the decline of the state of cultivable land makes the problems of the situation of agriculture discernible to the entire society.

In the first years of the transition quite significant changes took place in the branch's weight within the national economies, its contribution rate to GDP and its role in employment in given countries. Decline was particularly significant in Hungary, Czechia and Slovakia, while in Romania and Bulgaria the sector's share continued to increase. In Poland the agricultural sector's contribution to the GDP decreased, but has maintained its significant role in employment (Table 5.4).

The rapid increase in production that many expected never occurred. The transformation of the agricultural sector and the economies of rural areas have lagged behind the scale of changes in the economy as a whole. However, the results of reforms vary across countries (Csáki, 1998).

Table 5.3 Average holding size, 2000

	Average farm size (ha)
Czechia	151.0
Bulgaria	4.4
Hungary	4.7
Poland	6.6
Romania	3.1
Slovakia	50.2

Source: Eurostat.

Increased integration has characterised agricultural commerce in the CEE macro-region, which has become a player in global agricultural commerce. As such, the execution of competitiveness and efficiency factors in the branch has become important and unavoidable. This, however, has generated a contradictory situation regarding rural employment opportunities, with serious arguments taking place between rural developers and those representing the interests of the sector. The sector with spatial and rural development are both characterised by clashing interests and values, and the divergent argumentation and value systems of these two fields are difficult to reconcile.

From the point of view of the agriculture situation, the development of the non-agriculture segment of rural economies is of key importance. The problems of rural areas are not confined to the crisis in agriculture. Non-agricultural processes have also had significant negative effects on rural areas in CEE. Foremost, the industrial crisis bore down on rural areas, given that the so-called 'marginal employees' in industry were typically those who still resided in the countryside and who were most sensitive to drops in employment (given a lack of training and the costs of commuting). The collapse of non-agricultural activities in the 'side branches' of cooperatives also affected those village dwellers who did not work in agriculture. Illés (2002) has shown that in those countries where change to the agricultural structure – preceding structural change in industry – ensued quickly (e.g. Czechia), industry was able to incorporate workers arriving from the agricultural sector. Where the agricultural structure was slow to change or did not change (characteristically Romania, Bulgaria and to a degree Poland), crisis-stricken industry was unable to take on agricultural labourers and as such the ranks of the unemployed swelled. Table 5.4 offers an explanation for what transpired in Romania: the return of industrial workers to rural areas and the agricultural sector appeared to be a conscious method of managing the industrial crisis. In the majority of countries advanced reforms and the (diversifying) rural economy's upturn were able to counterbalance the decrease in the number of agricultural workers, and at the same time improve efficiency and competitiveness. Partly thanks to more favourable features and partly to relatively weak industrial output, the weight of the agricultural sector – along with the food industry attached to it – remained significant in all the countries of the macro-region (Csáki and Jámbor, 2012).

In the years following the regime change the fragmenting sector – which had been affected by unfavourable processes in the external environment and fractured by domestic economic policy mistakes – and rural areas that were facing ever graver issues arrived at EU accession in a

Table 5.4 Changes in the significance of agriculture in the first years of the transition, 1990–1995

	GDP (%)		Employment (%)	
	1990	1995	1990	1995
Czechia	6.3	3.1	9.9	4.7
Bulgaria	11.0	10.8	18.1	22.0
Hungary	15.6	6.4	17.9	8.5
Poland	11.8	6.1	26.4	25.0
Romania	13.7	21.6	28.2	35.7
Slovakia	9.3	5.3	12.1	9.4
Slovenia	4.4	4.4	11.8	10.4

Source: Author's construction based on OECD survey of agriculture in transition countries.

condition in which they did not have clear goals during negotiations and had not solved their own interest and value conflicts.

The macro-region's agricultural sector after EU accession

The accession of the macro-region's countries to the EU was accompanied by the introduction of the only true joint policy of integration in the CAP – even if these countries were temporarily unable to make use of all of CAP's advantages. The previous sections have clearly shown that the integration of the macro-region and its agricultural sectors was to take place in an environment with divergent relations, structures and levels of development, and as a result the new member states were able to utilise advantages and opportunity of membership at different levels. The dangers of membership also appeared at different degrees in given CEE countries. The success of preparatory processes for membership played a role as well. All in all it appears that divergent development paths led to the development of diversity that common policy could not moderate, and the structures, policies and features in the macro-region have further differentiated the sector and the situation of rural areas. Many are of the opinion that the countries that were able to develop more quickly after accession (e.g. Poland) were those where producer support stayed low before CAP and a significant portion of resources (under the programmes ISPA, PHARE and SPARAD) was spent on fortifying competitiveness (here enlargement had an inciting effect on production). The countries that became less successful in the mid-term (e.g. Hungary) were those where market support was characteristic. Partly independently and partly in agreement with Somai (2014) it can be stated that CAP's direct payments created unequal competition conditions between old and new member states. Another justified position holds that antecedents confirm that the form of CAP in CEE is incapable of recognising differences between producers and regional and structural disparities, and in fact does not indicate that it will be able to decrease agricultural developmental differences in Europe (Bazin, 2007).

As such, EU-accession put the macro-region's agricultural sector in a situation of intense competition while the national agricultural policy's autonomy shrunk to a significant degree (Meisel and Mohácsi, 1997). In all countries of the macro-region discussed in this study one of the most important tasks of the sector is to create the basis for competitiveness within the framework of integration and market economy conditions. The ways in which certain member states can meet this challenge is largely dependent on economic and agricultural policy, the quality of the political institutional system, and the degree to which the period of preparation for accession was successful. Divergences in the institutional background have generated significant differences and keep doing so in the developmental opportunities for agriculture and rural areas in CEE states. The causes of diverging development paces vary, including factors like historical heritage (most importantly), positions at the starting line determined by unique characteristics, the stock of tools and machinery (which was outstanding in Slovenia), infrastructural background, the labour force reserve (which was largest in Romania and Poland), ownership structure, etc. Output growth was biggest in Poland, with Polish results standing out from the significant covariance among the macro-region's countries (Table 5.5 and Figure 5.4).

Over a decade after EU-accession, it can be established that agricultural policies can change over time, and as a result so can the circles of countries that produce well or poorly. The performance indicators of various countries are differentiated not only by changes that can be measured during the preparation process, but also in differences in agricultural policy since accession. In the majority of states in the macro-region a dual structure has developed, the essence of which is a duality and coexistence of large concerns and smallholding enterprises. The operations of smallholdings can only be judged productive in Poland and Slovenia, while – as

Table 5.5 Changes in agricultural output since EU accession (2004 = 100)

	2005	2006	2007	2008	2009	2010	2011	2012	2013	2014
Hungary	94.0	91.7	102.0	119.6	89.0	93.4	118.2	114.4	119.0	120.9
Czechia	95.2	99.4	119.3	132.3	102.0	111.8	133.2	133.9	136.0	136.8
Slovakia	90.8	95.1	108.0	126.3	99.6	101.1	123.1	128.5	129.0	128.2
Slovenia	97.4	97.5	103.0	107.7	97.1	101.5	113.5	105.0	105.9	114.3
Romania	94.1	105.2	104.7	133.2	103.5	112.1	132.2	105.5	130.1	122.8
Bulgaria	96.9	100.2	95.7	129.8	110.0	110.3	125.8	127.7	126.9	124.2
Poland	105.8	113.6	141.6	153.5	122.8	139.0	159.9	163.1	167.1	162.0

Source: Author's construction based on Eurostat data.

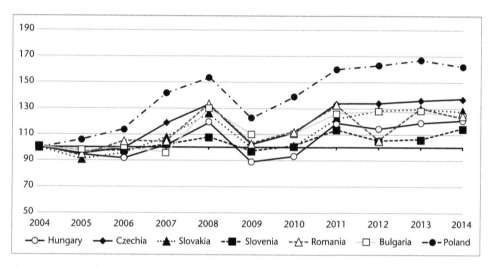

Figure 5.4 Development of agricultural output since EU accession (2004 = 100)
Source: Compiled by the author based on Eurostat data.

articulated by Csáki and Jámbor (2010) – the large concerns bear some elements inherited from the collectivised system, which are disadvantageous and result in low efficiency. Political motivations were more emphatically executed in agricultural policy than economic considerations since accession. A significant portion of structural problems can be traced back to this situation. The trouble with the dual-character system is that there is a lack of cooperation between large and small concerns working alongside one another, i.e. their necessary integration has not been carried out.

A rather interesting debate has emerged concerning desired ownership scales. There are those who would integrate smallholdings into larger concerns, and then those who would break up large-scale operations into smallholdings. An example of such is the divergent set of aspirations on the two sides of the Dráva River, which forms the border between Hungary and Croatia. While in the Slavonia region the inefficient, fractured smallholding structure is seen as the source of problems, and the goal of establishing larger concerns has appeared as a policy aim, on the

Hungarian side of the river concepts on the Ormánság subregion are just the opposite, where many smallholders hope for salvation from structures designed to provide them with opportunities. The situation on both sides of the border is rife with problems.

The Common Agricultural Policy has contributed an increase in prices and incomes, and many producers have been able to take advantage of the common market, while agricultural commerce has also expanded significantly. On the other hand, in competition with capital-rich, multinational commercial and processing industry companies, several CEE players have found it impossible to keep up. This has been positive for consumption, but negative for the majority of producers. The logic of CAP does not favour animal husbandry, which survived a severe crisis after the regime change but has not been very successful since. CAP instead supports crop production. The pork and poultry industries were excluded from the support system.

Actors in the agriculture and rural policy sectors often come into conflict with one another. We frequently hear that sectoral support draws air from the tyres of rural development (agricultural support makes for almost half of the European Agricultural Fund for Rural Development (EAFRD) resources). A characteristic demand holds that instead of supporting agriculture, resources should be directed toward supporting rural settlements, maintaining ecological values, etc. Not to underestimate the strength of the agricultural lobby in this process, but it should once again be emphasised that agriculture is the most important sector in rural areas (especially if we think in terms not only of basic agricultural activity, but also of the full scale of agro-business) and supporting it (to a very significant degree) is justified for several reasons. Foremost, the sector faces serious deficiencies in its technical level, and technical–technological development (Buday-Sántha, 2010) – especially compared to competitors. The gaps and lags experienced in this field can only be closed with intense technical development, although this is a very capital-intensive process. Buday-Sántha also points out 'the development level of agriculture determines what other activities can be developed[3] ... and what landscapes and cultural landscapes greet visitors' (p. 75); and emphasises that up-to-date agriculture can provide long-term employment. Support for the sector is thus justified from employment aspects, which, when they appear as rural development as opposed to sectoral issues, seem remarkably important. Agriculture can only be an essential employer in rural economies if it can live up to its production potential. This, however, necessitates significant reform to the production and ownership structures and the complex development of the sector (understood as agro-business). Industrialised agriculture is increasingly open: an expanding portion of its products is processed by the food industry, and it is increasingly dependent on the output of other sectors (e.g. chemicals, machinery). The term agro-business serves to describe this multifaceted system of relations. That is to say that the term agro-business is the grouping together of provision sectors, agricultural (raw material) production, food industry and food commerce. It seems obvious that from this point of view, the significance of agriculture is several times larger than is reflected in GDP production and its role in the employment structure. If we add to the above today's modern models based on high-tech (gene technology, organic farms operating in special ecological conditions), and markets that are transforming as a result of changing consumer attitudes, we can see that the significance of agro-business will continue to increase. To enhance competitiveness within the common market, it is essential to fortify the point of view of agro-business.

Although considerable changes took place in the ownership structure, the presence of microholdings still dominates in the region. Table 5.6 shows that significant ownership concentration ensued in the early 2000s in Bulgaria and Hungary, and that the exceptionally fractured Romanian ownership structure has remained almost untouched. Should these small family holdings be able to step outside the framework of self-sustenance, their integration is characteristically weak, as is their ability to generate income.

Table 5.6 Changes in ownership structure

	Average size of farms (ha)	
	2000	2012
Bulgaria	4.4	12.1
Czechia	151.0	152.4
Hungary	4.7	8.1
Poland	6.6	9.6
Romania	3.1	3.4
Slovakia	50.2	77.5
Slovenia	5.6	6.5

Source: Eurostat.

Alongside smallholders we find truly efficient large holdings operating on sizeable plots of land. Missing, however, is the mid-sized farming stratum, which is one of the target groups of the EU's agricultural and rural development policy. The number of mid-size (5–50 hectares) holdings is half that found in the EU15 on average. Slovenia and Poland lead in this category, with Slovakia and Hungary lagging somewhat behind, followed by Czechia, where such holdings are more rare.

The macro-region's agricultural sector was unable to completely overcome the shock of 1989–1990, and at the time of accession the productivity of its labour force was low, which, combined with a chronic shortage of capital, served as a brake on the sector and the successful integration of rural areas. In recent years capacity development has become more regular, but results have appeared largely in wheat production (in Hungary, Czechia and Slovakia), and less so in animal husbandry. EU accession has not been able to contribute to increasing agricultural productivity. Labour needs have decreased partly due to rapidly changing production systems (capital is replacing labour, especially in large concerns) (Halamska, 2011).

Accession has hardly assisted the macro-region's smallholdings, as EU agricultural policy has not truly offered/found solutions to their problems. One of the necessary consequences of this is that smallholders often give up the struggle and leave rural areas, where it is difficult to make a living. Smallholdings of insufficient scale continue to be characteristic in CEE rural areas.

Overall, accession has been mainly positive for the macro-region's agricultural sector, but has had a negative effect on the macro-region's rural areas. The latter statement is supported by the fact that the lag of rural space types has continued since accession. The success of European integration demands the redefinition of the most important tasks necessary to develop rural areas in the macro-region, and also the definition of conditions. In most countries this has not taken place or has proceeded insufficiently. In their analysis, Gorton et al. (2009) concluded that in itself CAP is incapable of solving the macro-region's rural development problems. European support is incoherent. The countries of CEE have received the least support – in per hectare, gross national product and output proportion terms (Beke et al., 2011). The same authors have shown that the region is under-supported, even though higher support received through CAP helps make use of potential.

The situation and development of rural areas

The most widely accepted criteria system for demarcating the countryside, or rural areas, was worked out by the Organisation for Economic Co-operation and Development (OECD).

Its essence lies in the fact that the proportion of employed persons working in agriculture has fallen in developed countries to a point where its magnitude cannot be definitive when defining rural areas. As a result we can only distinguish between urban and rural areas using population density indicators. The OECD drew this dividing line at 150 persons/sq.km.[4] There is a degree of fine tuning used when defining rural areas. The population density criterion is used at the level of settlements, and as such settlements with population density lower than 150 persons/sq.km are considered rural, while those above are urban. Only after making this distinction are the proportions of those living in settlements classified as urban or rural within specific regions calculated. These are placed in three categories:

- largely rural, if the proportion of the rural population is over 50 per cent;
- significantly rural, if this proportion lies between 15 and 50 per cent;
- largely urban, if the rural population is less than 15 per cent.

Some 90 per cent of land in CEE is rural. In certain regions, the proportion of the population living in rural areas exceeds 50 per cent. According to EU categorisation, Hungary has the highest proportion of land belonging to the rural category, along with the highest proportion of population in rural areas, whereas Poland and Czechia have the lowest numbers (Table 5.7).

Given that the level of progress and active development of rural areas cannot be separated from the agricultural sector, and considering its status as the most relevant sector, the situation of the macro-region's rural areas largely rests on the productivity of agriculture. However, although the agricultural sector has mostly recovered from the shock that followed the regime change, the same cannot be said for rural areas (or the vast majority of typical rural areas).

The rural population has declined significantly over the last decade, and there is a severe social and demographic crisis in the majority of villages. The most significant exodus from rural areas has taken place in border areas and mountainous areas (Mladenov and Ilieva, 2012).

Localisation, which is connected to globalisation, has led to an upsurge in the value of rural areas and a widening of opportunities through a new prominence of local values and the role of endogenous resources. Under the effects of the world economy, changes in the roles and functions of rural areas have been outlined mostly in countries with more developed market economies, while these processes are not stimulating localities in the countries of CEE. These changes (e.g. the depopulation of rural areas, commuting and social mobility, increasing demand

Table 5.7 Areas categorised as predominantly rural and the proportion of population living in them

	Predominantly rural	
	Area(%)	Population(%)
Bulgaria	54	37
Czechia	48	33
Hungary	66	47
Poland	51	36
Romania	60	45
Slovakia	59	50
Slovenia	59	44

Source: Author's construction based on Eurostat.

for recreational, natural preservation and environmental protection, the new spatial organisation of the economy) appeared in Western European countries in the 1970s and brought to the surface the *need for consistent rural development*. These concerns are further tied to questions of sustainable development, discussed in more detail in Chapter 17. The effects of economic and social changes pushed the emergence of new modes of utilising space (e.g. tourism, recreation, nature preservation, etc.), while rural areas became increasingly differentiated.

The creation of a multifunctional rural area requires success not only in specific functions, but also regarding the prosperity of the macro-region's rural areas. The following are agricultural functions (Gazdag, 2009):

- the safe and uniform-quality domestic provision of food;
- provision of an appropriate export goods base;
- providing means of living for rural populations;
- maintenance of the quality and quantity of agricultural land and the cultural state of landscapes.[5]

Rural areas are to a large degree affected by forces that threaten the fragile social–economic balance of such areas. Weakening economic output is strongly tied to the decline of agricultural activity, which results in the exodus of the young and the ageing, and ultimately in the shrinkage of the population. These in turn lead to the dismantling of services necessary for the provision of an appropriate standard of living. In rural areas ecological and social–cultural functions are known and recognised alongside those of agriculture. The functions of rural areas are as follows (Buday-Sántha, 2011, p. 18):

- *Economic (production) function*: provision of market-ready, income-generating agricultural production (foods, industrial raw materials), offering adequate means of income for producers, increasing productivity, technological development and innovation.
- *Ecological function*: which includes the following elements:
    - protection of natural elements (water, land, air);
    - protecting biodiversity, diversity of the living world, with an emphasis on genetic diversity and the functioning of the ecosystem;
    - maintaining the cultural state of the landscape;[6]
    - protection of wild animal and plant life through legal means and providing appropriate ecological conditions.
- *Socio-cultural function*: assistance of rural communities and populations through providing an appropriate standard of living for them in the interest of encouraging self-government, allowing locals to control their values and opportunities. Having local populations ready and able to maintain national, cultural values and traditions.

EU development policy has attempted to put a stronger emphasis on rural development. This is indicated by the fact that rural development (EAFRD) has been placed in the fund that serves agricultural development, which as an independent pillar aims to strengthen diversification in line with the functions discussed above. The increase in the emphasis on rural development came at the same time as Eastern expansion – although the two processes were independent of one another. As such, the agricultural and rural development policies of CEE countries within CAP serve sectoral, structural policy elements and territory-specific elements.

Rural development has a defined programme and regulation regime approved and supervised by the European Commission and the national governments. A targeted institutional system has been built to carry out this programme as the second pillar of agricultural policy. Despite significant resources having been put aside, rural development has access to considerably fewer resources than agricultural support to reach rural development goals. The resources for rural development actions (axes 3 and 4), unlike that of agricultural support, are not normative but competitive instead, meaning they are transferred to the final beneficiaries through a tendering system. The political goals and actions of rural development policy are articulated by the governments of member states in accordance with EU principles and directives.

Having analysed development strategies designed to access EU (rural development) resources and exclusively studying documents from various CEE countries, the macro-region's rural areas appear remarkably homogeneous, although the above have shown a significant degree of variation. For the most part there is no difference between the definitions of the most pressing problems and the demarcation of the most important goals. *Strategies following the logic of EU development policy appear one-size-fits-all. Everywhere the goal seems to be the improvement of living standards and employment opportunities of rural populations –characteristically through economic growth.* Demand for the support of 'sustainable development' is stated without defining its content (c.f. Chapter 17), and the same holds for improving the competitiveness of agriculture.

The situation, prospects, difficulties and opportunities of the macro-region's agriculture and rural areas are well summarised using SWOT – strengths, weaknesses, opportunities, threats – analysis (Table 5.8).

Jámbor (2014) evaluated ten years post-accession in a study. He weighed agricultural output and rural performance equally. His general conclusion held that everyone benefited from accession, but while Poland was a clear winner, Slovakia and Hungary were least able to profit from the opportunities at hand. Regarding rural performance, rural populations grew in Poland,

Table 5.8 The SWOT analysis of the macro-region's agriculture and rural areas

Strengths	*Weaknesses*
• Favourable, sometimes exceptional cultivation features and natural conditions in several countries • Absorbable labour force • Outstanding higher education base in some countries (esp. Hungary), high level of expertise • Low proportion of changes to the natural environment • Tourism potential • Cultural and natural values • Favourable organic farming conditions	• Underdeveloped social infrastructure • Emigration • State of transportation infrastructure • Significant regional differences • Lack of capital • Low mobility • Issues in ownership structure • Weak agro-business
Opportunities	*Threats*
• Geopolitical situation • Expansion of market for organic products • Single European Market • Income increase for agricultural concerns	• Increase in spatial differences • Climate change – increased production risks • Loss of position in Single Market • Cutbacks in EU development support • Dangerous impoverishment of rural areas

Source: Author's construction.

Czechia, Slovakia and Slovenia, while the largest decrease occurred in Hungary (14 per cent). According to Jámbor's figures rural employment decreased the most in Hungary, Slovenia and Czechia. To measure rural performance the author employed GDP differences between urban and rural areas, highway density and the proportion of early school leavers. Based on these measurements rural performance was best in Poland, Czechia and Slovakia, while Hungary's was poorest among the countries that joined the EU in 2004.

LEADER programmes in the countries of Central and Eastern Europe

The LEADER programme (from the French: Links between the Rural Economy and Development Actions) has been deemed successful by many. Evidence for such, for example, lies in the fact that since 2007, after three experimental programming phases, the programme was integrated into the EU's mainstream rural development policy, despite the fact that with its grass-roots construction and philosophy based on independent local decisions, it is difficult to harmonise with centralising government policies and EU programmes that try to supervise down to the finest details. Christopher Ray (2000) has called this rural development method and programme post-modern, or a kind of 'anarchist' programme, given that its philosophy does not conform to European and national administrative decisions. Instead, from planning to programme management, from tender to execution, decision-making rights are in the hands of the cooperating local communities. This approach contradicts the agricultural- and rural-development programmes' top-down, tightly regulated and supervised bureaucratic nature.[7] Given this, it is clear that this is a programme whose successful execution and targeted, efficient use of resources depend on frameworks constructed at the member-state level and the level of preparedness and creativity of local communities.

In their study of Poland, Czechia, Lithuania and Hungary, Chevalier and Maurel (2010) examined what social–economic and political context awaits the practice of EU rural development in rural areas in CEE, and how this modifies their effective mechanisms. Within this, the authors emphasise the nature of 'policy transfer and policy reception processes', claiming that their results depend on the national and local level 'institutional opportunity structure'. The policy transfer-reception cycle has four key players:

- process manager: the initiator of the policy model;
- operators: the key decision-makers;
- facilitators: those helping to spread the model;
- recipients: those utilising the model.

In the opinion of the authors, the regulation of EU rural development policy is based on 'soft' regulations, where member states and regions enjoy a high degree of relative autonomy and are somewhat free to modify the model. At the same time, they also show that the policy transfer-reception process is significantly affected by the given country's political system, institutions and normative framework. Common characteristics include a centralised nature, a lack of transparency and the relative slowness of both the evaluation of strategies recommended by local actors and financing. This observation is connected to the dilemma that affects the essence of the programme, whereby the low level of social capital in certain rural areas is accompanied by concerns over the ability of the LEADER approach to revive the existing capacities of local actors – and whether locals can construct the programme as initiators. *This discussion leads to the conclusion that a top-down policy that attempts to support endogenous models of development is rife with internal conflicts. It is important to consider whether, amongst such conditions, general guiding principles are sufficient, or whether these principles will be undermined, significantly compromising the effectiveness of these endogenous models.* Chevalier et al.'s (2012) CEE study[8] led to the conclusion that although the macro-region is very enthusiastic about taking on those projects that have proven successful in old EU member states, and have high expectations of them, this spiritedness is short-lived in most of the new member states. The beneficiaries of development and local communities feel too much centralisation and bureaucracy. The tendering mechanism is slow, lags are regular and regulations are too strict. Problems exist in all the examined CEE states, but not with the same severity. This is one of the reasons why the effectiveness of utilising resources varies among states. While Czechia is largely a positive example, Hungary, Romania and Bulgaria are negative. In these countries, disappointment is characteristic in the majority of local action groups, and as such this method leads to underperformance in making the most of the potential of important rural development programmes.

Summary

Agriculture is not the only source of rural welfare, but it is one of the most important and will remain so. The further modernisation of the sector, high-level management and real and measurable performance are necessary. Lacking these, there is little chance for the satisfactory development of rural areas. There is a need for innovations, the successful adaptation of the most developed technologies, bigger risk-taking (and an institutional system that supports this), integration and the establishment of cooperatives.

Output is increasingly shifting toward crop production (with large-holding cultivations modes), which is partly a natural result of the ownership structure. Should we expect agriculture to help develop rural areas, the structure of agriculture must be diversified. Opportunities must be provided for more labour-intensive branches (vegetables, fruits), animal husbandry and related processing industries must be developed and, above all, various integration programmes must be supported. By putting the agricultural structure and performance at the centre, the sector can be made capable of having a meaningful role in improving the quality of life in rural areas.

In the interest of developing rural areas there is a need to incorporate capital, from foreign sources as well as domestic. Through modernisation, technical development and lowering self-financed costs the opportunity for the agriculture, food industry plants to become competitive vis-à-vis imports can be secured. Beke et al. (2011) have established that – with meaningful capital injection – the agricultural sectors in Poland, Romania, Slovenia and Hungary have significant developmental potential.

Acknowledgement

The publication of this research has been supported by project #104985 of the Hungarian National Research, Development and Innovation Office.

Notes

1 There were significant differences even between the *Baltic States*, which had lost their independence and were annexed by the Soviet Union. They not only had differing agricultural product structures, but divergent heritages of ownership structure. Close to half of Estonia's land is forest, and the vast majority of its agricultural lands are arable lands. In Latvia, animal husbandry is definitive (milk products, eggs), while in Lithuania, arable land cultivation is outstanding, including wheat production. The land reform of 1920 in Latvia brought to life a viable middle landowning stratum, while the large estate structure in

Lithuania was largely unchanged. After Soviet-style collectivisation, the nationalisation of land and the establishment of *kolkhoz*, household operations were still worth mentioning.
2 In the years of transition, Romania's agricultural sector managed to grow somewhat, but this was basically a rebound from the catastrophic situation in previous years, and is not an indicator of the success of Romanian agriculture.
3 For example, processing of agricultural and forestry products, bioethanol and biodiesel production, biomass energy production, biogas development, etc.
4 Illés (2008) recognises the merits of the definition, but points out that it can hardly be utilised in areas outside the OECD, and its viability is limited in Central and South-Eastern Europe.
5 The sector can deliver all these functions when ownership relations, operations and organisational structures are appropriate and scales deliver economically efficient operations. Gazdag (2009) claims the appropriate framework for such is not small and mid-size firms, but competitive, appropriately scaled multifunctional operations capable of diversifying.
6 The cultural state of the landscape has a definitive role in the formation of a country's image, and for this reason the development of the state of rural areas is an important condition.
7 Naturally, to be able to experiment and for the programme to be able to diverge from main rural development programmes, it was necessary to have such decisions, institutions and regulations from European and national decision-making bodies that facilitate the operations of a decentralised programme. This also necessitates those local actors who can be addressed with such programmes. Evidently those regulations that concern the spending of public funds for development goals remain in force.
8 ALDETEC: Action Locale et Développement du Territoire en Europe Centrale.

References

Bazin, G., 2007. Politique agricole commune à l'Est. Premiers résultats. *Le courrier des Pays de l'Est*, 1063, pp. 4–11.

Beke J., Forgács, A. and Tarján, T., 2011. Európai uniós országcsoportok mezőgazdasági teljesítményének összehasonlító vizsgálata [Comparative study of agricultural performance in EU country groupings]. *Gazdálkodás*, 55(1), pp. 39–51.

Beluszky, P., 2010. Mi a helyzet falun [What is the situation in villages?]. In: A. Fábián, ed. *Párbeszéd és együttműködés: Területfejlesztési szabadegyetem, 2006–2010*. Sopron: Nyugat-magyarországi Egyetemi Kiadó, pp. 83–96.

Bourdeau-Lepage, L. and Bazin, G., 2011. L'agriculture dans les Pays d'Europe Centrale et Orientale: continuité et adaptation. *Économie rurale*, septembre–décembre. http://economierurale.revues.org/3214 [Accessed 10 January 2016].

Buday-Sántha, A., 2010. A vidék sorsa, a vidék helyzete, vidékfejlesztés [The fate of rural areas, the situation of rural areas, rural development]. In: A. Fábián, ed. *Párbeszéd és együttműködés: Területfejlesztési szabadegyetem, 2006–2010*. Sopron: Nyugat-magyarországi Egyetemi Kiadó, pp. 67–82.

Buday-Sántha, A., 2011. *Agrár- és vidékpolitika* [Agricultural and rural policy]. Budapest, Pécs: Dialóg Campus Kiadó.

Burgerné Gimes, A., 2001. *A közép-európai országok gazdaságának és mezőgazdaságának összehasonlító elemzése* [The comparative analysis of the economies and agricultural sectors of Central European countries]. Budapest: Századvég Kiadó.

Chevalier, P. and Maurel, M.-C., 2010. Policy Transfer of the Local Development Model. The Leader Program Implementation in Central European Countries. Paper presented at the Regional Studies Association Annual International Conference, Pécs, May.

Chevalier, P., Maurel, M. and Póla P., 2012. L'expérimentation de l'approche LEADER en Hongrie et en République Tcheque: Deux logiques politiques différentes. *Revue d'études comparatives Est-Ouest*, 43(3), pp. 91–143.

Csáki, Cs., 1998. Közép-Kelet-Európa és a volt Szovjetunió agrárgazdasága a kilencvenes évek második felében [The agricultural economy in CEE and the former Soviet Union in the second half of the 1990s]. *Közgazdasági Szemle*, 45(3), pp. 203–222.

Csáki, Cs. and Jámbor, A., 2010. Five Years of Accession: Impacts on Agriculture in the NMS. *EuroChoices*, 9(2), pp. 10–17.

Csáki, Cs. and Jámbor, A., 2012. Az európai integráció hatása a középkelet-európai országok mezőgazdaságára [The effect of European integration on the agriculture of CEE countries]. *Közgazdasági Szemle*, 59(7–8), pp. 892–910.

Darrot, C., 2008. Les paysans polonais à l'épreuve de la PAC, une analyse multi-disciplinaire d'un référentiel de politique publique. Rennes, thèse de doctorat, Agrocampus-Ouest.

Farkas, M. and Szabó, P., 2014. Regionális térszerkezeti sajátosságok Kelet-Közép-Európában és országaiban [Regional spatial structure uniqueness in CEE and its countries]. *Közép-európai Közlemények*, 7(3–4), pp. 120–132.

Gazdag, L., 2009. Magyarország útvesztése – a rendszerváltás közgazdaságtana [Hungary losing its way – economics of the transition]. Budapest: Mundus.

Gorton, M., Hubbard, C. and Hubbard, L., 2009. The Folly of European Union Policy Transfer: Why the Common Agricultural Policy (CAP) Does Not Fit Central and Eastern Europe. *Regional Studies*, (43)10, pp. 1305–1317.

Halamska, M., 2011. *The Polish Countryside in the Process of Transformation 1989–2009*. Polish Sociological Review, 1, pp. 35–54.

Horváth, Gy. ed., 2015. *Kelet- és Közép-Európa régióinak portréi* [Portraits of the regions of Eastern and Central Europe]. Budapest: Kossuth Kiadó.

Illés, I., 2002. *Közép- és Délkelet-Európa az ezredfordulón. Átalakulás, integráció, régiók* [Central and South Eastern Europe at the turn of the millennium. Transformation, integration and regions]. Budapest, Pécs: Dialóg Campus Kiadó.

Illés, I., 2008. *Regionális gazdaságtan – területfejlesztés* [Regional economic studies – spatial development]. Budapest: Typotex.

Jámbor, A., 2014. Tíz évvel a csatlakozás után: az új tagországok agrárteljesítményei [Ten years after accession: The agricultural performance of new member states]. *Gazdálkodás*, 58(6), pp. 508–518.

Káposzta, J., 2014. Területi különbségek kialakulásának főbb összefüggései [The key correlations of the development of spatial differences]. *Gazdálkodás*, 58(5), pp. 399–412.

Lieskovský, J., Bezák, P., Spulerova, J., Lieskovský, T., Koleda, P., Dobrovodska, M., et al., 2015. The Abandonment of Traditional Agricultural Landscape in Slovakia – Analysis of Extent and Driving Forces. *Journal of Rural Studies*, 37, pp. 75–84.

Maciulyte, J., 2004. La recomposition de l'espace rural lituanien dans la perspective de l'intégration européenne/The reconstitution of the Lithuanian rural space in light of the European integration. *Annales de Géographie*, 636, pp. 188–210.

Meisel, S. and Mohácsi, K., 1997. Az Európai Unióhoz való csatlakozás néhány élelmiszergazdasági összefüggése [Some food economy correlations of accession to the European Union]. *Közgazdasági Szemle*, 44(3), pp. 217–232.

Mladenov, C. and Ileiva, M., 2012. The Depopulation of the Bulgarian Villages. *Bulletin of Geography. Socio-economic Series*, 17(17), pp. 99–107.

Pospěch, P., 2014. Discursive No Man's Land: Analysing the Discourse of the Rural in the Transitional Czech Republic. *Journal of Rural Studies*, 34, pp. 96–107.

Ray, C., 2000. The EU LEADER Programme: Rural Development Laboratory. *Sociologia Ruralis*, 40(2), pp. 163–171.

Somai, M., 2014. Tíz éve az EU-ban: Az agrártermelés fejlődése [10 years in the EU: the development of agricultural production]. http://vilaggazdasagi.blog.hu/2014/11/04/10_eve_az_eu-ban_az_agrartermeles_fejlodese [Accessed 10 January 2016].

6
Factors influencing regional entrepreneurial activity in Central and Eastern Europe

Balázs Páger

Introduction

The importance of regional entrepreneurial activity has increased in the last few years, and many scholars have started to research different aspects of this topic. Regional-level investigation of entrepreneurial performance and the effects of entrepreneurship on regional development have quickly become popular research topics since the last decade. The shifted economic environment, global competition, new scientific results and new communication tools have all supported their growing importance. This does not mean that entrepreneurs were not important in the beginning of the last century. However, accelerated economic processes and competition among enterprises and countries now require flexible business units, which can respond to negative and positive externalities faster than large firms. Several relatively young industrial sectors have developed in the last few decades. These industries are based on new products and services, and a high number of fresh start-ups have come to life within their framework. These newcomers may also stimulate competition, the division of labour and the introduction of innovations (Glaeser et al., 1992; Ács and Armington, 2004). Hence, greater variety may have an indirect influence on regional development (Boschma, 2004; Fritsch, 2012). A new firm may inject diversity into the market and 'entrepreneurship is an important source of diversity by transforming knowledge into economic knowledge that otherwise would have remained uncommercialised' (Audretsch and Keilbach, 2004, p. 608). Therefore, new firms and enterprises may play a significant role in regional economies due to the knowledge and the novelties that they bring into the market. Audretsch and Thurik (2001) summarised the changes concerning the role of entrepreneurship in 14 trade-offs. These mark the difference between a 'managed economy' and an 'entrepreneurial economy'. The emerging concept of 'entrepreneurial economy' has been characterised by small and medium firms whose strategy is built on diversity and flexibility. The role of local policies and the focus on local and regional space became more significant than in a managed economy (Audretsch and Thurik, 2001; Audretsch, 2009).

In this chapter I investigate entrepreneurial activity in the Central and Eastern European regions. These territories were facing hard and significant challenges in the last decades while having to find their own development paths, and new firms have been instrumental in realising this development potential. Smallbone and Welter (2001) describe different courses for enterprises in transition economies, among them the Central and Eastern European countries. In the planned economies of these countries, a limited kind of entrepreneurial activity and mindset could already be observed. Since the regime change, regulations and the economic environment have changed to provide opportunities for individuals to start their own private businesses. Several people set up their own firms, thus an 'entrepreneurial boom' occurred in the early 1990s in the CEE countries (Smallbone and Welter, 2001). This 'boom' was characterised by an increasing rate of unemployment (thus people became self-employed), as well as by the emerging new business opportunities. The institutional framework for private firms was created more or less on a similar basis across the CEE countries. The EU accession process also played an important role in the evolution of these institutions (Smallbone and Welter, 2012). Since the transition period, many positive and negative effects have influenced the formation of new firms – such as Foreign Direct Investment (FDI), support from the European Union, legal barriers, negative opinions about entrepreneurs or the depressed economic heritage of specific regions. Some of the above are national factors (like acts or regulations), but there are regional components as well, e.g. the regional socioeconomic environment, which is able to create an adequate background for new businesses.

This study focuses on the regional factors that may influence the establishment of enterprises in the CEE regions. Although I focus on the number of new firms, I also claim that the qualitative factors of regional entrepreneurship are just as important (Szerb et al., 2013). Qualitative differences between entrepreneurial activities in the regions may influence the performance of the discussed enterprises. These aspects primarily depend on individual abilities and aspirations, but also on certain regional characteristics, according to which a person starts or operates a new business. In the next section I will review the theoretical background of this study. The data and methods I used are described in the third section, while the results of the investigation are reported in the fourth section. Conclusions and further research orientations will be summarised at the end of the chapter.

Theoretical background

The literature discussing entrepreneurial motivation, aspirations and the effects that new enterprises may have on economic development has broadened widely in the last few years. In this section I will summarise the most important theoretical concepts and empirical findings regarding entrepreneurial activity and performance.

Several recent studies have proved that new enterprises have a positive effect on economic growth in developed countries (see among others Ács and Audretsch, 1988; Ács and Varga, 2005; van Stel et al., 2005; Ács and Szerb, 2007). New firms have an impact on regional development over a longer time period as well (Audretsch and Fritsch, 2002; Fritsch and Müller, 2004), and they may also influence regional employment – but the latter also depends on regional productivity level (Fritsch and Müller, 2008). However, the relationship of entrepreneurial activity and economic growth is not unambiguously positive. Namely, a relatively high number of entrepreneurs (self-employers) can be observed in countries with relatively high unemployment rates. The probability of finding necessity-motivated firms is higher in these economies (Fernández-Serrano and Romero, 2013). As economic development and individual income increase, the number of new entrepreneurs will start to decrease, but after reaching a threshold it will rise again. However, these individuals are actually 'real entrepreneurs', who can positively affect economic development because of their opportunity-driven motivation (Wennekers and Thurik, 1999). Autio et al. (2014) note that evidence shows that quality also matters in entrepreneurship. The above-mentioned diversity of entrepreneurial motivations and the qualitative aspects of low-income and high-income countries also indicate that the context around the individual has an important role (Autio et al., 2014). Qualitative differences among entrepreneurial activities and regions may influence the performance of the enterprises. These aspects depend on individual abilities and aspirations, but also on certain regional characteristics, according to which a person starts or operates a new business. The Global Entrepreneurship Index (GEI) and the Regional Entrepreneurship and Development Index (REDI) were also developed with this in mind. These indicators argue that the effect of quantity-based entrepreneurial activity rates and index values on economic development is ambivalent (Ács et al., 2013, 2014; Szerb et al., 2012, 2013, 2014).

Entrepreneurial activity can be derived from different motivations, which influence entrepreneurial aspirations. According to Ács et al. (2008), three types of motivations can be distinguished: independence, wealth increase and necessity. Attitudes and entrepreneurial skills influence the creation of start-ups, as well as innovations introduced at the firm level. Thus they can influence competition, variety and also the establishment of further new enterprises (Wennekers and Thurik, 1999). Moreover, entrepreneurial behaviour and attitudes have an important role in creating new businesses on the regional level (Tamásy, 2006). An adequate entrepreneurial climate positively influences the establishment of new firms, thus regional policy-makers should focus on indirect tools (like improving regional entrepreneurial attitudes) in order to increase regional entrepreneurship rates (Bosma and Schutjens, 2011). Regional entrepreneurial culture also has a positive effect on regional entrepreneurial attitudes (Beugelsdijk, 2007; Fritsch and Wyrwich, 2014).

Individual attitudes are influenced by the business environment as well. Namely, different economic conditions may constitute disparities in the business environment in which local individuals and local enterprises are embedded. Objective regional attributes influence opportunity perception, and they also have an effect on entrepreneurial motivations and attitudes (Kibler, 2013; Stützer et al., 2014). If one looks at economic performance, competitiveness and the institutional sets of European countries, many differences can be observed between them and also within them. The institutional set may influence different (formal and informal) rules and has a crucial role in determining entrepreneurial attitudes and aspirations through these rules (Minniti, 2008). Regional circumstances (institutions, economic climate) and the individual's direct surroundings ('macro- and microsocial environment') will affect the personal decision to start a new enterprise (Feldman, 2001; Wagner and Sternberg, 2004). The microsocial environment involves among other things personal perception and entrepreneurial motivation. Macrosocial environmental factors include the cultural, social, political and financial conditions of the region. Social and cultural norms, the development of infrastructure and socioeconomic aspects (such as higher education, unemployment rates or social capital) also belong to the regional conditions that influence regional entrepreneurial activity (Wagner and Sternberg, 2004). According to Stam (2010, p. 141), 'entrepreneurship is the result of the interaction between individual attributes and the surrounding environment'. All of the components of macro- and microsocial environments are summarised in Figure 6.1.

I am investigating not only entrepreneurial activity in general, but also the same process in various industry sectors. This decision was supported by the assumption that entrepreneurial activity may depend on the given economic activity as well, since entry barriers are different in each sector (Brixy and Grotz, 2007; Kibler, 2013).

This study focuses on the macrosocial environment of the regions, especially on their socioeconomic features. The regional socioeconomic environment was examined from four aspects: regional economic performance; rate of persons with tertiary education; science-related factors (patents and human resources in science and technology sectors); and the entrepreneurial climate. Hypotheses were lined up along these dimensions.

The effect of *economic development* on new enterprises has been described by the relationship of increasing economic performance and necessity- and opportunity-driven entrepreneurship,

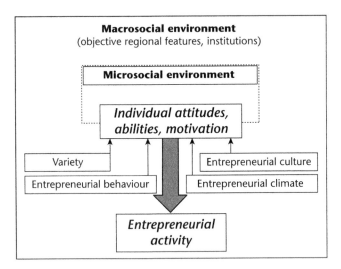

Figure 6.1 The concept of the study

Source: Author's construction.

respectively (Wennekers and Thurik, 1999; Bosma and Harding, 2007; Ács et al., 2008). My first hypothesis is the assumption that economic performance affects regional entrepreneurial activity in a positive way.

Next, I investigate the possible effect of persons with tertiary education. Graduates may have better skills and knowledge for leading a firm. Universities may be the source of new knowledge, and innovations may stimulate the establishment of enterprises (Feldman, 2001; Audretsch and Keilbach, 2004; Stam, 2010). The presence of universities is especially valid for new firms in high-tech manufacturing and other knowledge-intensive sectors (Baptista and Mendonça 2010; Baptista et al., 2011; Fritsch and Aamoucke, 2013). Therefore it is assumed that higher rates of graduates influence regional entrepreneurial activity positively.

My next assumption is about science-related factors. Two factors were used to measure this aspect: the number of patents and the rate of science and technology sector workers with graduate degrees. This assumption is based on the commercialisation of knowledge and the importance of innovations regarding new products that are new findings in the literature. However, the question of patents may also be influenced by other components (like protection and institutions) (Ács and Sanders, 2012). It is assumed that these science-related factors have a positive impact on entrepreneurial activity.

The density of existing firms refers to the entrepreneurial climate in a given region and economic sector. If this indicator shows high values, it points to an entrepreneur-friendly business environment (Wagner and Sternberg, 2004; Sternberg, 2009; Stam, 2010). Data were collected from all sectors, and also specifically from the industrial branches. Besides the factor of existing firms, it is assumed that regional concentration of the industrial sector (measured by location quotient (LQ) for industries) may also stimulate regional entrepreneurial activity in industrial branches. Therefore, the LQ indicator was applied in the industrial sector investigation (but not in the all-sector analysis).

Data and methods

In order to measure the number of new firms, I used business demography data published by Eurostat. Among other data, it contains regional statistics about the number of active, newly established and newly liquidated firms in a given year and economic sector. Data were available for 15 countries at the NUTS-3 level from 2008 to 2010. Since a significant amount of data was missing for 2010, this year was excluded from the investigation and I focused on data from 2008 and 2009. As mentioned in the previous section, industrial sector firms received special attention in the analysis. Thus, the analysis had two lines: all economic branches and the industrial sector. According to NACE2 classification, this involves classes from B to E. In order to compare regions and conduct an analysis, the number of established firms per 1,000 inhabitants (B1kPOP), and a similar variable for liquidated firms (D1kPOP) were created. In order to compute these two variables, I used the sums of established and liquidated firms for 2008 and 2009, as well as the average regional population in the mentioned years. These variables were calculated for all of the available NUTS-2 regions, and also for the involved NUTS-3 territories.

This study focuses on the Central and Eastern European regions, hence I included 190 NUTS-3 regions from seven countries (Bulgaria, Czechia, Hungary, Poland, Romania, Slovakia and Slovenia) in the investigation. The descriptive statistics of the variables were checked for extreme values and high skewness that might disfigure the results. If skewness values were out of the range [−1;1] the variable was altered. The calculated new variables indicated much higher positive skewness values, thus I applied logarithmic transformation to handle them (Table 6.1).

The regional socioeconomic context was described by factors related to regional economic performance, regional employment, higher education, patents and the density of previously existing firms. Data for these dimensions were collected for an earlier period than for the case of regional entrepreneurial activity indicators, because I assumed that the effect of these socioeconomic indicators might have had an impact on entrepreneurial activity later. Regional economic performance was described by the *average value of regional GDP per capita PPS* (NUTS-3 level) for the period 2003–2007, and by the *average value of disposable income per capita* (NUTS-2 level) for 2003–2007. The presence of graduates in a given region was measured by the *average rate of people with tertiary education* in the age group 25–64 years, from 2003 to 2007. This figure was available on the NUTS-2 level. There were two science-related factors: *patents per 1 million inhabitants* and *the rate of human resource: graduated from tertiary education and working in science and technology-related sectors* (2003–2007). The number of patents was summarised for the period of 2003–2007 on the NUTS-3 level. The density of existing firms was measured by the *number of local units per 1,000 inhabitants* on the NUTS-2 level for the period 2003–2007, first in all sectors, then in the industrial branches. The concentration of the industrial sector was described by *LQ* measures (for 2003–2008), where regional employment in industrial branches was compared to the whole examined territory, and not only to the country of the given region (Table 6.2).

As a first step, regions were clustered according to the number of established firms and liquidated firms. To determine the number of groups, hierarchical cluster analysis was used with the Ward method on the logarithmic values of the variables. This analysis suggested six groups for NUTS-3 level analysis. For identifying the different groups, k-means cluster analysis was used. In the case of the CEE regions, the analysis of variance (ANOVA) values of k-means cluster analysis showed that six groups may be a good solution, but the 7- and 5-group versions were also tested. In the end, the 5-group version was chosen. Similar results and numbers of groups were achieved in the case of the industrial sector analysis (Table 6.3).

As a next step, I repeated the classification to identify those socioeconomic factors that may have an influence on entrepreneurial activity. This was an important step in order to see which

Table 6.1 Descriptive statistics of the variables related to regional entrepreneurial activity

	N	Min	Max	Mean	Variance	Skewness		Kurtosis	
	Statistic	Statistic	Statistic	Statistic	Statistic	Statistic	S.E.	Statistic	S.E.
bper1kpop	190	2.19	39.90	12.25	33.852	.958	.176	2.76	.351
dper1kpop	190	2.53	27.87	9.60	14.603	1.07	.176	2.45	.351
LOG_bper1kpop	190	.34	1.60	1.03	.054	-.73	.176	.21	.351
LOG_dper1kpop	190	.40	1.45	.95	.031	-.35	.176	.39	.351
bper1kpop (ind)	190	.23	5.29	1.14	.570	2.07	.176	6.58	.351
dper1kpop (ind)	190	.30	5.00	1.05	.416	2.94	.176	11.48	.351
LOG_bper1kpop (ind)	190	-.64	.72	-.024	.068	.09	.176	-.26	.351
LOG_dper1kpop (ind)	190	-.52	.70	-.035	.042	.72	.176	1.38	.351

Source: Author's calculations and construction.

Table 6.2 Descriptive statistics of the regional socioeconomic factors

	N	Min	Max	Mean	St. Dev.	Skewness	Kurtosis
	Statistic	Statistic	Statistic	Statistic	Statistic	Statistic	Statistic
Patents per 1 million inhab.	190	.10*	577.30	31.04	70.28	4.839	27.598
GDP per capita PPS	190	3840	40220	10753	5442	2.180	7.111
Rate of higher educated pop.	190	.07	.32	.15	.05	1.067	1.449
Disposable income per capita	190	2800	12560	6183	2238	.523	-.159
Local units per 1,000 inhab. (all sectors)	190	5.79	150.62	34.78	20.17	1.709	5.637
Local units per 1,000 inhab. (industry sector)	190	.98	17.89	5.42	3.49	2.168	5.291
LQ of industry	190	.36	1.81	1.06	.31	.041	-.691
HRSTC**	190	.06	.22	.11	.04	1.020	1.127

Source: Author's calculations and construction.

Notes: * The original minimum value was zero, but we had to apply a positive technical minimum because of the logarithmic transformation.
** Measures the rate of persons who graduated from tertiary education and work in science and technology-related sectors.

Table 6.3 ANOVA results in the cluster analysis for the CEE regions

	Number of groups	Clustering variables	Cluster Mean Square	df	Error Mean Square	df	F	Sign.
All sectors	6 groups	LOGBper1Kpop	1.897	5	.004	184	440.04	.000
		LOGDper1Kpop	1.008	5	.004	184	229.49	.000
	5 groups	LOGBper1Kpop	2.367	4	.004	185	538.42	.000
		LOGDper1Kpop	1.234	4	.005	185	251.03	.000
Industry sector	6 groups	LOGBper1Kpop (ind)	2.349	5	.007	184	359.80	.000
		LOGDper1Kpop (ind)	1.339	5	.006	184	208.76	.000
	5 groups	LOGBper1Kpop (ind)	2.925	4	.007	185	435.53	.000
		LOGDper1Kpop (ind)	1.622	4	.008	185	215.84	.000

Source: Author's calculations and construction.

factors are related more strongly (and which more weakly) to entrepreneurial activities. Our assumption was that if a factor played a relatively significant role in the reclassifying, it might have a higher influence on regional entrepreneurial activity. I used discriminant analysis for this part of the investigation. The grouping variable was the classification of regions; it was created previously in the cluster analysis. The independent variables were the socioeconomic factors. Discriminant analysis has stricter assumptions for analysing categorical variables than logistic regression, and one of them is that the predictor variables have to be normally distributed. Thus, variables that indicated high positive or negative skewness had to be transformed (see Table 6.2). In this case, I applied logarithmic transformation, as in the case of the original grouping variables. There are two different methods (enter and stepwise) for running a discriminant analysis. Our investigation was carried out using both of them, but the stepwise method was applied using Wilks's lambda values. This method might help us distinguish the stronger and weaker predictor variables by creating different discriminant functions. The variance of dependent indicators is explained by the independent factors through these functions.

Results of the cluster analysis

Five groups of regions in the case of the all-sector analysis, and also five clusters regarding the industry branches were determined for the CEE countries. Since NUTS-3 regions have been analysed, we also attempted to discover the inner disparities within the regions. The clusters were ranked according to the number of established firms per 1,000 inhabitants, but we must note that the rate of activity (rate of established and liquidated firms) may differ from this ranking (Table 6.4).

Regarding all sectors, the regions performing best are the capital cities, the Polish metropolitan regions and certain areas with special characteristics (such as sea harbours on the Adriatic Sea or the Black Sea coasts). Czechia, Hungary, Poland, Slovakia and Slovenia show relatively good activity in general. Compared to these countries, less entrepreneurial activity can be observed in Bulgaria and Romania, but there are disparities within these countries as well. The worst-performing regions are located in Romania. Some counties – which boast more highly populated cities (Cluj-Napoca, Timișoara, Oradea or Brașov) – perform above the national average, but many Romanian territories have the lowest activity among all the CEE regions. These findings show notable parallels with urbanisation patterns (Chapter 8) and regional resilience (Chapter 15).

The results regarding the industrial branches are different from the all-sector investigation, although the main trends are similar. Hence, the Czech, Polish, Slovakian and Slovenian regions have the best performance in general. Compared to them, Bulgarian, Hungarian and Romanian territories have a lower average in measured activity. We observed that the entrepreneurial activity in industrial branches is also related to the industrial structure of the given region. Thus, the capital city and the most populated towns have a relatively moderate performance, since their agglomeration (surrounding counties) show higher entrepreneurial activity (Figure 6.2).

The results proved that more developed regions have higher entrepreneurial activity in the case of all sectors. However, the fact that the used data measured the situation in 2008 and 2009 should be taken into account. Therefore, it should be noted that the economic crisis might also affect the trends of entrepreneurial activity (e.g. higher concentration of population in more developed areas). In the case of the industrial branches, higher entrepreneurial activity can be observed in regions where these branches are concentrated. We must note that Slovakia shows relatively high activity in both cases, although some Slovak regions are not among the most developed CEE regions and do not have high concentrations of industry compared to others.

Table 6.4 Clusters within the CEE regions

	Description of cluster	Number of regions	Birth per 1,000 inhab (average)	Death per 1,000 inhab (average)	Birth/Death rate	Rank (Birth)	Rank (Death)	Rank (BDrate)
All sectors	Highest activity	30	21.55	15.90	1.36	1	1	1
	Active	78	13.72	10.35	1.33	2	2	2
	Moderate activity	47	9.74	7.66	1.27	3	3	3
	Less active	18	5.09	6.16	0.83	4	4	5
	Low activity	17	3.54	3.99	0.89	5	5	4
Industry sectors	Highest birth number	14	2.96	2.92	1.03	1	1	3
	Active	42	1.70	1.28	1.43	2	2	1
	Moderate activity	62	1.04	0.93	1.19	3	3	2
	Less active	48	0.62	0.73	0.95	4	4	4
	Low activity	24	0.37	0.50	0.83	5	5	5

Source: Author's calculations and construction.

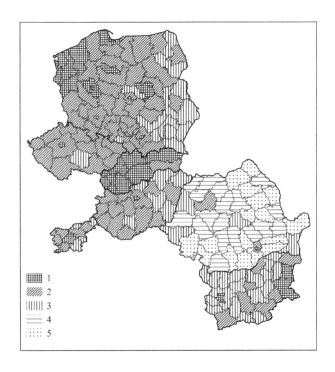

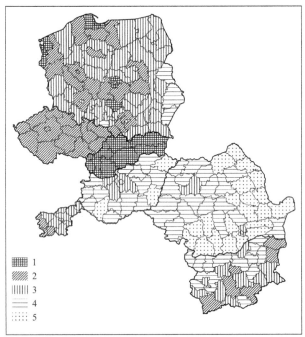

Figure 6.2 Regional entrepreneurial activity in the CEE regions

Source: Author's construction.

Notes: The higher the regional entrepreneurial activity, the darker the colour of the region.
1 – outstanding, 2 – above average, 3 – moderate, 4 – below average, 5 – underperforming.

Results of the discriminant analysis

The groups of the cluster analysis served as dependent variables in the discriminant analysis. This means that the clusters were used as categories here. The regions were reclassified with the regional socioeconomic factors serving as predictor variables. I applied the stepwise method. The Wilks's lambda values determined which variables were used in the analysis (see Appendix 6A). In the case of the all-sector analysis, five variables were involved: regional GDP per capita (log), disposable income per capita (log), rate of persons with tertiary education (log), local units per 1,000 inhabitants (all sectors) between 2003 and 2007 (log), and rate of graduated employees in science and technology-related sectors. The patents variable was removed from the analysis with the stepwise method. In the case of the industrial sector, LQ of industry was added to the analysis, and the local units per 1,000 inhabitants variable was applied for the industrial branches. In this part the LOGgdp and LOGpat indicators were excluded.

The number of discriminant functions depends on the number of determined groups. As we were working with five groups, four discriminant functions were created. The eigenvalue of the function serves as a kind of coefficient of determination (see also the covered percentage of variance). In the case of all sectors, the first two functions summarise 94.2 per cent of variance. The most important function is 'Function 1' which can explain 83.5 per cent of the variance according to the model. All the other functions have relatively low importance compared to the first. Similar values can be observed in the case of the industrial branches. 'Function 1' seems to be the most important: it explains about two-thirds (67.1 per cent) of variance. 'Function 2' has moderate importance, since it covers a bit more than 25 per cent of the whole variance. The other functions have very low effects (Table 6.5).

The structure matrix shows the correlation between the variables and functions. It also contains those variables which were excluded from the analysis. When investigating all of the sectors, all indicators have positive correlation values, with the dependent variable in the case of 'Function 1', and it is only the number of local units that shows a negative correlation in the case of 'Function 2'. The industrial branches analysis displays a similar pattern in the case of 'Function 1', but the correlation values between the dependent variable and explanatory variables are negative except LQ and local units measures in 'Function 2' (see Appendix 6A).

The canonical discriminant function coefficients have a role similar to regression coefficients (beta values). Therefore, they can be used to determine the importance of the regional socioeconomic factors. It can be observed that tertiary education plays an important role in both cases. Regional economic performance has ambivalent values: the income per capita of the earlier

Table 6.5 The main characteristics of discriminant functions

	Function	Eigenvalue	% of Variance	Cumulative %	Canonical correlation
All sectors	1	2.686	83.5	83.5	0.854
	2	.344	10.7	94.2	0.506
	3	.182	5.6	99.8	0.392
	4	.006	0.2	100	0.078
Industry sectors	1	1.766	67.1	67.1	0.799
	2	.713	27.1	94.1	0.645
	3	.100	3.8	97.9	0.302
	4	.055	2.1	100	0.228

Source: Author's calculations and construction.

Table 6.6 Canonical discriminant function coefficients

		Function			
		1	2	3	4
All sectors	LOGgdp	2.198	9.130	0.848	4.446
	LOGterteduc	18.648	.177	−12.414	2.067
	LOGdispincome	4.853	−6.767	3.093	−9.427
	LOGlu0307all	1.666	−2.582	1.637	3.592
	HRSTC_0307	−55.694	7.126	23.635	−2.920
	(Constant)	−7.393	−7.818	−30.602	14.519
Industry sectors	LOGterteduc	14.721	−5.061	6.405	2.750
	LOGdispincome	4.929	−9.559	−4.824	−0.177
	LOGlu0307ind	1.959	7.118	−1.021	−1.013
	LQind	0.830	0.165	1.331	3.236
	HRSTC_0307	−43.908	19.773	7.303	−14.328
	(Constant)	−3.295	24.536	21.991	1.859

Source: Author's calculations and construction.

Note: Unstandardised coefficients.

period positively affects both cases, and the regional GDP had an impact in the all-sector analysis. However, this indicator was excluded from the industrial sector analysis with the stepwise method. Previously established firms affect the groups created according to the entrepreneurial activity as well. Their coefficient is higher in the case of industrial sectors. The concentration of industry employment (LQ) also has a lower importance regarding new firms in industrial sectors. Patents were excluded in both cases. However, this does not mean that they do not have any impact on entrepreneurial activity. It was believed that if the establishment of firms in high-tech manufacturing or knowledge-intensive sectors would have been investigated, it would have a more significant impact (Audretsch and Keilbach, 2004; Ács and Sanders, 2012). The HRSTC variable has a negative effect in 'Function 1', but a positive effect in 'Function 2' in both cases. This may mean that the presence of workers in science and technology is not so important from the all-sectors viewpoint, and their importance emerged in the case of industrial sectors (Table 6.6).

As 'Function 1' plays the most important role in our analysis, the position of cluster centroids from the aspect of this function allows us to control the results of cluster analysis. Three main groups can be observed from the aspect of 'Function 1' in the case of the all-sector analysis: best (Group 1), moderate (Groups 2 and 3) and the lagging behind (Groups 4 and 5). The best cluster and the 'moderate' clusters are much closer to each other than the lagging behinds. From the aspect of 'Function 2' all the groups are around zero, only the best cluster has slightly better results. In the case of industrial sectors the first cluster (Group 1) does not have the best results from the aspect of 'Function 1'. This can explain why it has the highest establishment of firms number, but has only the third highest entrepreneurial activity (the establishment and liquidation of firms together). Groups 2 and 3 are close to each other and Group 4 is also situated closer to them than to Group 5. The entrepreneurial activity of Group 5 is unambiguously the lowest.

To summarise the results, the tertiary-educated persons indicator has a significant impact in both cases. The density of existing firms showed lower importance, but their effect is lower than the university indicator. Regional economic performance had ambivalent results, since disposable

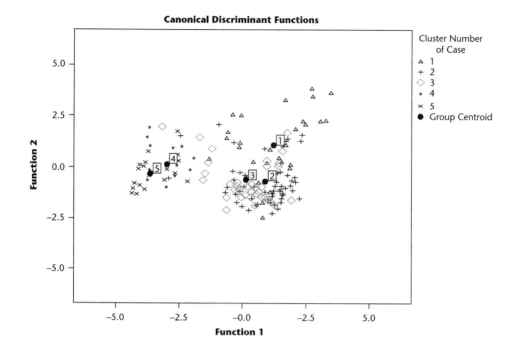

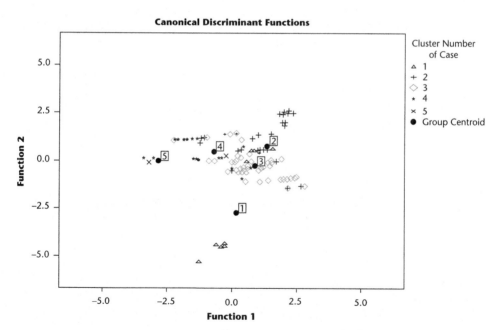

Figure 6.3 The situation of groups from the aspects of 'Function 1' and 'Function 2'

Source: Author's construction.

Note: The above diagram shows the situation of groups in the all-sector analysis, the below diagram concerns the industrial sectors.

income positively affects entrepreneurial activity, but the regional economic performance (regional GDP) of the former period had insignificant effects in the case of industrial sectors. The science-related indicators had low importance: patents were insignificant and the 'science workers' had a low impact on entrepreneurial activity.

Conclusion, limitations, further research directions

I investigated the regional socioeconomic factors that may influence regional entrepreneurial activity, especially the establishment of new enterprises. Four different dimensions (regional economic performance, tertiary education, science and regional entrepreneurial climate) were analysed by six indicators in the case of all sectors and seven measures in the case of industry sectors.

According to our results, among the mentioned factors, tertiary education has the most important role both in the case of all-sector and industrial sector analysis. This means that regional entrepreneurial activity may be higher in those regions where the rate of graduates is higher. Thus, it may be concluded that people may receive adequate skills, social capital or more opportunities in such regions. However, this result raises two new questions regarding the educational structure on the one hand and the concentration and migration of higher educated people on the other hand. Namely, is the presence of a university enough as a key to the higher entrepreneurial activity? And may the migration of highly qualified people to regions or countries that provide better economic environment and opportunities influence regional entrepreneurial activity in those regions where they graduated? Also, the education structure of universities may affect entrepreneurial activity, since universities that have wider scientific portfolios and opportunities and higher ranked research results may serve to produce adequate environments for founding firms based on new scientific results or innovative products. For example, Ďuricova and her colleagues (2014) tested whether the level and quality of entrepreneurial education has significant effects on the establishment of new firms among students.

The science-related indicators showed relatively low effects: patents showed insignificant impacts in both cases, and science workers have low impacts. The impact of science workers prevails primarily through 'Function 2'. It can be observed that 'Group 1' had more negative values than other clusters from the aspect of this function. This may refer to the fact that in the regions that have the highest establishment number in industrial sectors, new firms are founded mainly in low-tech industry sectors. But to prove this, a deeper investigation on the entrepreneurial activity in industrial sectors (especially manufacturing) in the CEE regions needs to be conducted. However, the fact that the analysis included all of the economic sectors and all of the industry sectors, respectively, should be taken into account. Therefore, I venture that these factors would have higher importance in an investigation about entrepreneurial activity of firms in knowledge-intensive sectors.

Regional economic performance showed ambivalent results. Disposable income per capita has significant effects in both all-sector and industrial sectors analyses. According to the discriminant analysis it had significant effects on the dependent variable, but this does not prove unambiguously that higher income per capita stimulates regional entrepreneurial activity. Disposable income correlates highly with the regional GDP per capita. This latter relation had significant effects in the case of all-sector analysis, but was excluded from the analysis in the case of industry sectors. Thus, this means that the more developed a region is, the higher entrepreneurial activity it has in general. But in the case of industrial sectors the highest establishment numbers were in those regions which traditionally had an industrial economic structure (except some Slovak regions). However, these regions do not have high economic performance compared to those regions that showed the highest entrepreneurial activity in the all-sector analysis.

This may explain why regional economic performance was excluded by the stepwise method in the case of industrial sectors.

The density of existing firms had a relatively important role in the classification. It strengthens the assertion that the formerly developed entrepreneurial climate matters in regional entrepreneurial activity. According to the analysis, this factor had a significant impact on the establishment of new enterprises in general in both cases. In the case of industrial sectors, the LQ measure also had an impact, but its importance was lower than that for existing firms.

Certainly, the study has its limitations. As mentioned in the data description, some factors could only be measured on the NUTS-2 level, and this may disfigure the results. Also, the use of time periods could be improved, since mostly average values of a given period were applied instead of annual values. Business demography data were only available for 2008 and 2009. These years were affected by a global economic crisis and could have had an impact on the results. However the data for NUTS-3 level are available in this way only. Therefore, a solution has to be found for expanding our data about entrepreneurial activity in future research. We must attempt to capture multiple factors like the education and entrepreneurial climate together. Some other regional socioeconomic factors could be also involved, for example the concentration of population within a given period (e.g. in the last 10 years), or field of study in universities. The above-mentioned limitations and questions may give us opportunities to improve this research, continue investigations and involve new dimensions in our research concerning regional macrosocial environments around entrepreneurs and future establishers.

Acknowledgement

This research has been supported by project #104985 of the Hungarian National Research, Development and Innovation Office.

References

Ács, Z.J. and Audretsch, D., 1988. Innovation in Large and Small Firms: An Empirical Analysis. *American Economic Review*, 78(4), pp. 678–690.
Ács, Z.J. and Armington, C., 2004. Employment Growth and Entrepreneurial Activity in Cities. *Regional Studies*, 38(8), pp. 911–927.
Ács, Z.J. and Varga, A., 2005. Entrepreneurship, Agglomeration and Technological Change. *Small Business Economics*, 24(3), pp. 323–334.
Ács, Z.J. and Szerb, L., 2007. Entrepreneurship, Economic Growth and Public Policy. *Small Business Economics*, 28(2), pp. 109–122.
Ács, Z.J. and Sanders, M., 2012. Patents, Knowledge Spillovers, and Entrepreneurship. *Small Business Economics*, 39(4), p. 801–817.
Ács, Z.J., Desai, S. and Hessels, J., 2008. Entrepreneurship, Economic Development and Institutions. *Small Business Economics*, 31(3), pp. 219–234.
Ács, Z., Szerb, L. and Autio, E., 2013. *Global Entrepreneurship and Development Index 2013*. Cheltenham: Edward Elgar.
Ács, Z.J., Autio, E. and Szerb, L., 2014. National Systems of Entrepreneurship: Measurement Issues and Policy Implications. *Research Policy*, 43(3), pp. 476–494.
Audretsch, D., 2009. The Entrepreneurial Society. *Journal of Technology Transfer*, 34(3), pp. 245–254.
Audretsch, D. and Thurik, R., 2001. What's New about the New Economy? Sources of Growth in the Managed and Entrepreneurial Economies. *Industrial and Corporate Change*, 10(1), pp. 267–315.
Audretsch, D. and Fritsch, M., 2002. Growth Regimes Over Time and Space. *Regional Studies*, 36(2), pp. 113–124.
Audretsch, D. and Keilbach, M., 2004. Entrepreneurship and Regional Growth: An Evolutionary Interpretation. *Journal of Evolutionary Economics*, 14(5), pp. 605–616.
Autio, E., Kenney, M., Mustard, P., Siegele, D. and Wright, M., 2014. Entrepreneurial Innovation: The Importance of Context. *Research Policy*, 43(7), pp. 1097–1108.

Baptista, R. and Mendonça, J., 2010. Proximity to Knowledge Sources and the Location of Knowledge-based Start-ups. *Annals of Regional Science*, 45(1), pp. 5–29.

Baptista, R., Lima, F. and Mendonça, J., 2011. Establishment of Higher Education Institutions and New Firm Entry. *Research Policy*, 40(5), pp. 751–760.

Beugelsdijk, S., 2007. Entrepreneurial Culture, Regional Innovativeness and Economic Growth. *Journal of Evolutionary Economics*, 17(2), pp. 187–210.

Boschma, R., 2004. Competitiveness of Regions from an Evolutionary Perspective. *Regional Studies*, 38(9), pp. 1001–1014.

Bosma, N. and Harding, R., 2007. *Global Entrepreneurship. GEM 2006 Summary Results*. Babson, MA and London: Babson College and London Business School.

Bosma, N. and Schutjens, V., 2011. Understanding Regional Variation in Entrepreneurial Activity and Entrepreneurial Attitude in Europe. *Annals of Regional Science*, 47(3), pp. 711–742.

Brixy, U. and Grotz, R., 2007. Regional Patterns and Determinants of Birth and Survival of New Firms in Western Germany. *Entrepreneurship & Regional Development*, 19(4), pp. 293–312.

Ďuricova, V., Grünhagen, M., Bischoff, K., Varabei, T. and Blahovec, R., 2014. *University-based Entrepreneurial Ecosystems: Regional Specifics in Eastern and Western Europe*. In: P. Nijkamp, K. Kourtit, M. Buček and O. Hudec, eds. *5th Central European Conference in Regional Science Conference Proceedings*. Košice: Technical University of Košice, pp. 186–196.

Feldman, M.P., 2001. The Entrepreneurial Event Revisited: Firm Formation in a Regional Context. *Industrial and Corporate Change*, 10(4), pp. 861–891.

Fernández-Serrano, J. and Romero, I., 2013. Entrepreneurial Quality and Regional Development: Characterizing SME Sectors in Low Income Areas. *Papers in Regional Science*, 92(3), pp. 495–513.

Fritsch, M., 2012. Methods of Analyzing the Relationship between New Business Formation and Regional Development. *Jena Economic Research Papers*, 2012–064.

Fritsch, M. and Müller, P., 2004. Effects of New Business Formation on Regional Development Over Time. *Regional Studies*, 38(8), pp. 961–975.

Fritsch, M. and Müller, P., 2008. The Effect of New Business Formation in Regional Development Over Time: The Case of Germany. *Small Business Economics*, 30(1), pp. 15–29.

Fritsch, M. and Aamoucke, R., 2013. Regional Public Research, Higher Education, and Innovative Start-ups: An Empirical Investigation. *Small Business Economics*, 41(4), pp. 865–885.

Fritsch, M. and Wyrwich, M., 2014. The Effect of Regional Entrepreneurship Culture on Economic Development – Evidence for Germany. *Jena Economic Research Papers*, 2014–014.

Glaeser, E.L., Kallal, H.D., Scheinkman, J.A. and Shleifer, A., 1992. Growth in Cities. *Journal of Political Economy*, 100(6), pp. 1126–1152.

Kibler, E., 2013. Formation of Entrepreneurial Intentions in a Regional Context. *Entrepreneurship & Regional Development: An International Journal*, 25(3–4), pp. 293–323.

Minniti, M., 2008. The Role of Government Policy on Entrepreneurial Activity: Productive, Unproductive, or Destructive? *Entrepreneurship Theory and Practice*, 32(5), pp. 779–790.

Smallbone, D. and Welter, F., 2001. The Distinctiveness of Entrepreneurship in Transition Economies. *Small Business Economics*, 16(4), pp. 249–262.

Smallbone, D. and Welter, F., 2012. Entrepreneurship and Institutional Change in Transition Economies: The Commonwealth of Independent States, Central and Eastern Europe and China Compared. *Entrepreneurship & Regional Development: An International Journal*, 24 (3–4), pp. 215–233.

Stam, E., 2010. Entrepreneurship, Evolution and Geography. In: R. Boschma and R. Martin, eds. *The Handbook of Evolutionary Economic Geography*. Cheltenham: Edward Elgar, pp. 139–161.

van Stel, A., Carree, M. and Thurik, R., 2005. The Effect of Entrepreneurial Activity on National Economic Growth. *Small Business Economics*, 24(3), pp. 311–321.

Sternberg, R., 2009. Regional Dimensions of Entrepreneurship. *Foundations and Trends in Entrepreneurship*, 5, pp. 211–340.

Stützer, M., Obschonka, M., Brixy, U., Sternberg, R. and Cantner, U., 2014. Regional Characteristics, Opportunity Perception and Entrepreneurial Activities. *Small Business Economics*, 42(2), pp. 221–244.

Szerb, L., Aidis, R. and Ács, Z.J., 2012. *A Comparative Analysis of Hungary's Entrepreneurial Performance in the 2006–2010 Time Period Based in the GEM and the GEDI Methodologies*. Pécs: PTE-KTK.

Szerb, L., Ács, Z.J. and Autio, E., 2013. Entrepreneurship and Policy: The National System of Entrepreneurship in the European Union and in Its Member Countries. *Entrepreneurship Research Journal*, 3(1), pp. 9–34.

Szerb, L., Ács, Z.J. and Autio, E., Ortega-Argilés, R., Komlósi, É., 2014. *REDI: The Regional Entrepreneurship and Development Index – Measuring Regional Entrepreneurship*. Report for the European Commission

Directorate-General Regional and Urban Policy under contract number NO 2012.CE.16.BAT.057. Luxembourg: European Union.

Tamásy, C., 2006. Determinants of Regional Entrepreneurship Dynamics in Contemporary Germany: A Conceptual and Empirical Analysis. *Regional Studies*, 40(4), pp. 365–384.

Wagner, J. and Sternberg, R., 2004. Start-up Activities, Individual Characteristics, and the Regional Milieu: Lessons for Entrepreneurship Support Policies from German Micro Data. *Annals of Regional Science*, 38(2), pp. 219–240.

Wennekers, S. and Thurik, R., 1999. Linking Entrepreneurship and Economic Growth. *Small Business Economics*, 13(1), pp. 27–55.

Appendix 6A

Table 6A.1 Structure matrix in the all-sector analysis

	Function			
	1	2	3	4
LOGdispincome	.629*	.113	.554	−.473
LOGgdp	.465	.653*	.590	−.019
LOGterteduc	.511	.079	−.654*	.096
HRSTC_0307	.393	.172	−.410*	.097
LOGpat**	.270	.305	.361*	−.233
LOGlu0307all	.398	−.393	.442	.671*
LOGdispincome	.629*	.113	.554	−.473
LOGgdp	.465	.653*	.590	−.019

Source: Author's calculations.

Notes: Pooled within-group correlations between discriminating variables and standardised canonical discriminant functions. Variables ordered by absolute size of correlation within function.
 * Largest absolute correlation between each variable and any discriminant function.
 ** This variable was not used in the analysis.

Table 6A.2 Variables involved in the all-sector analysis

	Function			
	1	2	3	4
LOGgdp	.309	1.282	.119	.624
LOGterteduc	2.015	.019	−1.341	.223
LOGdispincome	.545	−.760	.347	−1.058
LOGlu0307all	.344	−.534	.338	.742
HRSTC_0307	−1.663	.213	.706	−.087

Source: Author's calculations.

Table 6A.3 The functions at group centroids in the all-sector analysis

	Function			
	1	2	3	4
1	1.250	1.235	−.060	−.031
2	.913	−.368	.353	−.007
3	.142	−.311	−.679	.029
4	−2.983	.395	.341	.172
5	−3.629	−.051	.001	−.174

Source: Author's calculations.

Note: Unstandardised canonical discriminant functions evaluated at group means.

Table 6A.4 Structure matrix in the industrial sector analysis

	Function			
	1	2	3	4
LOGdispincome	.708*	−.249	−.438	−.063
LOGlu0307ind	.683*	.569	−.368	−.092
LOGpat**	.429*	−.128	−.290	−.032
LOGgdp**	.375*	−.123	−.363	−.038
LOGterteduc	.492	−.089	.729*	−.444
HRSTC_0307	.386	−.058	.639*	−.435
LQind	.152	.100	−.023	.954*
LOGdispincome	.708*	−.249	−.438	−.063

Source: Author's calculations.

Notes: Pooled within-group correlations between discriminating variables and standardised canonical discriminant functions. Variables ordered by absolute size of correlation within function.
* Largest absolute correlation between each variable and any discriminant function.
** This variable was not used in the analysis.

Table 6A.5 The Standardised Canonical – Discriminant Function Coefficients in the industrial sector analysis

	Function			
	1	2	3	4
LOGterteduc	1.735	−.596	.755	.324
LOGdispincome	.578	−1.120	−.565	−.021
LOGlu0307ind	.338	1.230	−.176	−.175
LQind	.250	.050	.401	.976
HRSTC_0307	−1.378	.621	.229	−.450

Source: Author's calculations.

Table 6A.6 The functions at group centroids in the industrial sector analysis

Clusters	Function			
	1	2	3	4
1	.135	−2.605	−.234	.343
2	1.238	.731	−.401	.106
3	.814	−.273	.194	−.251
4	−.754	.450	.344	.244
5	−2.840	.045	−.352	−.225

Source: Author's calculations.

Note: Unstandardised canonical discriminant functions evaluated at group means.

7
Creativity and culture in reproducing uneven development across Central and Eastern Europe

Márton Czirfusz

Introduction

Creativity and culture have become key sectors in raising the competitiveness of cities and regions across Central and Eastern Europe as 'new urban rationalities' (Oosterlynck and González 2013) penetrate policy-makers' and scholars' mindsets in the macro-region. Richard Florida was one of the keynote speakers of Brain Bar Budapest in 2015, a festival organised by the Design Terminal state agency supporting the creative economy. Hódmezővásárhely (Hungary) and Mizil (Romania) were participating in the Creative Clusters network of URBACT, aiming to extend the 'creative city model' to low-density urban areas. The research project ACRE (Accommodating Creative Knowledge), funded by the sixth framework programme of the EU, selected three cities from the macro-region (Budapest, Poznań and Sofia) for understanding the role of creative economy in CEE urban development. Transnational companies also entered the scene: IBM commissioned a smart city study about Hungarian cities (Lados and Horváthné Barsi 2011), the company's Smarter City Challenge reached Bucharest, Katowice, Łódź and Plzeň with recommendations about how to make the cities better; and Siemens issued a leaflet about intelligent technologies for Budapest (Siemens Zrt. et al. 2011).

Research on cultural and creativity-based policy mobilities and policy transfers revealed how ideas are adaptable (González 2011; Peck and Theodore 2012; Prince 2012): they seem to work as off-the-shelf means without taking into account local socio-spatial contexts (Clifton et al. 2015), but in the end are quite different in each locality (Dzudzek and Lindner 2015). Despite their differences, the critique regarding the ambivalent nature of creativity and culture-led redevelopment policies, based on cases from the core countries (Krätke 2010; Peck 2005; Sternberg 2013; Tochterman 2012), is largely valid in the CEE context, exacerbated by the relative lack of capital on Europe's periphery.

This chapter discusses creativity and culture-based development in CEE. Comparative analyses in the CEE context are rare, and a significant blind spot of existing research is how creativity and culture-led spatial policies reproduce geographically uneven development. A proper theorisation of this question would help us to understand why regional development theories and policies failed in CEE (cf. Hadjimichalis and Hudson 2014), and how our spatial concepts might be revisited about the role of creativity and culture in urban and regional development. This offers preliminary remarks for such a comparative study with the use of illustrative examples from the macro-region. First, I theorise which new modes of capital accumulation have gained importance in the region with creativity and culture in their discursive framings. The second section focuses on the governance of large-scale cultural redevelopment and cultural mega-events. The last section discusses new social movements which appeared following the growing use of creativity and culture in national, regional and local policies.

Creativity, culture and new rounds of capital accumulation

After the collapse of state socialism in CEE one of the most striking effects of the economic and political change was a rapid de-industrialisation. Since then, there has been continuous pressure to carve out new ways of economic development on different geographical scales. Apart from reindustrialisation strategies covered by Chapter 3, tertiarisation (including the creative economy) seemed a viable alternative. The 'side-effects' (or rather, intrinsic characteristics) of many of these new development paths, however, have been growing regional inequalities and new urban–rural differentiations, feeding into and resulting from new rounds of capital accumulation.

Although knowledge-intensive and creativity-based regional and local development strategies (Chapter 14) show several commonalities with those in core countries, there are some significant differences from the CEE macro-region. As Mayer (2015) observes, the creative economy is assumed to be the only sector bringing growth in core countries, after the recent relocation of manufacturing to other parts of the world. This is perhaps less true in CEE, as manufacturing-based reindustrialisation exercises less pressure on creativity-based strategies. Nevertheless, the blossoming creative economy is a powerful discourse to show how we left behind the CEE way of capital accumulation and urbanisation, associated with a strong industrial base (Pobłocki 2013), and the sector is promoted as an opportunity of further economic development (Musterd and Kovács 2013), without discussion about the failures of the policies and the reproduction of geographical inequalities.

The knowledge-intensive and cultural economy and their role in regional development are on the agenda of all European countries. Business service offshoring to CEE has been a general trend, in which EU enlargement, competitive infrastructure costs and the strong education system have been considered as important location factors (Gál and Sass 2009). Apart from the capital cities Gál and Sass (2009) identified almost 30 second-tier cities as service offshoring locations in the four Visegrad countries, Romania and Bulgaria. Knowledge-intensive service activities have played a significant role in reproducing uneven development within countries (Nagy and Nagy 2009), resulting in unemployment and further proletarianisation of the peripheries across CEE (Blažek and Netrdová 2012; Petrovici 2013).

The future of these regional development paths might be questionable. Although there are some examples of upgrading in value chains, such as described by Glückler (2008) on the case of Budapest, other examples refer to the dependent nature of these developments and lock-in effects in less knowledge-intensive parts of global production networks (Blažek and Csank 2015). Gál (2015) as well as Chapter 4 emphasise that the economic crisis resulted in diverging paths of cities, regions and countries in international financing market integration. In another case study, Sokol (2013) shows how the attraction of IT service firms to Košice might be interpreted as a 'coping strategy' with neoliberalism on the regional scale. Sokol also emphasises that knowledge-based development cannot take place in a free-market environment, as it is always dependent on policy measures at different geographical scales, and on local elites (see also Nagy and Nagy 2009).

The spatial patterns and trends described above are echoed by studies which measured creativity, territorial capital or the strength of the cultural economy across regions of CEE. One of the most important issues shared by the studies is the urban–rural dichotomy with regard to the 'concentration of creativity'. Firms in the creative economy of Poland represented 7.8 per cent of all economic entities in 2010, and location quotients of the sector were higher than 1 only in 6.1 per cent of LAU-2 local administrative units (Środa-Murawska and Szymańska 2013). Similar patterns are observed in Hungary: 30 per cent of creative occupations are in the capital city of Budapest, and a further 10 per cent in the surrounding NUTS-3 unit, exacerbated by a general increase (and thereby an assumed growth in unevenness) of the share of creative occupations (Lengyel and Ságvári 2011). The creativity index is higher in those Slovakian regions in which urbanisation rates are higher, replicating to some extent existing patterns of uneven development (Hudec and Klasová 2016). The top league of cities in the creative economy of Romania simply reproduce former unevenness in the manufacturing sector: the top five cities apart from the capital (Cluj, Timişoara, Braşov, Mediaş, Sibiu) have been important hubs in manufacturing production (Becuţ 2016).

From this point of view, it is not surprising that regional creativity policies are to a large extent a reformulation of former cluster strategies (with a heyday in the second half of the 1990s and early 2000s in the macro-region). Musterd and Kovács (2013) bring forward a general theoretical argument to revisit Richard Florida's much-debated ideas on creativity policies (Florida 2005) with the inclusion of cluster theory. This might be appealing for policy-makers in the macro-region who are seeking to restore the fizz to local and regional economies after the failures of conventional industrial cluster policies, although there are serious concerns that creative cluster policies will only reproduce geographically uneven development across CEE.

This is certainly true for development within the macro-regions' cities. Creativity and culture-based redevelopment of cities across the world are 'smoke screens behind which agendas of privatisation, modernisation of public services or tertiarisation of the economy can be implemented' (González 2011, p. 1412). In spite of the fact that these processes are generally the same across CEE, how they have unfolded throughout the macro-region is specific in each case. First, the tertiarisaion of the economy had a strong impact on culture and consumption-based brownfield redevelopment and capital reinvestment in cities of CEE. Second, privatisation played a different role in the macro-region in fuelling redevelopment. Last but not least, changes in public services (and I will limit myself to the provision of culture) also show some differences.

The tertiarisation of the economy has run parallel with the abandonment of former industrial sites in CEE's cities. The excellent geographical positions of these areas led to a closure of the rent gap of the neighbourhood, through public or private investment. Many of the former industrial sites have been flagship projects of cultural redevelopment in CEE cities. It was the case with the renovation of the Ganz electric works premises in Budapest in which the area was converted to a mix of cultural attractions and green areas, now managed by a state-owned company (Barta et al. 2006). The heavily debated, mixed-use Millennium City Centre at the Danube waterfront in Budapest on a former railway site was developed by a Hungarian billionaire; the Palace of Arts with cultural functions is an exemplary of the controversies of early 2000s public–private partnerships (Berki 2012). The redevelopment of the Zsolnay factory in Pécs, one of the second-tier cities in Hungary, was linked to the European Capital of Culture programme. One part of the Dolní Vítkovice iron works in Ostrava (Czechia) underwent a cultural redevelopment with the aim to rebrand the industrial image of the city (Sucháček and Herot 2015). In Poland, the European Regional Development Fund was used to convert former industrial sites to new cultural centres, such as in Katowice (Baranyai and Lux 2014), or the EC1 former power plant in Łódź. For a mixed-use cultural redevelopment one of the most striking examples is Manufaktura in Łódź: the 27-hectare area used by a textile factory was redeveloped for new forms of capital accumulation, with the abandonment of residential buildings, coupled with evictions of tenants (Sowińska-Heim 2014). Apart from the cultural functions represented by

Creativity and culture

Figure 7.1 Cultural–commercial redevelopment of a former industrial site: Manufaktura in Łódź
Source: Photo by the author.

museums, a large shopping centre, a luxury hotel and several other forms of entertainment dominate the cityscape (Figure 7.1).

For the creativity and culture-led redevelopment of cities, small-scale capital is as important as large-scale public or private capital mentioned in the previous cases. Small enterprises – as in textbook cases of gentrification – help transform blighted neighbourhoods (Mayer 2015), sometimes with the help of municipalities. One of the examples from Budapest is to propagate the reuse of vacant retail spaces. Apart from the Budapest municipality – which advertised the initiative with successful young Hungarian creative entrepreneurs smiling to passers-by from posters placed in large windows of empty shops – the Hungarian Contemporary Architecture Centre (KÉK, a non-profit organisation) and the URBACT network joined the initiative with festivals and a database of empty shops (Majorné Venn 2014). The project was correct in assuming that many enterprises in the creative sector were working in precarious conditions and that helping them might be a cheap and promising publicly-funded programme for the middle class, but other examples – such as the Prague story analysed by Pixová (2013, p. 222) – shed light on the fact that non-mainstream cultural or creative initiatives are in an insecure position precisely 'due to the appropriation, displacement, and destruction of urban space'. Also, this use of vacant sites by the middle class – such as in Budapest's 'ruin bars' (Lugosi et al. 2010) – are only a temporary phenomenon paving the way for further gentrification and displacing urban pioneers which are celebrated as the new creative class by local decision-makers.

How alternative cultural milieus move within the city and how this is linked to privatisation (combined with the restitution of properties in some CEE countries after 1990), local policies and violence, was traced by Pixová (2013) in Prague and Czirfusz (2014) in Budapest. New cultural initiatives claiming their right to the city open up in former smaller industrial sites, schools or daycare facilities abandoned because of the falling population and economic decline in the inner city. Some of them have been in existence for a longer time, and some of

109

them institutionalised or embedded themselves in the urban fabric over the years – such as the Trafó cultural centre in one of Budapest's gentrifying neighbourhoods (Keresztély 2005) or the Off Piotrkowska cultural complex in Łódź's main pedestrian axis. Official places of culture are also threatened by privatisation. A large-scale example is Warsaw where the tallest building in Poland, Palace of Culture and Science, built in the 1950s, is currently 'counterbalanced' with new commercial, residential and office high-rises, representative of the growing amount of real estate capital since the 1990s.

In the larger urban scale, commercialising cultural heritage has been a recurring topic in CEE cities. Jewish quarters across the macro-region were among the first to be gentrified after 1990. The redevelopment of the Jewish quarter in Prague (located in the historical city centre) was backed by the heavy inflow of tourists after 1990, and by the rising corporate and state control over cultural production (Pixová 2013). Kraków's Kazimierz neighbourhood has also witnessed a massive revival, in which different visions and interests of the local residents, the private sector and the municipality are hard to be reconciled (Murzyn 2006). Market-led gentrification is also present in Budapest's Jewish quarter, now well known and advertised in tourist guides listing synagogues and ruin-pubs, leading to skyrocketing property prices, conflicts with local residents and further privatisation of the municipal social housing stock. Although these neighbourhoods are trendy and nurture the 'local creative class', empirical case studies show how this kind of cultural and creativity-led development increases gentrification and displacement, raises socio-spatial inequalities and starts new rounds of capital accumulation within the city.

As a counterpoint to this kind of development in urban environments, culture is also an indicator of new rounds of capital accumulation in rural areas of CEE. The general assumption about cultural institutions in the countryside is that with the roll-out of neoliberalism the state is not capable anymore of providing public funding for cultural facilities, and culture is less prevalent in these areas (Fábián and Tóth 2014).

Institutions for civic cultivation have been partaking in building the nation-state and a competitive labour force across CEE since the second half of the nineteenth century. Cultural sites became widespread throughout the macro-region during socialism, being important centres of reproduction. Cultural institutions underwent liberalisation following the reforms of socialism in the late 1960s which also meant a diminishing significance of offering equal access to culture for everyone within the country (Szarvas 2016). These were trends exacerbated by a harsher marketisation after 1990 (Becuț 2016): as public money decreased, private functions appeared in houses of culture (from labour recruitment of large multinational companies or temporary staffing agencies to multilevel marketing events), changing their roles in capital accumulation from the tertiary to the primary circuit of capital. A differentiation and change of functions takes place following a neoliberal turn of cultural policy: there is a growing quest for additional financial resources, be it for EU-funded renovation or cultural programmes, or national money to secure the jobs of cultural workers. For instance, Hungarian houses of culture employ half of their staff, almost 6,000 people, in precarious labour conditions within the nation-state's public works programme, in which workers are not covered by the Labour Code and earn less than the minimum wage (Belügyminisztérium 2016). Thereby, market logics, labour policies and cultural politics reproduce not only 'internal' unevenness within the cultural sector, but also more general patterns of uneven development.[1]

Cultural projects, mega-events and re-scaling governance

After the initial boom of FDI-driven investment during the 1990s and early 2000s (Chapters 2 and 3), culture and creativity have become sectors of further capital accumulation throughout

CEE. This has not only led to a fossilisation of existing patterns of uneven geographical development, but also to changing governance structures. Different geographical scales have repositioned themselves vis-à-vis others as culture and creativity gained importance in regional and urban political economies.

This section discusses re-scaling governance in CEE contexts related to large-scale cultural projects and cultural mega-events. These mega-events have been 'an important actor in the reconstruction and repositioning of emerging economies' (Müller and Pickles 2015, p. 122). Most of the literature has focused on the interrelationship of culture, cultural infrastructure and mega-events in building nationhood (Koch 2013; Müller 2011) or the reproduction of Eastern European paternalistic political culture (Pálné Kovács and Grünhut 2015). This section shows how the dependent, peripheral role of CEE in European political economies (see for example Wallerstein 1976) influences new governance structures related to culture.

The discussion starts with the role of global and European scale of governance. The next scale to be considered is the nation-state, and the section ends with an analysis of urban entrepreneurialism (Harvey 1989). The illustrative examples from CEE are European Capitals of Culture programmes (as one of the European flagships of cultural redevelopment), sports mega-events (Olympic Games and UEFA European Championship) and universal expositions.

Large cultural events are often regarded as symbolic actions which aim to reposition cities, urban regions, even whole countries in global or European hierarchies. The idea fits well with CEE realities, as the past 150 years marked a failure of CEE to catch up with developed countries through mainstream economic and trade policies (Gál and Lux 2014). Culture appears to offer a relatively cheap and quick way to 'do the trick' and represent the region as equivalent to other developed, democratic countries.

This aspect has been self-evident in recent cultural mega-events across the macro-region. The first of such large cultural ventures after 1990 was the Expo 1996 with a joint candidacy of Budapest and Vienna. The leading politicians in Hungary would have 'project[ed] Hungary onto the world stage' with the Expo (Keresztély 2002, p. 32), but economic realities – i.e. the lack of public money to organise the event in the turmoil of the transition of the early 1990s – led to the event being called off by both cities. Europeanness has also been a recurring theme in the European Capitals of Culture events, such as in Košice (Slovakia, 2013) aiming to put the city back on the map of Europe (Hudec and Džupka 2016), or in Pécs (Hungary, 2010) which ventured to become a gateway to the Balkans with the help of the cultural parade (Pálné Kovács and Grünhut 2015). Sports events have taken on the same veneer: the Sarajevo Olympics in 1984 was an early example; Poland aimed to join the global family of sport and entertainment with the UEFA European Championship in 2012 (Cope 2015); and the (now retracted) Budapest Olympics 2024 bid underlined that the Olympics 'represent an unprecedented opportunity to build new global economic connections' (Budapest2024 2016, p. 58).

Many of these events are economically designed to establish spaces for global capital accumulation, thereby reproducing uneven geographical development. Mega-events such as the UEFA Euro 2012 or the Olympic Games are functioning as 'special economic zones' in the primary circuit of capital: UEFA and the International Olympic Committee (IOC) are exempt from paying taxes, they extract monopoly rent (Cope 2015) and strengthen uneven geographical development before and during the events. Furthermore, local economic actors are structurally excluded from the profits and works related to the organisation of these events because of the existing contracts of UEFA or the IOC with sponsors and suppliers from core countries (Woźniak 2013).

Cultural mega-events are also a well-established means for transnational class formation, as these events are deeply embedded in global capitalist social relations (Apeldoorn 2004). The local

Márton Czirfusz

comprador service class capitalises on the events: branches of large consultancy firms write feasibility studies and offer other services for cities and national governments. This was the case during the UEFA Euro 2012 in Poland in which the big four consultancy companies (Deloitte, Ernst & Young, PWC, KPMG) were all involved (Cope 2015); the Budapest Olympic Games 2024 feasibility study was written by PWC; and the Swiss Event Knowledge Services was involved in the Kraków Winter Olympics 2022 bid (Kozłowska 2015). Moreover, through the international policy transfer, political and economic elites travel across the world to get first-hand experiences about supposedly best practices of organising such events (González 2011; Oanca 2010).

Large cultural events represent a means to modernise selected, largely prosperous cities and regions – thereby increasing socio-spatial unevenness within countries. Infrastructural development in the secondary circuit of capital related to cultural events are strong labels which render any critique about the quality or necessity almost impossible (Woźniak 2013). Nation-states welcome these monopoly events 'to increase [their] competitiveness in the international tourism and investment market' (Cope 2015, p. 167).

These modernisation projects are generally financed and governed by the nation-state (Müller and Pickles 2015). This was the case in the Expo 1996 in Budapest: although the world fair was cancelled, the central government finished a large amount of the originally planned investments (Keresztély 2002); in the Sibiu European Capital of Culture project 56 per cent of the budget was covered by the state (Oanca 2010). The UEFA Euro 2012 included €20–22 billion of infrastructural projects advertised as an important element of developing Poland (Cope 2015), the modernisation discourse also being important in the Kraków Olympics bid (Kozłowska 2015). The Budapest 2024 Olympics bid book expected that games-related projects would be responsible for 8.7 per cent of government-funded capital expenditure of Hungary in the next years. In the case that additional financing comes from EU cohesion funds (as in the UEFA Euro 2012 or in the Košice European Capital of Culture project), the nation-state and its bodies often function as fulcrums in the process (Pálné Kovács and Grünhut 2015).

Although not fully comparable, European Capitals of Culture budgets are interesting cases about how money is largely spent on infrastructural investments, rather than the operation of the programmes, thereby literally solidifying uneven geographical development (see Figure 7.2). Surpluses, also in the form of new debt for infrastructural projects, are absorbed in secondary circuits of capital spatially unevenly, as in many other cases of development at the European periphery (López and Rodríguez 2011).

Cope (2015) notes that there are competing roles with the state acting as both an investor and a regulator in mega-events: it generates investments which are financed by itself. Statehood intermingles with business interests: modernisation projects related to cultural events not only mobilise the transnational service class, but are also a means of strengthening national elites, such as in the construction industry. Although for the UEFA Euro 2012 Poland's infrastructural developments were predominantly implemented by international consortia (Cope 2015), in the current Hungarian stadium-development national programme (32 stadiums with an expected budget of about €700 million, financed by the state) it is reported that medium-sized national companies based in the countryside were winning most of the bids (Csepregi 2016). In this case, rather than leading to more even geographies of capital accumulation, building the national bourgeoisie results in rising inequalities across classes and within localities.

The urban (local) scale plays an important role in governing cultural mega-events as well. Large cultural events maintain and even increase intra-urban competition and existing inequalities in the urban hierarchy in the age of urban entrepreneurialism (González 2011; Oanca 2010), with the changing importance of the local state. In the case of the CEE local state, lacking resources

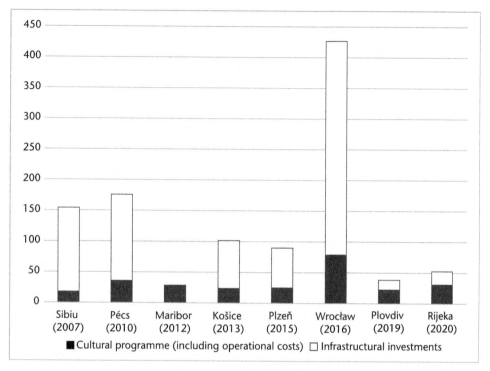

Figure 7.2 Budgets of European Capitals of Culture in CEE (€m)

Source: Ex-post evaluations available at https://ec.europa.eu/programmes/creative-europe/node/150_en and bid books of the cities.

Note: No data were available for Maribor's infrastructural investments.

after 1990 has meant either that other resources have to be brought into large-scale cultural projects or redevelopment, or that the local state has become the driver in conceptualisations of these projects (cf. Oanca 2010). However, as Pálné Kovács and Grünhut (2015) emphasise, this often leads to controversial authority issues, such as local cultural events during the European Capitals of Culture programmes being managed by a cultural agency of the nation-state.

A central issue in understanding the local consequences of culturally led redevelopment is the pro-market imaginary embraced by all of the examples mentioned before. The European Capitals of Culture is often regarded as an economic tool in redeveloping de-industrialised cities across CEE, following the global 'cultural turn' of urban redevelopment (Cope 2015; Hudec and Džupka 2016; Oanca 2010; Pálné Kovács and Grünhut 2015). The general aim is to continue economic growth or attract new investors to local economies – either into traditional FDI-led manufacturing sectors (Sibiu) or by promoting a post-industrial urban image based on the creative economy (Košice) (Hudec and Džupka 2016; Oanca 2010). During the Warsaw stadium redevelopment for the UEFA Euro 2012, 'non-capitalist' economic relations (a semi-formal marketplace was operating in the stadium) were simply erased from the urban fabric (Sulima 2012), informality was commodified by current urban policies (cf. Krivý 2016). Later on, redeveloped urban neighbourhoods might become focal points for private sector capital investment – a part of the abandoned Budapest Expo 1996 site was developed by an international real estate company as Infopark Budapest, attracting foreign companies from mostly IT industries.

Cities and different groups of populations are not necessarily winners in cultural redevelopment of neighbourhoods. Money flowing into secondary circuits of capital in the form of infrastructural investment is putting new burdens on local governments: high operating costs of cultural institutions are typical postscripts to cultural events (Pálné Kovács and Grünhut 2015). They do not solve structural challenges of common issues of semi-peripheral economic development across CEE either: in spite of feasibility studies and official discourses, population numbers or number of jobs will not automatically increase in the wake of mega-events and urban redevelopment. Gentrification and politics of exclusion are common 'side-effects'. The population of Sibiu living in the historic city centre was a loser of the European Capital of Culture project. Certain independent local cultural or artistic groups or institutions have often been excluded from the events (Oanca 2010; Pálné Kovács and Grünhut 2015), even if there have been efforts of polycentric development – as in Košice where local cultural centres in the urban peripheries were established in the European Capital of Culture programme (Hudec and Džupka 2016). Citizens are not regarded as social actors; this is only ascribed to visitors and consumers of the events (Woźniak 2013). One of the consequences of these rising socio-spatial inequalities is the emerging social movements against culturally led redevelopment in CEE. The last section of this chapter discusses some of these.

New social movements against creativity-led and cultural redevelopment

Mobilisations and social movements against creativity policies and cultural redevelopment of cities have been a recurring topic of scholars in core countries. Questions might be asked as to whether there are similar movements emerging in CEE, and what are the common and specific features of this macro-region.

Social movement studies refer to a 'crisis of democratic capitalism' in analysing current social movements. But as states and class dynamics are different in CEE from those in core countries (some aspects were shown in the previous sections), forms of social mobilisations are also different in the macro-region (Gagyi 2015b). Middle classes in CEE are systematically more dependent on state institutions in securing their standards of living and reproduction; grassroots movements might be more easily captured by political parties' interests (Gagyi 2015b). Nevertheless, they largely deploy narratives and political agendas which are simply taken from Western activist groups, without questioning the basic assumptions from a CEE experience (Gagyi 2015a). This might also explain why academics largely remain silent about how culture-led large-scale redevelopment projects reproduce socio-spatial inequalities (Woźniak 2013).

With the crisis of Fordism also in CEE, creative subjectivities come into the foreground of the futures of cities and regions – although with the internal contradiction that freedom associated with creativity is very much based on internalising the capitalist logic (Dzudzek 2013). As Pixová (2013, p. 234) observes in the Prague alternative cultural scene, most of these alternative cultural groups 'prefer to comply with the status quo' and do not challenge neoliberal urbanisation in general. As Mayer (2015) notes, cities seeking to boost their economies with the creative city branding need the creatives. That is why official low-cost creativity strategies after the crisis often incorporate some artist groups, subcultures and movements (by marginalising other groups and 'non-creative' labour). In Budapest, for example, one of the founders of the formerly mentioned Hungarian Contemporary Architecture Centre (KÉK) was appointed as the chief-architect of the city in 2012. Despite the chief-architect's weak institutional role in the municipal decision-making structure, some initiatives (mainly in the creative–cultural urban policies) were actively involving non-governmental organisations (NGOs) and 'independent' groups of artists, architects and creative people in developing small-scale creativity-based redevelopment in the Hungarian capital city. Mayer (2015) notes the drawbacks of these initiatives, by referring to how art activism might easily turn into art-based place-marketing, which was the case to some extent in Budapest as well.

Another movement against cultural redevelopment is activism against the demolition of historical buildings in urban areas. Krivý (2011) reminds us that these movements resist the logic of land speculation and strengthen conservation discourses which are in fact about how to make old urban structures profitable. Kauko (2015, p. 15) shows how growing spatial inequalities because of the growing 'new economy' in CEE cities (see also Krivý 2016) leads to movements 'protecting some specific elements of the original urban environment'. This controversial interplay was present in the internal workings of the civil movement Óvás! in Budapest's Jewish quarter since the early 2000s. The movement on the one hand was successful in not letting the local municipality demolish listed buildings and sell the empty plots to investors, but on the other hand implicitly strengthened a possibility of culture-based gentrification of the district (Földi 2006, chap. 7.2.3 offers a thorough analysis of actors and their interests).

Neoliberal urbanism tends to eliminate affordable inner-city workplaces for creative professionals: places valorised by cultural capital are appropriated later by more powerful economic capital (Mayer 2013). Disappearance of affordable inner-city workplaces for creative workers in precarious conditions was followed by Pixová (2013) in Prague, it is also present in the changing urban structure of inner-city Budapest (Czirfusz 2014). Cultural activities have been important building blocks of the Polish squatting scene in their struggles about housing and the right to the city in general (Polanska and Piotrowski 2015). Disappearance of some of the local movements is thereby a direct consequence of broader market forces.

Struggle over representative democracy is another crucial element in new social movements, as power and decision-making have shifted to specific institutions established for delivering certain (mostly infrastructural) projects in the city (Mayer 2013). The changing (and diminishing) role of the local state is mirrored by the success in recent years of the national network of right to the city movements in Poland, calling for more democratisation and decommercialisation (Domaradzka 2015). Democratic decision-making is also immanent in struggles against mega-events. Successful initiatives include the popular initiative for a local referendum about the 2024 Budapest and the 2022 Kraków Olympic Games candidacy. In Kraków, after pressure from social movements and the general public, a referendum took place with 70 per cent voting against the bid. During the process, residents secured rights over direct decision-making about projects with significant public resources and large scope of urban development (Pasquinelli 2015). Another, less successful case until now is the opposition to a cultural megaproject in Budapest with scattered mobilisations. The Városliget public park (home also to some important museums in the city, with buildings erected at the end of the nineteenth century) is being redeveloped as a museums quarter, with three new museum buildings and one extension, by considerably reducing the green area. The mobilisations have taken place largely because of environmental concerns (i.e. how many trees will be cut down), but the activists have also been questioning the scale and costs of the project, as well as the institutional, non-democratic decision-making structures (a state-owned company is coordinating the project, the affected municipal and district governments have minimal influence on the development, as local jurisdiction was overwritten by the interests of the nation-state).

The challenges of these new social movements are manifold. First, many of them organise themselves around narrow material and professional interests (Mayer 2015) or struggle for the right to the city for temporary use (Pixová 2013). Although several redevelopment projects in the region stopped in the wake of the economic crisis, we certainly need to go beyond offering more time for temporal use by the 'creative class' and cultural initiatives. The more general agenda would be to contest neoliberal urbanism which necessitates shifting our analysis on the repoliticisation of creative subjects, and not on their current labour prospects during crisis (Dzudzek 2013). Taking a more optimistic stance, 'resistance along with longstanding critiques of creative and neoliberal cities by urban scholars and activists presents a ripe opportunity to confront current neoliberal and entrepreneurial urban order' (Holgersen 2015, p. 702). To be successful, an important move is to deploy alternative interpretations about creativity, culture and their role on regional and urban development, as well as new spatial concepts about how creativity and culture influence political economies on different geographical scales. In line with the ideas of Mayer (2013) and Dzudzek (2013) this also necessitates linking these movements to larger right to the city debates, and uniting local struggles across CEE and SEE, taking into consideration the common historical–geographical structures of capitalism in the macro-region.

Acknowledgement

This research has been supported by project #115870 of the Hungarian National Research, Development and Innovation Office.

Notes

1 I thank Márton Szarvas for helping me to clarify the argument on cultural politics in the countryside.

References

Apeldoorn, Bastiaan van. 2004. Theorizing the Transnational: A Historical Materialist Approach. *Journal of International Relations and Development* 7 (2): 142–176.

Baranyai, Nóra, and Gábor Lux. 2014. Upper Silesia: The Revival of a Traditional Industrial Region in Poland. *Regional Statistics* 4 (2): 126–144.

Barta, Györgyi, Pál Beluszky, Márton Czirfusz, Róbert Győri, and György Kukely. 2006. *Rehabilitating the Brownfield Zones of Budapest*. Vol. 51. Discussion Papers. Pécs: Centre for Regional Studies of the Hungarian Academy of Sciences.

Becuț, Anda Georgiana. 2016. Dynamics of Creative Industries in a Post-Communist Society. The Development of Creative Sector in Romanian Cities. *City, Culture and Society* 7 (2): 63–68.

Belügyminisztérium. 2016. *Havi Tájékoztató a Közfoglalkoztatás Alakulásáról. 2016. Május*. Budapest: Belügyminisztérium, Közfoglalkoztatási és Vízügyi Helyettes Államtitkárság. http://kozfoglalkoztatas.kormany.hu/download/7/ac/71000/Havi%20t%C3%A1j%C3%A9koztat%C3%B3_2016_m%C3%A1j.pdf

Berki, Márton. 2012. Post-Socialist Transformation of Former Industrial Areas: A Case Study of Soroksári Road, Budapest. In *Metropolitan Regions in Europe*, edited by Viktória Szirmai and Heinz Fassmann, 83–99. Budapest: Austrian–Hungarian Action Fund.

Blažek, Jiří, and Pavlína Netrdová. 2012. Regional Unemployment Impacts of the Global Financial Crisis in the New Member States of the EU in Central and Eastern Europe. *European Urban and Regional Studies* 19 (1): 42–61.

Blažek, Jiří, and Pavel Csank. 2016. Can Emerging Regional Innovation Strategies in Less Developed European Regions Bridge the Main Gaps in the Innovation Process? Environment and Planning C: Government and Policy, 34 (6): 1095–1114.

Budapest2024. 2016. Candidature File Stage 1. Budapest 2024 Olympic and Paralympic Games. www.budapest2024bid.com/Budapest_2024_Candidature_File_Stage_1_EN.pdf

Clifton, Nick, Roberta Comunian, and Caroline Chapain. 2015. Creative Regions in Europe: Challenges and Opportunities for Policy. *European Planning Studies* 23 (12): 2331–2335.

Cope, Benjamin. 2015. Euro 2012 in Poland: Recalibrations of Statehood in Eastern Europe. *European Urban and Regional Studies* 22 (2): 161–175.

Csepregi, Botond. 2016. Stadionláz, 3. Rész: Minden Út Felcsútra Vezet – a Stadionépítő Cégek Toplistája. *Atlatszo.hu*, June 30. https://atlatszo.hu/2016/06/30/stadionlaz-3-resz-minden-ut-felcsutra-vezet-a-stadionepito-cegek-toplistaja/

Czirfusz, Márton. 2014. Obliterating Creative Capital? Urban Governance of Creative Industries in Post-Socialist Budapest. *Europa XXI* 26: 85–96.

Domaradzka, Anna. 2015. Changing the Rules of the Game: Impact of the Urban Movement on Public Administration Practices. In *Civil Society and Innovative Public Administration*, edited by Matthias Freise, Friedrich Paulsen, and Andrea Walter, 188–217. European Civil Society 16. Baden-Baden: Nomos Verlagsgesellschaft.

Dzudzek, Iris. 2013. Unternehmen oder Unvernehmen? – Über die Krise des Kreativsubjekts und darüber Hinaus. *Geographica Helvetica* 68 (3): 181–189.

Dzudzek, Iris, and Peter Lindner. 2015. Performing the Creative-Economy Script: Contradicting Urban Rationalities at Work. *Regional Studies* 49 (3): 388–403.

Fábián, Attila, and Balázs István Tóth. 2014. Which Attributes of Culture Can We Measure? The Case of Hungarian Micro-Regions. *International Journal of Global Environmental Issues* 13 (2–4): 294–307.

Florida, Richard L. 2005. *Cities and the Creative Class*. London: Routledge.

Földi, Zs., 2006. *Neighbourhood Dynamics in Inner-Budapest: A Realist Approach*. (Netherlands Geographical Studies 350.) Utrecht: Koninklijk Nederlands Aardrijkskundig Genootschap.

Gagyi, Ágnes. 2015a. Why Don't East European Movements Address Inequalities the Way Western European Movements Do? A Review Essay on the Availability of Movement-Relevant Research. *Interface: A Journal for and about Social Movements* 7 (2): 15–26.

Gagyi, Ágnes. 2015b. Social Movement Studies for East Central Europe? The Challenge of a Time–Space Bias on Postwar Western Societies. *Intersections. East European Journal of Society and Politics* 1 (3): 16–36

Gál, Zoltán. 2015. Development of International Financial Centres in Central and Eastern Europe during Transition Period and Crisis. The Case of Budapest. *Studia Regionalne I Lokalne* 60 (2): 53–80.

Gál, Zoltán, and Magdolna Sass. 2009. Emerging New Locations of Business Services: Offshoring in Central and Eastern Europe. *Regions Magazine* 274 (1): 18–22.

Gál, Zoltán, and Gábor Lux. 2014. ET 2050 – Territorial Scenarios and Visions for Europe. Vol. 8. Territorial Scenarios and Visions for Central and Eastern Europe. Central and Eastern European Impacts of Scenarios. Report. ESPON (European Spatial Planning Observation Network). www.regscience.hu:8080/xmlui/handle/11155/753

Glückler, Johannes. 2008. 'Service Offshoring'. Globale Arbeitsteilung Und Regionale Entwicklungschancen. *Geographische Rundschau* 60 (9): 36–42.

González, Sara. 2011. Bilbao and Barcelona 'in Motion'. How Urban Regeneration 'Models' Travel and Mutate in the Global Flows of Policy Tourism. *Urban Studies* 48 (7): 1397–1418.

Hadjimichalis, Costis, and Ray Hudson. 2014. Contemporary Crisis across Europe and the Crisis of Regional Development Theories. *Regional Studies* 48 (1): 208–218.

Harvey, David. 1989. From Managerialism to Entrepreneurialism: The Transformation in Urban Governance in Late Capitalism. *Geografiska Annaler. Series B, Human Geography* 71 (1): 3–17.

Holgersen, Ståle. 2015. Economic Crisis, (Creative) Destruction, and the Current Urban Condition. *Antipode* 47 (3): 689–707.

Hudec, Oto, and Peter Džupka. 2016. Culture-Led Regeneration through the Young Generation: Košice as the European Capital of Culture. *European Urban and Regional Studies* 23 (3): 531–538.

Hudec, Oto, and Slávka Klasová. 2016. Slovak Creativity Index – A PCA Based Approach. *European Spatial Research and Policy* 23 (1): 47–64.

Kauko, Tom. 2015. Territorial Competition in the New Economy. Different Strategies in Different Urban Settings. *Urbanism. Architecture. Constructions* 6 (1): 5–28.

Keresztély, Krisztina. 2002. *The Role of the State in the Urban Development of Budapest*. Discussion Papers 39. Pécs: Centre for Regional Studies of the Hungarian Academy of Sciences. http://discussionpapers.rkk.hu/index.php/DP/article/view/2199

Keresztély, Krisztina. 2005. Cultural Investments in a City in Transition: The Budapest Case. In *Hungarian Spaces and Places: Patterns of Transition*, edited by Györgyi Barta, Éva G. Fekete, Irén Szörényiné Kukorelli, and Judit Timár, 449–465. Pécs: Centre for Regional Studies.

Koch, Natalie. 2013. Sport and Soft Authoritarian Nation-Building. *Political Geography* 32 (January): 42–51.

Kozłowska, Małgorzata. 2015. Future of Winter Olympic Games in the Context of Ongoing Application Process to Host Winter Olympic Games 2022: Study Case: Cracow Bid. *Current Issues of Tourism Research* 4 (2): 27–37.

Krätke, Stefan. 2010. 'Creative Cities' and the Rise of the Dealer Class: A Critique of Richard Florida's Approach to Urban Theory. *International Journal of Urban and Regional Research* 34 (4): 835–853.

Krivý, Maroš. 2011. Speculative Redevelopment and Conservation: The Signifying Role of Architecture. *City* 15 (1): 42–62.

Krivý, Maroš. 2016. Towards a Critique of Cybernetic Urbanism: The Smart City and the Society of Control. *Planning Theory*, doi:10.1177/1473095216645631.

Lados, Mihály, and Boglárka Horváthné Barsi. 2011. 'Smart Cities' tanulmány. Győr: MTA Regionális Kutatások Központja Nyugat-magyarországi Tudományos Intézet. www.rkk.hu/rkk/news/2011/smart_cities_tanulmany_IBM_RKK.pdf

Lengyel, Balázs, and Bence Ságvári. 2011. Creative Occupations and Regional Development in Hungary: Mobility of Talent in a One-Centred Transition Economy. *European Planning Studies* 19 (12): 2073–2093.

López, Isidro, and Emmanuel Rodríguez. 2011. The Spanish Model. *New Left Review* 2 (69): 5–29.

Lugosi, Peter, David Bell, and Krisztina Lugosi. 2010. Hospitality, Culture and Regeneration: Urban Decay, Entrepreneurship and the 'Ruin' Bars of Budapest. *Urban Studies* 47 (14): 3079–3101.

Majorné Venn, Mariann. 2014. Open! Community Utilisation of Vacant Retail Space. *URBACT Blog. For Cities, by Cities, about Cities*. November 7. www.blog.urbact.eu/2014/11/open-community-utilisation-of-vacant-retail-space/

Mayer, Margit. 2013. Urbane soziale Bewegungen in der neoliberalisierenden Stadt. *sub\urban. Zeitschrift für kritische Stadtforschung* 1 (1): 155–168.

Mayer, Margit. 2015. Afterword: Creative City Policy and Social Resistance. In *Making Cultural Cities in Asia: Mobility, Assemblage, and the Politics of Aspirational Urbanism*, edited by June Wang, Tim Oakes, and Yang Yang, 234–250. Abingdon: Routledge.

Müller, Martin. 2011. State Dirigisme in Megaprojects: Governing the 2014 Winter Olympics in Sochi. *Environment and Planning A* 43 (9): 2091–2108.

Müller, Martin, and John Pickles. 2015. Global Games, Local Rules: Mega-Events in the Post-Socialist World. *European Urban and Regional Studies* 22 (2): 121–127.

Murzyn, Monika A. 2006. 'Winners' and 'Losers' in the Game: The Social Dimension of Urban Regeneration in the Kazimierz Quarter in Krakow. In *Social Changes and Social Sustainability in Historical Urban Centres. The Case of Central Europe*, edited by György Enyedi and Zoltán Kovács, 81–106. Discussion Papers, Special. Pécs: Centre for Regional Studies of the Hungarian Academy of Sciences.

Musterd, Sako, and Zoltán Kovács. 2013. Tailored – Context-Sensitive – Urban Policies for Creative Knowledge Cities. In *Place-Making and Policies for Competitive Cities*, edited by Sako Musterd and Zoltán Kovács, 315–327. Oxford: John Wiley.

Nagy, Erika, and Gábor Nagy. 2009. Changing Spaces of Knowledge-Based Business Services in Hungary. *Hungarian Geographical Bulletin* 58 (2): 101–120.

Oanca, Alexandra. 2010. Governing the European Capital of Culture and Urban Regimes in Sibiu. MA thesis, Budapest: Central European University. www.etd.ceu.hu/2010/oanca_alexandra.pdf

Oosterlynck, Stijn, and Sara González. 2013. 'Don't Waste a Crisis': Opening Up the City Yet Again for Neoliberal Experimentation. *International Journal of Urban and Regional Research* 37 (3): 1075–1082.

Pálné Kovács, Ilona, and Zoltán Grünhut. 2015. The 'European Capital of Culture – Pécs': Territorial Governance Challenges within a Centralised Context. In *Territorial Governance across Europe: Pathways, Practices and Prospects*, edited by Peter Schmitt and Lisa Van Well, 81–94. Abingdon: Routledge.

Pasquinelli, Cecilia. 2015. The Olympics Bidding Process: A Matter of Branding? *CritCom: A Forum for Research & Commentary on Europe*, May 22. http://councilforeuropeanstudies.org/critcom/the-olympics-bidding-process-a-matter-of-branding/

Peck, Jamie. 2005. Struggling with the Creative Class. *International Journal of Urban and Regional Research* 29 (4): 740–770.

Peck, Jamie, and Nik Theodore. 2012. Follow the Policy: A Distended Case Approach. *Environment and Planning A* 44 (1): 21–30.

Petrovici, Norbert. 2013. Neoliberal Proletarization along the Urban–Rural Divide in Postsocialist Romania. *Studia Sociologia* 58 (2): 23–54.

Pixová, Michaela. 2013. Spaces of Alternative Culture in Prague in a Time of Political–Economic Changes of the City. *Geografie* 118 (3): 221–242.

Pobłocki, Kacper. 2013. Neither West nor South: Colour and Vernacular Cosmopolitanism in Urban Poland. *Przegląd Kulturoznawczy* 3 (17): 205–215.

Polanska, Dominika V., and Grzegorz Piotrowski. 2015. The Transformative Power of Cooperation between Social Movements: Squatting and Tenants' Movements in Poland. *City* 19 (2–3): 274–296.

Prince, Russell. 2012. Metaphors of Policy Mobility: Fluid Spaces of 'Creativity' Policy. *Geografiska Annaler: Series B, Human Geography* 94 (4): 317–331.

Siemens Zrt., Studio Metropolitana, and HAS Centre for Regional Studies, Central and North Hungarian Research Institute. 2011. *A Good Place to Live – Budapest. Smart Local Solutions for an Emerging European Metropolis*. Budapest: Siemens Zrt.

Sokol, Martin. 2013. Silicon Valley in Eastern Slovakia? Neoliberalism, Post-Socialism and the Knowledge Economy. *Europe-Asia Studies* 65 (7): 1324–1343.

Sowińska-Heim, Lulia. 2014. Margins and Marginalizations in a Postsocialist Urban Area. The Case of Łódź. *Art Inquiry. Recherches Sur Les Arts* 16 (25): 297–312.

Środa-Murawska, Stefania, and Daniela Szymańska. 2013. The Concentration of the Creative Sector Firms as a Potential Basis for the Formation of Creative Clusters in Poland. *Bulletin of Geography. Socio-Economic Series* 20 (20): 85–93.

Sternberg, Rolf. 2013. Learning from the Past? Why 'Creative Industries' Can Hardly Be Created by Local/Regional Government Policies. *Die Erde* 143 (4): 293–315.

Suchacek, Jan, and Pavel Herot. 2015. Case F: The City of Ostrava – From Industrial Image to Industrial Image 2.0. In *Harnessing Place Branding through Cultural Entrepreneurship*, edited by Frank M. Go, Arja Lemmetyinen, and Ulla Hakala, 191–210. Basingstoke: Palgrave Macmillan.

Sulima, Roch. 2012. The Laboratory of Polish Postmodernity: An Ethnographic Report from the Stadium-Bazaar. In *Chasing Warsaw. Socio-Material Dynamics of Urban Change since 1990*, edited by Monika Grubbauer and Joanna Kusiak, 241–268. Frankfurt am Main: Campus Verlag.

Szarvas, Márton. 2016. Orfeo's Maoist Utopia: The Emergence of the Cultural Critique of Existing Socialism. CEU Sociology Department Master's thesis, 2016/20. Budapest: Central European University. www.etd.ceu.hu/2016/szarvas_marton.pdf

Tochterman, Brian. 2012. Theorizing Neoliberal Urban Development: A Genealogy from Richard Florida to Jane Jacobs. *Radical History Review* (112): 65–87.

Wallerstein, Immanuel. 1976. Semi-Peripheral Countries and the Contemporary World Crisis. *Theory and Society* 3 (4): 461–483.

Woźniak, Wojciech. 2013. Sport Mega Events and the Need for Critical Sociological Research: The Case of Euro 2012. *Przegląd Socjologiczny* 62 (3): 31–50.

Part II
Social structures and governance

8
Changing settlement networks in Central and Eastern Europe with special regard to urban networks

Zoltán Hajdú, Réka Horeczki and Szilárd Rácz

Introduction

The end of the Cold War led to the collapse of the Soviet Union and its European system of alliance, which was followed by the self-dissolution of the one-time superpower. The ensuing systemic changes fundamentally altered the social, economic, political life as well as the spatial structure of the countries concerned. In some of the 'post-socialist'/'post-communist' countries even the structure and the territory of the state was changed (Illés, 2002; Cho, 2013). In this sense, we can speak about 'post-Soviet', 'post-Yugoslav' or 'post-Czechoslovak' countries.

Settlement networks are rigid systems that can only be transformed in the long run. Their structures at the end of the state socialist era reflected the 'results' of the earlier periods, both historical and those coming from the state socialist era with its purposeful urban development and ideologically politically hostile attitude towards villages. From the 1950s, a new institutional system for developing settlements and their networks was established in every socialist country. The 1971 Hungarian settlement development concept and the disreputable Romanian 'systematisation' in the 1980s received international attention. Significant differences developed among settlements and also among their various groups in all these countries. There is abundant literature on the transformation processes in post-socialist countries, including post-socialist towns, from both within the region and outside it (see e.g. Musil, 1992; Enyedi, 1998; Andrusz et al., 1996; Horváth, 2000, 2015b; Kovács, 2014).

Naturally, systemic changes were not about settlement networks (albeit the Romanian opposition to systematisation almost directly turned into revolution). At the same time, the social, economic, political, legal, etc. elements of transition were directly or indirectly noticeable in the everyday life of the settlements and the changes in network relations (Slavík, 1997; Kuttor, 2009).

The establishment of new states and capitals, the opening of the western state borders, the improved permeability of the former Soviet western border (the so-called 'Eastern Iron Curtain'), and finally the abolition of border checks within the European Union have all been significant developments in the life of settlement networks after the systemic change.

In this study we shall focus our attention on the Visegrad Four and the South-Eastern European countries. Similarities as well as significant differences between these two macro-regions appeared already in the course of transition. An earlier volume (Horváth, 2015a) 'drawing the portrait' of the NUTS-2 regions analysed the settlement and town networks of the respective regions 'inwardly', detailing the specificities of these regions, micro-regions and in some cases even the settlements. In this volume we are attempting to describe the general characteristics, and relying on previous research in this area. Previous studies have been carried out at several levels: global (UN Habitat, OECD), all-European settlement networks, municipalities, functional, systematising studies (the European Spatial Planning Observation Network (ESPON)), Central and Eastern European settlement networks, the Visegrad Four, South-Eastern Europe, inter-state (transnational, transborder), national, within country (settlement, micro-, meso-, macro-regional level). As we are focusing our attention here on the common characteristics in the macro-region as a whole, only the comparative studies are included among the references.

Settlement systems in the Visegrad countries

The Visegrad countries (also called Visegrad Four or Visegrad Group) constitute a formal group of four countries (Czechia, Hungary, Poland, Slovakia) since 1991, aiming to cooperate in several fields, among others, in harmonising territorial and settlement network development. Due to occasional disagreement and strained relations between some of the countries, this co-operation is not always smooth (Máliková et al., 2015; Balázova et al., 2012; Novotny et al., 2015; Radics and Pénzes 2014).

The social, economic, political systemic changes also affected these countries' settlement networks as well as their various settlements themselves (from capital cities through regional centres and territorial administrative units to rural communities and suburban areas). We shall focus on urban networks and the towns themselves as they are decisive in the space-shaping processes. The EU's new regional and cohesion policies emphasise the role of large cities and urban regions (AEBR, 2009; Dijkstra and Poelman, 2012; European Commission, 2012, 2014).

A settlement network consists of settlements of various sizes and with various functions, usually having as a node a town fulfilling important social, economic and territorial roles. In the past decades, the central economic development projects had also served settlement policy and settlement network development aims. Large state investments sometimes intentionally and significantly modified the centres of historically developed settlement networks, thus transforming their inter-settlement relations.

Investors are primarily interested in the returns and profits of their capital, and in the short run the settlement network is just a stable background for them. It is not their task to plan or purposefully intervene in the settlement system. At the same time, the central government and the local self-governments (either at settlement or county level) might work out their settlement policies, development aims and strategies, and they can accordingly influence investments. Investments and development usually proceed within the coordinated system of these two approaches.

Settlement is one of the most complicated categories of space. The various scientific disciplines (dealing with administration, regional studies, statistics, regional geography) use it with different meanings, as they have worked out their own categories. Perhaps the most problematic terms are the 'village' ('*falu*' in Hungarian) which is a natural settlement category and the 'rural community' (Hungarian '*község*') which is an 'artificial', administrative one, but both are applied to the same place (Csapó and Balogh, 2012). There are minor problems with the term 'functional town' versus the 'administrative town'.

Table 8.1 Major settlement network characteristics in the Visegrad countries, 1990 and 2011

		Hungary	Slovakia	Czechia	Poland	V4 countries
Number of settlements	1990	3,093	2,669	6,590	2,880	15,232
	2011	3,154	2,890	6,251	2,479	14,774
Number of towns	1990	164	136	457	830	1,587
	2011	328	138	602	908	1,976
Population (1,000)	1990	10,374	5,297	10,302	37,849	63,823
	2011	9,986	5,404	10,486	38,533	64,410
Urban population (1,000)	1990	6,837	2,993	7,399	23,366	40,597
	2011	6,952	2,937	7,732	23,351	40,973
Urban population (%)*	1990	65.91	56.50	71.82	61.73	63.60
	2011	69.62	54.35	73.5	60.60	63.61

Source: Authors' calculations and construction based on census data (* based on data by UN).

In 1990 there were more than 15,000 basic administrative units[1] (Table 8.1) in the four Visegrad countries (although Czechia and Slovakia were at the time parts of the federative Czechoslovakia, their statistics were already separated). The number of settlements with city rank was close to 1,600[2] with an urbanisation rate of 64 per cent, showing great internal differences. Towns are organic nodes of settlement networks. Urban networks as systems, especially those of functional towns, are stable in the short run and can only be changed over a long period of time. However, the legal definition of a town or its administrative functions can be radically altered in a short time (Devecseri et al., 2008; Gorzelak, 2006; Gorzelak et al., 2010; Hamilton et al., 2005).

If we calculate the average settlement and urban populations for each country based on the data in Table 8.1, we get low values – with notable differences – in international comparison. When we take the four countries together, the average settlement population is 4,190, whereas the average city population is 25,922.

The national capitals play an outstanding role in regional administration, and they are also the most populous and most developed cities. The combined population of the four Visegrad capitals is over 5.3 million, but their ratio in the country's population differs greatly (see Table 8.5; Borsdorf and Zembri, 2004; Institute for Spatial Development, 2014). That of Budapest is the highest (24.9 per cent) and Warsaw's is the lowest (7 per cent).

Regions and 'official' regional centres appeared in the administrative organisational system of two countries only. Regional centres have been decisive in the urban networks of Poland and Slovakia. County seats have important positions in three countries, not only because of their population (half of the urban population lives there in Hungary and Slovakia), but also due to their economic, social, educational and health services. Smaller towns are of key importance in the micro-regions. Most of them have been the nodes of the services provided by the state. They have played the most important role in the Czechia.

Due to the differences in administrative structures, any further comparison is limited. There were no administrative regions in two countries, so the second largest cities after the capital functioned as county seats. Altogether in the four countries there are about 100 cities playing important political, economic and cultural roles.

In the early period of transition, there were countries where many people left the cities. Some had lost their jobs when large industrial plants closed down, others stopped commuting as

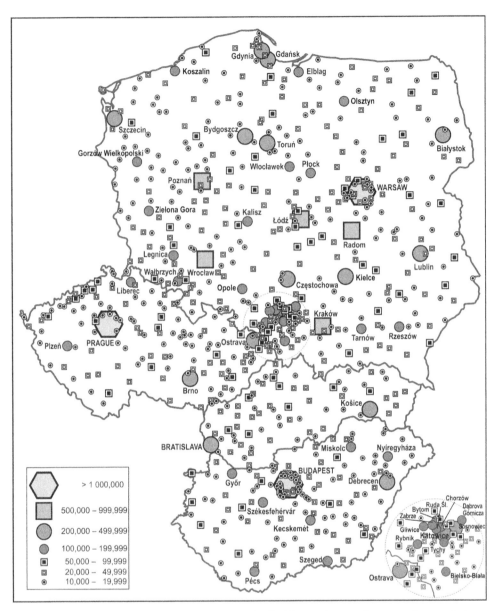

Figure 8.1 Towns with population over 10,000 in the Visegrad countries according to their size, 2011.

Source: Constructed by the authors.

transportation costs were raised. Furthermore, people and families who still had their rural connections also left the cities in the hope of having an easier life in the villages. On the whole, however, land privatisation and returning to agriculture remained rather modest.

At the beginning of the systemic changes there were large peripheral areas without towns in the settlement system of each country (Pénzes, 2013). The territory of these areas had diminished remarkably by 2000, but these changes were not fundamental.

The transformation processes of two decades

On evaluating the main processes and changes between 1990 and 2011, the following points should be stressed (see Table 8.2):

- The overall number of settlements in the four countries decreased significantly, although differently in each country.
- At both the national and the Visegrad group level, there was notable urbanisation, the number of settlements with town rank rose in all four countries (see Chapter 9 for the connected administrative issues).
- While the total number of population slightly increased, both the number and the proportion of urban population dropped remarkably (suburbanisation processes played a great role in this, many people moved from the cities to nearby villages).
- From a demographic aspect, the weight of the capitals developed differently (population decrease in Budapest is especially conspicuous, mainly due to suburbanisation), but on the whole it increased slightly within the settlement system.
- The capitals retained their dominant position and were winners of the massive changes in the financial, banking and insurance sector as well as in respect of Foreign Direct Investment

Table 8.2 Towns according to administrative functions, 1990 and 2011

		Hungary	Slovakia	Czechia	Poland	Visegrad countries
Capital city	1990	1	1	1	1	4
	2011	1	1	1	1	4
Regional (NUTS-2) centres (No.)*	1990		4		8	12
	2011	7	4	8	16	35
County seats (NUTS-3) (No.)	1990	19	8	6	49	82
	2011	19	8	13	66	106
Other cities, towns (No.)	1990	144	127	451	775	1,497
	2011	308	129	588	841	1,866
Population of the capital city	1990	2,016,681	442,197	1,214,174	1,644,515	5,317,567
	2011	1,733,685	608,287	1,241,664	1,740,119	5,323,755
Population of the regional centres	1990		848,730		7,070,370	7,919,100
	2011	2,670,992	2,736,172	2,383,345	7,396,629	15,187,138
Population of the county seats	1990	2,079,409	1,145,758	2,307,580	6,870,000	12,402,747
	2011	3,512,195	1,111,258	2,836,930	6,391,358	13,851,741
Population of other cities, towns	1990	3,194,321	1,405,279	3,877,447	9,426,454	17,903,501
	2011	1,706,563	1,217,968	2,969,107	9,158,839	15,052,477

Source: Authors' calculations and construction.

Note: * Statistical regions.

- (Chapters 2 and 4); their prestige in both society and market economy increased (see Table 8.5); their high real estate prices clearly indicate their distinguished economic position.
- Regions were not established in every country, but where they appeared, the population of their centres tended to grow considerably. The administrative and functional position of the NUTS-2 regional centres was not completely clear in 2011 either. There are no explicit, formalised regional centres in Hungary; neither the NUTS-2 regional centres nor the respective functional cities are clearly designated as such. Altogether, 35 cities and towns function as centres of large regions. Their central role is not always 'legitimised' by the 'rising county seats', their regional competitors, and there is rivalry between them in many cases. These cities and towns are in a special situation as they are 'known' in Brussels, and several regions and cities have representatives there.
- As a result of the administrative reforms, the number of county seats (NUTS-3 regional centres) increased, as did their population. The settlement network position of the over 100 virtual county seats did not change essentially during the administrative reforms; most of them play a key role within the network both administratively and functionally.
- The number of other (mostly small) towns grew considerably, but their population declined. Looking at the territorial distribution of the towns with a population over 10,000 (see Figure 8.1), we can see that areas without towns practically ceased to exist in the Visegrad Four; a small town can be reached from most of the villages within 30 minutes.

Functional characteristics of the settlement systems' dominant elements

After 1990, besides analysing the formalised hierarchy of the administrative system, attempts were made to delimit the most dynamic elements and functional spaces of the settlement systems in each of the four countries (in Hungary, for instance, the Central Statistical Office defined and then continually redefined the agglomerations and settlement groups[3]).

Functional settlement types and groups have also been delimited in some of the European comparative studies.[4] Due to different theoretical and methodological approaches and demographic limits, the categories used in this research covered different settlement groups in these four countries.

OECD has also worked out functional types and defined territorial groups[5] for the whole world in general, and for its members in detail. On defining territorial structures, OECD emphasised not only functional relations, but also population limits. The following categories were used: small urban areas with a population between 50,000 and 200,000; medium-sized areas with a population of 200,000–500,000; metropolitan areas with a population between 500,000 and 1.5 million; and large metropolitan areas with a population above 1.5 million.

In Czechia, 16 functional urban units have been delimited, of which Prague is a large metropolitan area; Brno, Ostrava, Plzen, Olomouc are medium-sized urban areas; and the remaining 11 are small urban areas. In Hungary, Budapest is a large urban area, while there are 7 medium-sized and 2 small urban areas. In Poland, Warsaw and Katowice belong to the first category, 6 areas to the second, and another 6 to the third. In Slovakia, Bratislava is in the second, Košice in the third, and a further 6 centres in the small urban area category (OECD, 2013, Slavík et al., 2005).

In the case of urban centres with a population of over 10,000, it is possible to speak about functional urban areas (FUAs). These actually form bottom-up functional and space hierarchy.

ESPON project 1.4.3 from 2007 revealed the differences among FUAs and cities as regards their area and population. It also categorised urban areas into five groups: small cities, medium

Table 8.3 The difference between functional urban areas (FUA) and morphological urban areas (MUA) population

	FUAs	Population	MUAs	Population	FUAs	Population	MUAs	Population
		Czechia				Poland		
Metropolitan*	2	3,187	10	2,267	13	12,817	43	9,729
Large city	1	352	1	165	9	3,178	9	2,130
Medium cities	14	2,302	17	1,228	28	3,920	29	2,821
Small cities	9	736	9	368	41	2,630	42	2,353
		Hungary				Slovakia		
Metropolis	1	2,523	5	2,232	1	711	1	444
Large city	2	580	2	393	1	343	1	239
Medium cities	13	2,021	17	1,333	8	1,080	9	612
Small cities	26	1,813	27	988	14	918	14	492

Source: Authors' construction based on ESPON Project 1.4.3.

Notes: * Metropolitan and polynuclear metropolitan areas (FUAs/MUAs: number of cities, Population: thousand people).

cities, large cities, metropolis, and metropolitan and polynuclear metropolitan areas (with a population over 2 million). Among the Visegrad Four, the smallest differences were found in Slovakia, the largest in Poland. Differences between the functional and the morphological urban areas (MUAs) were the greatest in the uppermost categories (metropolis and metropolitan areas) (see Table 8.3).

The four countries greatly diverge as regards their polycentric character:

- Slovakia has two centres; Bratislava and Košice belong to the same group as regards their size, albeit the capital has a larger population. Košice is the most populous city and a pivotal functional centre in Eastern Slovakia.
- Czechia is monocentric, as there is nothing in the next category 'under' Prague. Brno and Ostrava play a regional role in the eastern part of the country. While Brno is a kind of 'correctional' macro-regional centre between Bratislava and Prague, Ostrava is the Czech centre of the huge Silesian urban agglomeration, and has significant transborder relations.
- Hungary is doubtlessly Budapest-centred, almost monocentric. The agglomeration of Budapest is a pivotal core of the country in every respect. None of the informal centres of the NUTS-2 regions can compete with this populous and powerful economic and intellectual centre.
- From among the four countries, Poland has the only genuinely polycentric settlement network. The five most populous large cities following Warsaw in rank function as real macro-regional centres (Gorzelak et al., 2009; Szymańska, 2005). Their prestige is not much different from that of Warsaw.

After the western borders were opened, and especially following EU accession, the role of Berlin increased for the western part of Poland, while the same happened to Vienna for the southern Czech and the western Slovakian and Hungarian areas. The functional effects of the large-city networks increasingly transcend the borders. The eastern borders 'opened', too, but lacking substantial transborder relations, these areas and towns find it difficult to realise their potential. This is partly due also to Russian and Ukrainian domestic uncertainties, and the special Belarus situation.

The attraction zones of the county seats and the regional centres usually cover those of the small towns, which thus get into a very unfavourable position. Most of the small towns are in the 'hinterland' of a large city, connecting it to the rural areas. At the same time, small towns as centres of rural areas can provide such advantages which the big cities cannot because of their size. The role of small towns as local centres is greatly influenced by the settlements there, since the small towns themselves depend on the resources of the micro-region (districts, sections) surrounding them. The small towns in the Visegrad countries have a historical tradition as well as innovative capabilities. It has been increasingly emphasised that sustainable development and the future of the rural areas depend upon the small towns. In our globalising world regional and local centres – including small towns – can play a significant role as the centres of the hinterlands and the peripheries (see Chapter 5 on the subject of their role in rural development).

Factors shaping the urban networks in South-Eastern Europe

Due to size limitations, we are not considering here the concepts and delimitations of the Balkans (Zeune, 1808) and South-Eastern Europe (Fischer, 1893). In what follows, South-Eastern Europe will mean the Yugoslav successor states and Albania, Bulgaria and Romania. For historical reasons, we consider Greece to belong to Southern Europe, the Mediterranean, so we shall not deal with it here. In the case of the Balkans and the Balkan Peninsula we accept their most widely used – albeit far from being perfect (Todorova, 1997) – interpretation originating from physical geography. Studying this area is an interesting task as some of these countries lie on the border between the Balkans and CEE. That is, Western and CEE meet South-Eastern Europe here, and not only from a geographical point of view.

Settlement networks as systems develop over a long period and are fairly stable. In order to understand their deep-rooted structures, one must know their history. Their development has been affected by three groups of factors: (1) their geographical conditions, (2) where they are situated (how empires changed their borders and centres of gravity), and (3) their ethnic and cultural characteristics.

Physical–geographical conditions can promote or hinder inter-settlement relations. While mountains made it difficult to build relations between settlements on the shore and those in the inner territories, the Adriatic and the Black Sea almost predestined the coastal settlements to develop relations and trade with each other. Defended areas like valleys and bays were favourable places for social and economic relations, which then grew into larger settlements. Coastal settlements with road and railway connections had good potential for development. Among the inland towns, for those in valleys, it was primarily the settlements built on important crossroads that were able to keep their position over the course of history. The Danube and its tributaries as well as the roads running parallel to them were also suitable places to establish settlements (Erdősi, 2005). Almost all states and regions have an inner axis related to rivers. Most of the important towns can be found on the edge of the Balkans, at the coast or along rivers. This holds true even when we examine the peninsula taken in a broad context (Thessaloniki, Athens, Istanbul).

Due to their topography, the central areas of South-Eastern Europe are less suitable for the development of big cities with a wide area of influence. It is understandable, as they cannot provide work for a large population and they cannot extend their functions because of the fragmented spatial structure there. At the same time, conditions for autonomous (small) town

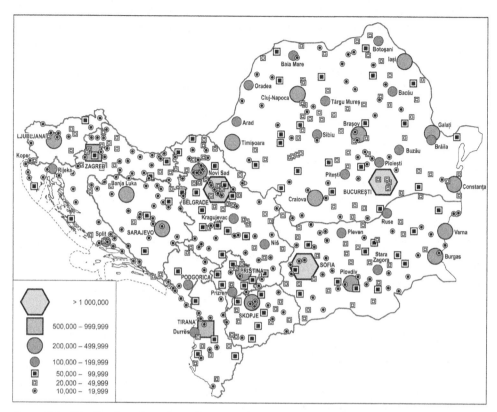

Figure 8.2 Distribution of towns and cities with population over 10,000 in South-Eastern Europe, 2011

Source: Rácz, 2014, p. 51.

development and for the integration of the closer surrounding (basins, plateaus, valleys) are favourable here. As a result, we can find central settlements everywhere, but their size is small even in a CEE comparison (Dimou and Schaffar, 2009). Urbanisation started here later than in CEE, and the urban network is rather sparse (BBR, 2006; CSIS-EKEM, 2010; ÖIR, 2006; Petrakos and Economou, 2002; RePUS, 2007). The level of urbanisation in the Balkans is still lagging behind that in Europe (see Figure 8.3).

The Balkans have always been a periphery in the past centuries; this area has never been a development centre of a large territory (Hajdú et al., 2007; Horváth and Hajdú, 2011; Illés, 2002; Kocsis, 2007). The main specificity influencing urban network development here was the often changing state structure: once there were small independent states (periods of disintegration), and at other times this area belonged to large empires (periods of integration). This political instability resulted in cyclical demographic, economic and administrative changes (see Table 8.4, and Rácz, 2011). The urban network and the settlements had to adapt to the given situations by changing their functions according to their new roles. As a consequence of these continuous changes, there have been no cities with uninterrupted development, and there are no large cities dominant over the entire South-Eastern European territory. There have been changes within the town hierarchy from time to time. As the classical (long-lasting, organic) conditions for urban development have been missing, there are no large cities in the Balkans in the Western European

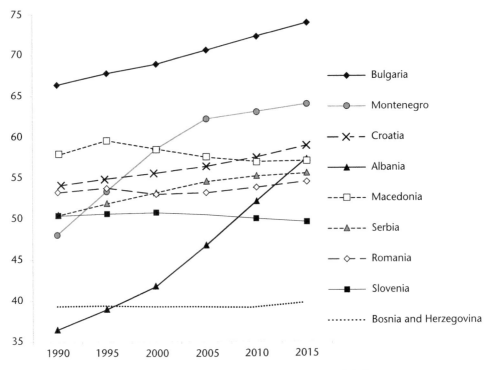

Figure 8.3 Urban population in South-Eastern Europe, 1990–2015 (%)

Source: Constructed by Sz. Rácz on the basis of WUP, 2010.

Table 8.4 The effects of cyclical development on the urban network and space structure in South-Eastern Europe

	Integrating (imperial) period	*Disintegrating (in-between, small-state) period*
Cooperation	territorial	ethnic
Borders	easy to cross (connecting)	ethnic borders are difficult to cross
Ethnic space	mixed, more heterogeneous	homogenising
Settlement network development	even, structurally and territorially balanced	fragmenting 'subsets', increasing differences
'Deficits' in the network	developed small towns	regional centres with rich functions
Large town functions	inter-regional, macro-regional	national (state)
Imperial capital(s)	strong development, rich functionality	relative power and social decline
Historical centres	relative recession, becoming peripheral	'selective' development
Ethnic centres	stagnation, limited growth	outstanding development
Gateway cities	mostly monopolistic	gateway cities nationally developed

Source: Constructed by Sz. Rácz.

sense; that is, there are no traditionally strong metropolises (Taylor, 2010). Having undergone significant growth in the last century, Belgrade, Bucharest, Ljubljana, Sofia and Zagreb are potential metropolises. Istanbul is of global importance, although it lies on the edge of the Balkan Peninsula, and we cannot take it as an unambiguous part of South-Eastern Europe.

The continually changing borders also entailed changes in the cities' area of influence. For instance, there were multiple attempts to ease ethnic separation in the Southern Slavic country (the so-called Banat system in 1929, the economic areas of the second Yugoslav state, etc.), but the new constitution of 1974 put an end to these efforts. Widespread decentralisation – even in the field of economic management that had been centralised earlier – led to strengthening the interests of the member republics and the fragmentation of the settlement network on a national basis. The cities' sphere of influence crossed the subnational borders only where ethnic interests made it necessary. Thus, after Yugoslavia broke up and these borders became international, only these areas were notably affected (Reményi, 2009). This transborder ethnic pattern[6] involves several issues: for instance, provision from the mother country (e.g. pension), or attempts to create independent institutions (e.g. own university) (Hajdú and Rácz, 2011).

Cities and their relations were influenced not only by the state they belonged to, but also by different cultures and their use of space. As it is known, the ethnic and religious composition of the Balkans is varied and mosaic-like (Horváth and Hajdú, 2011; Kobolka and Pap, 2011; Kocsis, 2007). Such multiculturalism cannot be found anywhere else in Europe. In the different cultural zones of the Balkans, different types of towns and cities developed (Mediterranean, Turkish, CEE, etc.), and in the 'transition' zones we can see their characteristics combined.

Changing urban network positions in the post-Yugoslav territory

The transformation of the post-Yugoslav territory is special in at least two respects. First, the federal member republics were quick to become independent states, and this affected the settlement system in a particular way. Second, the armed conflicts have made these transformation processes unusual and in part, tragic. That is why we are dealing with the post-Yugoslav territory in more detail.

Yugoslavia was one of the most complex peripheral areas of Europe from several aspects. The cyclic alternation of integration and disintegration was the main factor shaping the territorial and settlement processes here. Today, both the states and the settlement networks have achieved their most disintegrated condition; they have never been as fragmented as they are now. This structural change has not ended yet, and the establishment of new states can be expected in the case of Kosovo and Bosnia-Herzegovina. Further tension can be generated by the idea of Greater Albania and the issue of the Sandžak. New integration is hindered by unsolved problems like territorial disputes, unsettled conflicts, etc. Except for Slovenia, the other new states in the region carry ethnic and viability risks. Being separated on a national/ethnic basis, the urban network of the area is not connected; it is a system of 'monocentric' subsets with the capital cities in the centres.

There have been winners and losers among the cities of this area as a result of transformation. We might break down this transformation process into four phases which can only partially be separated from each other: (1) the break-up of Yugoslavia, (2) the armed conflict, (3) building nation states, and (4) Euro-Atlantic and global integration.

The dissolution of Yugoslavia and the way it took place had an essential effect on the settlements and their relations in this area. For a long time in the last two decades, the impact of this disintegration was stronger than that of globalisation and European integration. The individual countries were affected to varying degrees. If we compare transition in this area to that of the Visegrad countries, and also Romania and Bulgaria, we can find that transition was peaceful and integration faster in the latter countries. Yugoslavia's break-up was disadvantageous primarily to the federal capital Belgrade (losing federal resources and territorial power), and Rijeka was also a loser, just like some of the settlements that found themselves on the periphery at the new border areas. All the other settlements were winners, because as the former regions turned into independent states, their towns moved upwards in the settlement hierarchy. The Yugoslav settlement system, which had developed organically for almost seven decades, can also be considered a loser, as the new (ethnic) state borders made it fall into its 'subsets'. This fragmentation was further increased by building the new networks within the national borders, neglecting the earlier inter-regional relations. Settlement network 'subsets' do not necessarily adapt to frontiers. Where the ethnic patterns need it, towns may have a transborder scope of influence. This, of course, cannot be regarded as an international function.

We cannot speak of winners in the case of the bloody civil war. Even the settlements and nearby areas not involved in the armed battles and not burdened by waves of refugees were affected by the indirect consequences of the war. Lack of stability and trust, security risks and the deliberate termination of relations with the Yugoslav system led to loss of investors, cessation of traditional cooperations and declining economic performance. The human and material losses of the towns involved in the armed conflict were different. The various international statistics report different figures on this: the number of people killed and missing was between 150,000 and 300,000; the number of refugees having fled from the area was 600,000–800,000; the number of those leaving their homes temporarily was between 3.7 and 4.3 million. The worst devastation took place in those areas where opposing forces had previously lived together; which is why four-fifths of the death toll was concentrated in Bosnia-Herzegovina.

Besides occupying some territories, the aim of the war was to secure the loyalty of the population and the 'exclusiveness' of the ethnic group having military power over the given territory. To achieve ethnically homogeneous areas needed for redrawing the borders of the new nation states, ethnic minorities had to be ousted (or killed). The settlements thus 'cleansed' were then populated by people belonging to the same ethnic group who were either refugees or moved there from other territories. Accordingly, there were three parties fighting one another in the territories with mixed populations, not to speak of the intra-Bosnian fights. Following the war, most of the settlements, town districts and regions had ethnically homogeneous populations, even in those parts of the countries which were not involved in the war. In Bosnia the front lines became administrative borders (Reményi, 2009).

There is neither a possibility of (owing to destroyed, occupied homes) nor genuine requests for (through lack of personal or material motivation) restoring the original ethnic situation via the refugees' return, albeit the international community continues to urge this, and it is also communicated in several countries in the region as a requirement for EU integration. In the aftermath of the conflict, there are also examples of increasing ethnic diversity. For instance, in the Croatian Krajina the Serbs lost their majority as many of them fled. But losing the majority also happened in areas where many refugees arrived (e.g. in the northern part of Macedonia, or to the Hungarian areas of Vojvodina). Peaceful factors (e.g. assimilation) also contributed to the shift in the proportions of ethnic groups.

The break-up of Yugoslavia 'decreased' ethnic diversity as the ethnic category 'Yugoslav' ceased to exist (in 1991 almost 700,000 people declared themselves to be Yugoslav). On the other hand, when Montenegro become independent in 2006, it 'increased' ethnic variety, because its thus far uniform Montenegrin population got split: those in favour of a federal state declared themselves to be Serbs, while the people supporting separation, Montenegrins. The natural increase of the Albanian population and their immigration seemingly increased ethnic heterogeneity in Macedonia, but practically – at regional level – the situation is just the opposite: the area inhabited by Albanians is continuous and homogeneous. As regards the demographic effects of the conflict (e.g. changes in population), there are also 'loser' and 'winner' settlements both in absolute and in relative (compared to the national average) terms.

The successor states have strengthened their new territory and their urban network. As a result of the break-up of Yugoslavia, Belgrade suffered a loss of functions and sphere of influence. It is just a slight 'consolation' for its heavy losses that the new second-rank cities (Novi Sad, Niš) are much weaker than its former 'counterweights' (Zagreb, Ljubljana) were. The new capitals have been developed the most in every respect. Getting rid of the shadow of Belgrade, they have become the primary centres of their countries, and they have been concentrating a growing proportion of national resources. The macro-regional centres are relatively weak, or cannot even be interpreted as real 'counterweights', which is understandable, as these countries are small and have scant resources that are enough to 'create' only one large city. In spite of this, the 'large towns' following the capital in hierarchy have also grown appreciably. Even without widening functions or new administrative ranks, they 'rose' to the second level in the new country from their relatively lower position in the former Yugoslav settlement network. With the new international borders, the former attraction zones have changed and the number of towns fulfilling regional functions has grown; the new centres have integrated the areas that lost their earlier centres. It was important for the successor states to have direct access to the sea and to develop a national port-town. The new administrative systems introduced have broadened the functions of several settlements. The most conspicuous among them was the multi-tier decentralisation in Bosnia-Herzegovina, but the Croatian and the Macedonian administration reforms also increased the number of settlements having central functions. Irrespective of the administrative system, ethnicity-based institutional development (e.g. the establishment of universities for ethnic minorities) has increased the importance of some towns within the network.

Some of these processes (e.g. ethnic separation, mass emigration, decreasing rural population) would have most probably taken place without wars, too, but much more slowly and perhaps on a smaller scale. The interrelations within this area have changed essentially with the break-up of Yugoslavia, and the war and the establishment of nation states entailed their cessation or neglect. At the same time, mostly ethnic transborder cooperation has intensified, as have relations with the great powers, with various orientations in the individual countries. In the long run, a new epoch of integration might start at both European and global levels. In this process, not only whole countries but also their settlements will compete for Foreign Direct Investment. So far, only few countries and some cities have been successful in this integration, for the rest of them this is just an opportunity.

We assume that transition in this area will be similar to the experiences of the Visegrad countries. Namely, modernisation (integration) will be a top-down process from northwest to southeast along the European corridors. The capital cities entering this process first will have an advantage, further increasing their relative importance (Papadaskalopoulos et al., 2005). In general, gateway cities and large cities will be the winners. The loser group will include, among others, the settlements and areas close to the new ethnic borders, which have thus gotten into a peripheral situation due to the change in orientation. The new urban relations will not necessarily build on the earlier structures, as the European interests and scales basically differ from those in the Yugoslav era.

Depopulation is characteristic of the whole Balkan Peninsula, of the countries involved in the Southern Slav war and in Albania, Bulgaria and Romania alike. Rural population decline is a great problem in this area. The number of depopulated settlements has been growing in the last two decades. At the time of the last censuses, there were about a thousand depopulated settlements which will be administratively incorporated into one of the neighbouring villages, since without a permanent population, a settlement practically does not exist.

Albania is the most peculiar country in the region from the aspect of urbanisation. It is the 'most rural' country in Europe. Before the systemic change, urbanisation was restricted administratively and as a result, in 2006, 58 per cent of the employed population worked in the primary sector. It was only in the 1990s that people were allowed to move into towns. At present, still over 40 per cent are employed in agriculture. The intensive population movements and the outstanding natural increase of the population (in European comparison) have together led to the increased growth of the urban population. Jobs in industry, commerce and tourism as well as better living conditions attract masses of the population from the highlands to the capital and the coastal urban settlements; but there is also emigration present. Over three-quarters of these people move to the areas at Tirana and Durrës, and about half of them are younger than 30. The population of these two cities grew by 37 per cent, 260,000 people, between 2001 and 2010. Now 40 per cent of Albania's population live in this region. This concentrated and uncontrolled urbanisation has a disadvantageous effect on both the natural environment and the quality of life in these cities. The economic and social costs of the Albanian urban population explosion are progressively growing. Despite the negative experiences, this spontaneous and chaotic urbanisation is an inevitable concomitant of Albania's development, which also has its positive outcomes. However, it is imperative to plan and control these processes.

Central and Eastern European capitals and their specificities

The macro-region has had a very dynamic period in the past two decades. Previously, FDI had been unknown, but in two decades, investors discovered the countries of the region as profitable fields of investment. This did not happen to all regions simultaneously. Generally, investors first focused on the capital cities, on ports and on regions bordering EU member states. That is, investments were concentrated in a few cities and regions, while most regions remained without substantial FDI for several years, in many cases until now. Consequently, capital cities and their regions acted as – sometimes the only – hubs of economic growth, increasing their dominance (Table 8.5). In a wider context it also means that in the period between 1995 and 2000, peripheral countries and especially their capital cities grew more rapidly than other regions and cities in the EU. The global financial and economic crisis affected this trend, but the long-term trends have not changed: regional disparities between the centres and the peripheries have been increasing. Capital city regions have lower unemployment and their growth rate has declined less than the other regions'. The main driving force of economic growth continues to be the service sector being concentrated in the capital cities (Chapters 2, 4). National and EU policies also strongly influence the territorial development of countries, cities and regions.

The unambiguous winners, however, are again the same regions, exploiting their role as metropolitan growth areas. All of them have improved their relative positions, some substantially (București–Ilfov), while a number of non-central regions have experienced a drop in their relative development level. There are no non-capital regions that have experienced significant improvement in their development ranking. This also underlines the highly metropolitan character of high-technology manufacturing and knowledge-intensive services, as well as the intentions of economic, financial and political control, in which the higher tiers of the globally organised urban network predominate, while the competitive positions of functional urban areas lacking a critical mass are much less advantageous. The distribution of advanced business services

Table 8.5 Share of capital city regions (in percentage of the country)

	Population	Employed persons	GDP	Students in higher education
Bulgaria	18	33	36	47
Croatia	18	19	31	53
Czechia*	11.8	13	24.9	36.2
Hungary	29.8	28	49	48
Macedonia	30	34	48	37
Montenegro	23	33	n/a	67
Poland	13.8	7	22.3	18.9
Romania	11	12	21	35
Serbia	22	31	35	52
Slovakia	11.3	13.7	27.6	40.3
Slovenia	25	36	36	60

Source: Authors' calculations and construction.

* *Note*: Prague without the Central Bohemian Region.

corresponds to the urban network: it exceeds 60 per cent in the central regions of Croatia, Slovenia, Bulgaria, Hungary and Slovakia, and is only below 40 per cent in Poland, where a more polycentric urban network is present (Chapters 1, 4).

It is a general phenomenon in the Balkan states that the vast majority of economic activities are concentrated in the capital cities; they are mostly services such as banking, finance, trade, research, higher education, etc. The capital cities are by far the most important centres of transport, too; it is much more difficult and slower to access the other cities by means of transport. There are crucial differences between the capital cities and the second largest cities (Rácz, 2014).

Summary

Transition in CEE from the late 1980s has changed not only the social, economic and political structures, but also the external and internal circumstances of the settlements, the settlement groups and the whole settlement network. One of the essential elements of this transformation was the increasing role of private initiatives, of domestic as well as foreign capital in both decision-making and employment, replacing the earlier, centralised party and state decisions.

The state continued to play its role in settlement and territorial administration, in organising the provision of public services, and in shaping the big infrastructural networks. Many of the related decisions have already been made based on compromises between settlement and territorial interests. Political benefits (favouring larger groups to get more votes) have emerged as a new element in considering the various interests.

Social differences (e.g. unemployment) have been increasing both among the various settlement groups and within settlements. The role of the state in mitigating this has evidently remained. Granting preferences to the territories and settlement groups that turned out to be losers of the transformation has become an important element in providing support. However, this has led to only moderate economic results.

After a transformation of a quarter of a century, the settlement network of this area faces new challenges. The basic question is how the new type of inter-settlement competition evolving now within the European Union will affect the towns and the urban networks which have different functions and are of different sizes and in different situations (e.g. some are in member states, some in candidate countries).

Most of the poorest NUTS-2 regions in the EU can be found in this area. If bridging their national, macro-regional social and economic gaps is unsuccessful, the intra-settlement and intra-network relations in these countries might become even more strained and discrepancies might become more aggravated.

Acknowledgement

The publication of this research has been supported by project #104985 of the Hungarian National Research, Development and Innovation Office.

Notes

1 The differences in the interpretation of the terms 'settlement' and 'administrative unit' at the European level are also clear from the UN Habitat II (1996) report. For instance, 'nearly 58,000 settlements, including 853 towns' are mentioned for Poland, that is, all of the settlement groups, while in the other three countries the number of administrative units were treated as settlements. If we are calculating them on a common basis, there were almost 100,000 settlements in the four countries.
2 There are different approaches to towns in the various disciplines: historical (in social and economic studies), demographic (as the habitat of people), social (as used in sociology or ethnography), legal–administrative, nature-related, or the approach of urbanists (Enyedi, 2012).
3 It was in 1970 that the Hungarian Central Statistical Office first delimited Budapest's agglomeration. In 1985, it delimited 37 various functional settlement groups countrywide: 3 agglomerations (Budapest, Miskolc, Pécs), 5 agglomeration areas, 3 urbanising areas, 8 settlement groups of cities, 6 smaller settlement groups, 7 smaller groups of towns, and 5 groups of towns. These delimitations are specific to the Hungarian settlement network, and many of them could not be applied even to the Visegrad countries. Since 1997, these functional units have been delimited several times, but they have not been used in development planning. The practice of delimiting settlements functionally has also been used in the other three countries.
4 ESPON carries out research into every element of the European settlement system, from rural and urban spaces through the functional relations between towns and villages to the polycentric formation of urban systems.
5 OECD list of urban areas by country: www.oecd.org/gov/regional/measuringurban
6 *Albanians* in Northwest Macedonia, Preševo Valley (Serbia), Montenegro; *Bosnians* in Sandžak; *Croats* in Herzegovina; *Hungarians* in Vojvodina (Serbia), Partium, Székelyland (Romania); *Serbs* in Syrmia, Krajina (Croatia), along the banks of the rivers Sava and Drina (Bosnia-Herzegovina), in Mitrovica (Kosovo) and Montenegro; *Turks* in Kardzhali, Razgrad and Shumen (Bulgaria).

References

AEBR (Statement of the Association of European Border Region), 2009. Green Book on Territorial Cohesion. www.aebr.eu/en/publications/ [Accessed 14 March 2015].
Andrusz, G., Harloe, M. and Szelényi, I. eds., 1996. *Cities after Socialism. Urban and Regional Change and Conflict in Post-Socialist Societies*. Oxford: Blackwell.
Balázova, E., et al., 2012. *Benchmarking of the Public Services Provided by Municipalities in the V4 Countries*. Nitra: Slovak University of Agriculture in Nitra.
BBR, 2006. *Outlining Central and South East Europe. Report on Spatial Development in CenSE*. Bonn: Federal Office for Building and Regional Planning.
Borsdorf, A. and Zembri, P. eds., 2004. *European Cities Structures. Inside on Outskirts*. Paris: METL/PUCA.
Cho, Gyoujin, 2013. *Global Review of Human Settlements*. A Support Paper for Habitat: United Nations Conference on Human Settlements. Amersterdam: Elsevier.
Csapó, T. and Balogh, A. eds., 2012. *Development of the Settlement Network in the Central European Countries: Past, Present and Future*. Heidelberg: Springer Verlag.

CSIS-EKEM, 2010. *Re-linking the Western Balkans. The Transportation Dimension*. Policy Paper. Centre for Strategic and International Studies, Hellenic Centre for European Studies, Athens, February.

Devecseri, A., et al., 2008. *Presentation of the Hungarian Spatial Development Policy and Glossary of Terms*. (Preparatory Study for the Joint Spatial Development Document of the V4 Countries.) Budapest: VÁTI.

Dijkstra L. and Poelman, H., 2012. Cities in Europe. The New OECD-EC Definition. European Commission.

Dimou, M. and Schaffar, A., 2009. Urban Hierarchies and City Growth in the Balkans. *Urban Studies*, 46(13), pp. 2891–2906.

Enyedi, Gy., 1998. *Social Change and Urban Restructuring in Central Europe*. Budapest: Akadémiai Kiadó.

Enyedi, Gy., 2012. *Városi világ* [Urban world]. Budapest: Akadémiai Kiadó.

Erdősi, F., 2005. *A Balkán közlekedésének főbb földrajzi jellemzői* [Major geographical characteristics of transportation in the Balkans]. Pécs: Pécsi Tudományegyetem Természettudományi Kar Földrajzi Intézet Kelet-Mediterrán és Balkán Tanulmányok Központja. (Balkán Füzetek, 3).

ESPON, 2007. ESPON Project 1.4.3. Study on Urban Function. www.espon.eu/ [Accessed 12 March 2015].

European Commission, 2012. *Eurostat Regional Yearbook 2012*. ec.europa.eu/eurostat/ [Accessed 20 April 2015].

European Commission, 2014. Investment for Jobs and Growth. Promoting Development and Good Governance in EU Regions and Cities. ec.europa.eu/regional_policy/ [Accessed 12 March 2015].

Fischer, T., 1893. Die Südosteuropäischen Halbinsel. In: Kirchoff, A., ed. *Unser Wissen von der Erde. Allgemeine Erdkunde und Länderkunde*. Länderkunde von Europa. II. Vienna.

Gorzelak, G., 2006. Poland's Regional Policy and Disparities in the Polish Space. *Studia Regionalne i Lokalne*, Special Issue, pp. 39–74.

Gorzelak G., Smętkowski, M. and Jałowiecki, B., 2009. Metropolitan Areas in Poland – Diagnosis and Recommendations. *Studia Regionalne i Lokalne*, Special Issue, pp. 37–58.

Gorzelak G., Bachtler, J. and Smętkowski, M. eds., 2010. *Regional Development in Central and Eastern Europe. Development Processes and Policy Challenges*. New York: Routledge.

Hajdú, Z., Illés, I. and Raffay, Z. eds., 2007. *Southeast-Europe: State Borders, Cross-Border Relations, Spatial Structures*. Pécs: Centre for Regional Studies, Hungarian Academy of Sciences.

Hajdú, Z. and Rácz, Sz., 2011. Urbanisation, State Formation Processes and New Capital Cities in the Western Balkans. *AUPO Geographica*, 42(2), pp. 63–77.

Hamilton, F.E., Andrews, K.D. and Pichler-Milanovic, N., eds., 2005. *Transformation of Cities in Central and Eastern Europe: Towards Globalization*. Tokyo: United Nations University Press.

Horváth, Gy. ed., 2000. *Regions and Cities in the Global World*. Pécs: Centre for Regional Studies.

Horváth, Gy. ed., 2015a. *Kelet- és Közép-Európa régióinak portréi* [Portraits of the regions of Eastern and Central Europe]. Budapest: Kossuth Kiadó.

Horváth, Gy., 2015b. *Spaces and Places in Central and Eastern Europe: Historical Trends and Perspectives of Regional Development*. London: Routledge.

Horváth, Gy. and Hajdú, Z. eds., 2011. *Regional Transformation Processes in the Western Balkan Countries*. Pécs: Centre for Regional Studies of the Hungarian Academy of Sciences.

Illés, I., 2002. *Közép- és Délkelet-Európa az ezredfordulón. Átalakulás, integráció, régiók* [Central and South Eastern Europe at the turn of the Millennium. Transformation, integration and regions]. Budapest, Pécs: Dialóg Campus Kiadó.

Institute for Spatial Development, ed., 2014. *Common Spatial Development Strategy of the V4+2 Countries*. Brno.

Kobolka, I. and Pap, N., eds., 2011. *The Western Balkans: European Perspective and Tradition*. Budapest: Ministry of Foreign Affairs of the Republic of Hungary.

Kocsis K., ed., 2007. *South Eastern Europe in Maps*. Budapest: Geographical Research Institute Hungarian Academy of Sciences.

Kovács, Z., ed., 2014. Special Issue on Post-Socialist Cities. *Hungarian Geographical Bulletin*, 63(3), pp. 231–352.

Kuttor, D., 2009. Territorial Inequalities in Central Europe – Spatial Analysis of the Visegrad Countries. *Romanian Review of Regional Studies*, 5(1), pp. 25–36.

Máliková, L., Klobucník, M., Bacik, V. and Spislak, P., 2015. Socio-economic Changes in the Borderlands of the Visegrad Group (V4) Countries. *Moravian Geographical Reports*, 23(2), pp. 26–37.

Musil, J., 1992. Changing Urban Systems in Post-communist Societies in Central Europe. Analysis and Prediction. In: Gy. Enyedi, ed. *Social Transition and Urban Restructuring in Central Europe*. Budapest: European Science Foundation, pp. 69–83.

Novotny, L., Hruska, V., Egedy, T. and Mazur, M., 2015. Defining Rural Areas of Visegrad Countries. *Studia ObszarówWiejskich*, 39, pp. 21–34.
OECD, 2013. Definition of Functional Urban Areas (FUA) for the OECD Metropolitan Database.
ÖIR, 2006. *Metropolitan Networking in CenSE Backed by North–South Rail Corridors. Final Report of the Pilot Projects*. Wien: Österreichisches Institut für Raumplanung.
Papadaskalopoulos, A., Karaganis, A. and Christofakis, M., 2005. The Spatial Impact of EU Pan-European Transport Axes: City Clusters Formation in the Balkan Area Developmental Perspectives. *Transport Policy*, 12(6), pp. 488–499.
Pénzes, J., 2013. The Dimension of Peripheral Areas and Their Restructuring in Central Europe. *Hungarian Geographical Bulletin*, 62(4), pp. 373–386.
Petrakos, G. and Economou, D., 2002. The Spatial Aspects of Development in South-eastern Europe. *Spatium*, 8. pp. 1–13.
Rácz, Sz., 2011. Déli szomszédsági városhálózatok és hatások [Urban networks in the southern neighbourhood and their impacts]. In: A. Cieger, ed. *Terek, tervek, történetek. Az identitás történetének térbeli keretei 2*. Budapest: Atelier, pp. 211–237.
Rácz, Sz. 2014. New Integration Period? Changing Tendencies of the Urban Network in South East Europe. *Deturope*, 6(2), pp. 46–63.
Radics, Zs. and Pénzes, J. eds., 2014. *Enhancing the Competitiveness of V4 Historic Cities to Develop Tourism*. (Spatial-economic Cohesion and Competitiveness in the Context of Tourism.) Debrecen: Visegrad Fund.
Reményi, P., 2009. Jugoszlávia felbomlásának politikai földrajzi következményei és a térség városhálózatára gyakorolt hatásai [Political geographical consequences of the break-up of Yugoslavia and its impact on the urban system of the region.] PhD thesis. Pécs: Pécsi Tudományegyetem Földtudományok Doktori Iskola.
RePUS, 2007. Regional Polycentric Urban System – Strategy for a Regional Polycentric Urban System in Central-Eastern Europe Economic Integration Zone. Final Report.
Slavík, V., 1997. The Settlement Structure of the Slovak Republic during the Process of Transition. *Acta Universitatis Carolinae*, pp. 235–243.
Slavík, V., Kožuch, M. and Bačík, V., 2005. Big Cities in Slovakia: Development, Planning, Contemporary Transformation. *European Spatial Research and Policy*, 12(2), pp. 47–69.
Szymańska, D., 2005. New Towns in Settlement Systems in Central-Eastern European Countries. In: Iraj Etessam, Reza Karbalaei Nouri, and Faraneh Riahi Dehkordi, eds. New Towns Development Corporation (NTDC), Ministry of Housing and Urban Development, Tehran, Iran, pp. 187–219.
Taylor, P.J., 2010. Measuring the World City Network: New Developments and Results. Loughborough. *GaWC Research Bulletin*, 300.
Todorova, M., 1997. *Imagining the Balkans*. New York: Oxford University Press.
UN Habitat II Report, 1996. Report of the United Nations Conference on Human Settlements (Habitat II) (Istanbul, 3–14 June). United Nations.
WUP, 2010. World Urbanization Prospects, the 2009 Revision. New York, United Nations Department of Economic and Social Affairs Population Division.
Zeune, A., 1808. *Gea. Versuch einer wissenschaftlichen Erdbeschreibung*. Berlin: Wittich.

9

The development of regional governance in Central and Eastern Europe:

Trends and perspectives

Ilona Pálné Kovács

Shared roots and common starting point in a unified direction

When analysing regional governance systems in Central and Eastern European countries, the customary starting point is usually their common past – all the while declaring that regional government reforms were first launched at the time of the regime changes, and mainly follow a Western model. It is an indisputable fact that the governance efforts of these countries were clearly influenced by the intention of breaking with their (Eastern) pasts and trying to catch up with the West. Despite their common past, aspirations and patterns, these countries no longer form a homogeneous group, which manifests itself in the diverse content and form of the development of their regional governance systems. Reasons leading to a certain measure of divergence lie in their pre-socialist history, cultural characteristics and divergent geopolitical contexts, and it is accepted that convergence is not the only possibility regarding the future (Henderson et al., 2012). National traditions overlap the socialist phase: Napoleonic, Prussian and Scandinavian influences also need to be taken into account (Hendriks et al., 2011; Jordan, 2011). Nevertheless, the shared socialist heritage proved to be a strong cultural and structural determinant when examining regional governance, especially – as we will shortly see – in explaining the traditionally strong role of medium-level governance and regularly arising centralisation efforts.

The regime change meant a radical break with previous traditions and the directly inherited centralised Soviet-type council system. The very model of regional governance had to be drastically rethought, while almost in parallel efforts were made to consolidate territorial structure, to improve efficiency and to execute modernisation reforms. Today's balance sheet shows that this reform wave was only partially successful (Hendriks et al., 2011). Scruples were already raised at the time of the regime changes, claiming that these hasty and purportedly final reforms did not treat the period of transition as a separate context – making the very reforms obstacles to further adaptation and hindering the introduction of a long-term model (Bennett, 1994).

As a symbol of political change, new democracies in CEE dedicated a special role to local governments during the regime change. For example – save for a few exceptions such as Poland, Bulgaria and Romania – the fragmentation of municipalities and decentralisation to the local level evidently became the prevailing trend (Swianiewicz, 2010). The lack of trust in medium-level governments was also a common characteristic, in most cases referring to the prevalence of the autonomy of local governments. Almost every transition country under-financed its local government system, and it is generally accepted that fiscal decentralisation fell far short of the decentralisation of competences in the CEE new democracies (Owens, 2002; Šević, 2008).

When these democracies reached the 'consolidation phase' of local government reforms around the turn of the millennium, uncertainty regarding following models was already palpable. The 'European' standard manifests itself as a structural, operational, institutional and policy model, partly through compulsory rules, and partly via informal learning processes facilitated by cooperation between member states (Hughes et al., 2004). Although these standards leave room to adopt national characteristics, a certain asymmetry in power took shape between 'Western' experts conveying 'European' expectations and the political and professional elites of the 'learner' states in question. 'Recommended' Western policies were chiefly structural and institutional in their nature (or only transmitted general principles), and they offered little help in shaping the reform process in the new democracies. They could not be used to solve the technical, political and cultural problems of execution.

Pressure from the European Union to create regions was somewhat stronger. Every new CEE member state found it a fundamental challenge to gain access to Structural and Cohesion Funds of the EU. In order to achieve such access, separated so-called NUTS-2 regions were drawn. The EU was able to implement its will in a more decisive manner in the newly acceded post-communist countries using the so-called second-wave 'conditionalism' (Hughes et al., 2004; also see Chapter 10 for the power relations and dynamics of this process). Candidate states had to adapt to a moving target. On one hand, the expectations of European cohesion policy can only be articulated in a framework manner, as the Commission and member state governments have a joint responsibility in managing Structural Funds. For example, including regions as partners in developmental policy planning and resource allocation can only be voiced as a recommendation from the EU. Moreover, the institutional and spatial focal points of the cohesion policy change in each programming period. When the new countries gained accession, the old member states were preparing to implement the model of the programming period that was to start in 2007. Expectations concerning the legal status of the NUTS-2 regions (whether they should be public, administrative or regional in their nature) were difficult to articulate, given that the shaping of territorial administration lies exclusively in the discretional power of the member states. It is common knowledge that there was no universal standard to turn to when managing cohesion policy at the national level. Member states employed very different institutional tools, and institutionalising the NUTS-2 regions as local governments did not become a leading trend. Administrative units below the government and the NUTS-2 regions are different in the large majority of the 28 member states. This particularly applies to countries in the accession phase, which proves that the regime of the Structural Funds – although a strong influence in the preparation period – did not determine regional administrative reforms in a noteworthy manner.

Despite the above, the annual country reviews expressed considerable criticism regarding the administration of the states wanting to join the EU; however, accession clearly cannot be denied on these grounds (Moxon-Browne, 2005). It was mostly medium-level governance that sparked complaints when reviewing the candidate countries' administration systems, not only because it lacked clear professional competences, but because this level holds the most potential for foreseeable structural or even social resistance (COR, 2001).

The idea of 'New Regionalism' which reigned in Western Europe in the 1990s has long passed the administrational or cohesion policy directives, and instead expressed a new, more democratic and more plural governance paradigm which extended beyond the concept of the 'state' and spatially crossed administrative borders. In contrast, one of the specialities of CEE regionalisms was that in several countries they were tightly tied together with the process of nation-building leading to centralisation. The communist past also facilitated the fragmentation of former regional ties and networks in these states (Scott, 2009). It is obvious that the creation of new states and the redefining of the framework of state sovereignty put spatial power structures in a whole new context – consider Czechoslovakia or the successor states of the disintegrating Yugoslavia. Regional reforms concerning medium-level governance in many CEE countries were mostly regarded as 'greenfield investments', resulting in a situation whereby – save for a few exceptions – these NUTS-2 units have no common regional ethnic identity, economic cohesion or historical traditions whatsoever.

EU accession has introduced a new era where indirect expectations are less prominent, and only serve to motivate when the intention to learn is real (Bouckaert et al., 2011). New member states have quickly realised that they could still gain access to cohesion sources without real regional decentralisation. Moreover – despite the efforts toward regionalisation and institution-building prior to accession – it is the central governments that succeeded in ascertaining strong positions in the management of EU funds. The regional approach did not become a strong factor in development plans either, as most acceding countries are lacking independent Regional Operational Programmes (ROPs): not only after accession, but during the first, and now the second, programming periods as well. The reason for this is not restricted to the decreased pressure for adapting and learning after a country becomes a member state. It is also due to the new countries' recognition that the European Commission is actually mostly wary of regions. Forced fund absorption and the 'need' to concentrate administrative capacities also hindered further regionalisation (Bachtler et al., 2013).

Apart from a few exceptions (e.g. Poland), regional decentralisation in transition countries in CEE has come to a halt, and a general phase delay has ensued. Local and regional governments' task management models, proportions of sharing responsibilities between central and regional units, and even geographical borders have not been stabilised. In this field, the real regime change is only actually happening now – suffocated by financial restrictions and lacking the euphoria of political transition. The economic crisis and other, previously manifesting factors have led to strong central interference and the expansion of state power in order to consolidate local governments and service public utilities. We have to note that the dichotomy becomes less apparent when studying the behaviour of Western and Eastern countries during the crisis (Gorzelak and Goh, 2010), because centralisation as a trend has also appeared in other regions of Europe (Göymen and Sazak, 2014). It is not surprising that international organisations congregating local governments have expressed their worries when viewing current events, clearly stating that the now-general financial restrictions have a negative impact on regional governance autonomy, even despite reforms formally aiming for decentralisation (CEMR, 2013). The fact that regionalisation reforms have come to a halt is also palpable (although there are a few exceptions, such as France, which decided in 2015 to decrease the number of its regions substantially, but to give them more autonomy in turn). Seemingly, it is mostly the varieties with separatist motivations that have less trouble staying upright (e.g. Scottish referendum, Catalonian movements). The fact that spatial priorities of the EU cohesion policy have shifted toward cities and macro-regional cooperation, and the proportion of developed regions is getting larger compared to convergence regions, is also a factor. To summarise, the position of regionalism has weakened in Europe, and its role as a valid model or motivation in the eyes of the new CEE member states or countries waiting for accession has deteriorated.

Bouckaert remarks that reforms in CEE states have not been successful in many cases because only the frame of models was adopted, but content was not (2009, p. 102). Standards and 'best practices' originating from old member states often incited only superficial changes in new member countries, which lacked the necessary capacities for proper implementation (Stead and Nadin, 2011).

The history of regional governance reforms in CEE confirms that uniform decentralisation, one that is based on external expectations and implements a foreign model, is usually not successful. The following section will discuss the characteristics and the development process of regional governance systems in each country. Our aim is to demonstrate that in spite of their shared roots and common starting circumstances, the regional governments display substantial differences in this part of Europe, and have to face varying conditions for regional convergence and servicing community tasks.

Divergences: country reviews regarding regional governance and reforms

In *Bulgaria*, the regime change did not directly induce a substantial model change in regional governance – in fact, the administration and its personnel persisted throughout the transition. The unstable political situation was not ideal for the introduction of in-depth reforms (Dimitrova, 2007). A new constitution was adopted in 1991, one which recognised the concepts of rural communities and regional governance, and also discussed the role of the state in regional politics. However, it declared that creating autonomous regional units within the country was not possible, even though the exact definition of these was left unclear (Nikolova, 2011). Researchers claim that the true political transition and change of the elites started only in 1997, with the socialists losing power that year (Ellison, 2007). The first regulation left the existing, highly integrated regional government system untouched (Drumeva, 2001). There were 262 rural municipalities (Bulgarian: *obshtini*) in the beginning, and their number is almost the same now (264). At the same time, almost 5,400 settlements (Bulgarian: *naseleni mesta*) – which belong to approximately 1,600 smaller organisational units – have elected mayors and hold a certain sphere of authority (Kemkers, 2010).

Even today, there are no elected regional governments in Bulgaria. In 1998, 28 provinces (Bulgarian: *oblasti*) were set up in lieu of the 9 provinces which were dissolved 10 years prior, but these serve exclusively as administrative divisions with appointed governors leading them. Fragmentation characterises medium-level governance not only on a territorial basis, but on structural grounds too. Parallel to the administration of provincial governments, there are also decentralised sectoral units operating in the *oblasti* (Brusis, 2010). The legislation on regional development was adopted in 1999, but it did not take into account the consequences of joining the EU: conditions for accessing Structural Funds were left unexplained in both the planning phase and the institutional system (Bankova, 2014). Six NUTS-2 units were set up, but these were given no institutional framework, and the 28 provinces were classified at the NUTS-3 level. Development councils have been working in an ad hoc way in the NUTS-2 regions, mostly established to approve regional development plans (Yanakiev, 2010). EU accession did indeed speed up institutional reforms, but the system has stayed strongly centralised, and in 2008 the Minister of Regional Development acquired the right to allocate funds (Yanakiev, 2010). The participation of NGOs, non-state actors and rural communities is generally very limited, and the low capacity for administration related to any substantial development activity is still a problem on both local and regional levels (Nikolova, 2011). It is no coincidence that strong concerns have been voiced regarding the country's absorption performance (Gaydardzhieva, 2013). Systematic cross-compliance of development documentation is yet to be settled between the EU and national authorities – the very need for it only arose in the programming period 2007–2013. The National Concept for Spatial Development 2013–2025 includes the possibility of setting up special 'informal' regions, but this only underlines that there is no clear long-term concept regarding the regional division of the country (for both development policy and administrative purposes) (Velikova and Mihaleva, 2014).

The basics of a regional governance system in *Croatia* were already laid down by the constitution adopted in 1990, but their implementation was hindered by the war following the declaration of Croatian independence. Legislation regulating local self-governance was only introduced at the end of 2002, when the country basically returned to its pre-1998 status (Rácz, 2014). The lack of trust toward the autonomous regional governments that characterised this first decade is usually explained with the values of the party ruling at the time (Perko-Šeparović, 2001). The most important steps took place between 1999 and 2001, generally under the scope of decentralisation (Balázs, 2005). The Croatian state was highly centralised up to 2001, and truly substantial changes took place only in the past decade (Bajo and Jakir-Bajo, 2008).

The 1994 law regulating administrative divisions increased the number of rural municipalities (Croatian: *općine*) fivefold, compared to the former integrated system. The average population of the approximately 420 communities is 3,600 people (Ivanišević et al., 2001). The number of towns and cities (Croatian: *gradovi*)[1] was also doubled. This basic-level organisational system currently rules more than 6,700 settlements (Croatian: *naselja*), facilitating their operational efficiency by promoting partnerships (Kopajtich-Škrles, 2008).

At first, county governments (20 *županije* + capital) were not given much freedom; meanwhile, the public administration led by county governors had significant authority (Koprić, 2007).

The decentralisation process in Croatia took place in multiple phases, and decision-making was further weighed down by the lack of sufficient organised and analysed information related to the actual situation (Koprić, 2003). Substantial reforms were launched after the amendment of the constitution in 2001. Thanks to the 2004–2007 decentralisation package, the role of the counties and local governments has been significantly upgraded, but their resources remained scarce (Koprić, 2007). The 2005 amendment to the local self-governance law strengthened the competence of *gradovi* but, on the whole, central tutelage remained a strong factor in the Croatian public sector (Maslyukivska, 2006). The reform process is counteracted by the politicisation of the system (Bajo and Jakir-Bajo, 2008), which projects future problems regarding the absorption of EU funds.

The institutional framework of regional development has been regulated several times. Counties were classified as NUTS-3 regions, and in the end – after many debates – three NUTS-2 regions were marked in 2007 (Kopajtich-Škrles, 2008; Rácz, 2014). Part of the new development institutions were created in a centralised way (e.g. the council responsible for national development strategies), while development agencies were set up separately in the counties.

Institutions were created using pre-accession funds, and the process of EU planning has begun. The number of NUTS-2 regions was reduced to two in 2012, primarily in order to ensure that the most developed capital region could stay supported in the 2014–2020 programming period. The fact that the EU member Croatia has no Regional Operational Programme (and its management remains centralised) is due to the joint effect of the artificial nature of the regions and the fact that public administration has remained very centralised.

The disintegration of local-level administration can be viewed as the by-product of the regime change in *Czechia*: small rural communities that were unified under former rule, once again reclaimed their autonomous status after the transition. By the end of 2012, the already fragmented local governance system counted 6,250 self-governing municipalities (Czech: *obce*), substituting the approximately 4,000 local councils operating under the previous regime. The fact that in 1950 almost twice as many settlements had rights to self-government is worth noting (Illner, 2013). The constitution adopted in 1992 – following the political transition but before independent Czech statehood (1993) – guaranteed local self-governance on both levels.

The next step was the introduction of separate legislation regulating local and regional governance, but this process was ongoing until 2011 (Illner, 2011).

Similar to other cases, the arising concept of regionalisation in Czechia was connected to EU expectations of cohesion policy before the country's accession. The idea was that using Structural Funds would seem more legitimate in the eyes of the European Commission if there were democratically elected regions within the country (Baun and Marek, 2006). However, the 14 self-governing regions thus created in 1999 (including the capital) are not equal to the NUTS classification, as the EU considered these regions (Czech: *kraje*) too small. There are currently eight NUTS-2 regions in Czechia.

Since the regions were created, initiatives to readjust their borders have been launched repeatedly (Halász, 2000), and instability was further increased by the reform plan concerning districts (Czech: *okresy*) (Blažek, 2001). Currently there are 76 districts. One of the unique aspects of Czech territorial administration is that instead of the usual dual structure, public (state-level) and regional administration belong to the same organisational framework, although certain decentralised bodies of the central public administration are still not classified under the unified territorial system (Illner, 2011). The role of Czech regional governments is not strong in any case; they are especially limited in their financial capacities and lack the power of assessment. This fact partially explains the change that took place in public opinion regarding regions after the country joined the EU. Regional development councils and regional development plans needed to be created in the eight NUTS-2 regions (so-called 'cohesion regions'), but the cooperation between *kraje* has often been difficult in these cohesion regions. Their collaboration is often characterised by rivalry, which generates significant disadvantage compared to *kraje* that form NUTS-2 regions on their own. It is no coincidence that only one integrated Regional Operational Programme received EU support, in which the opinion of the European Commission was clearly weighted (Baun and Marek, 2006). As the next EU programming period draws nearer, debates over the issue of ROPs are starting to resurface. The strengthened regions succeeded in having every NUTS-2 region establish their own Operational Programmes with which to start the previous 7-year cycle. Nevertheless, regions have argued in vain for the continuation of this practice after 2014 (Ferry and McMaster, 2013), and the partnership agreement with the Commission once again lists an integrated Regional Programme for the Czech regions in the current programming period.

Centralisation is the thousand-year-old legacy of the *Hungarian* governance system. The law regulating local self-governments, which was adopted at the time of the regime change, clearly broke that tradition. At the beginning of the 1990s, decentralisation was further supported by the political atmosphere, as ruling political elites then viewed autonomous local governance as a normative value. Legislation specifically declared the elementary settlement level as the structural base of the Hungarian regional governance system, doubling the number of decision-making municipalities from one day to the next to approximately 3,200. Every other level or unit of integration was assigned a secondary role. Due to the fragmented nature of the self-governance of these municipalities, the institutional system of local-level public services could not escape fragmentation either. Partnerships have not come to characterise to the Hungarian system, despite the fact that in 1997 new legislation was adopted specifically regulating partnerships. Towns and cities (Hungarian: *városok*),[2] particularly in the beginning, did not strive to cooperate with rural communities (Hungarian: *községek*) – their participation in partnerships mostly depended on potential financial benefits. The town network expanded exponentially over time, due to the forced elevation of villages to town status (Chapter 8). Today there are more than 300 towns, but a maximum 200 of these can be considered actual towns based on their functions. Experts claim that the development of the remaining 100 'towns' in order to reach the level of real towns is an unrealistic objective (Beluszky and Győri, 2006).

Traditionally strong county governments (19 plus the capital) experienced a significant loss of authority when new legislation was introduced regulating their role. The political and legal status of these counties (Hungarian: *megyék*) was crippled along with their legitimacy, due to their weakening function and the uncertain terminology of the regulation itself. Also, the indirect election system did not make it possible for the counties to form a closer relationship with their population. Their situation was further weighed down by the introduction of the new model of 'cities with county rank' (Hungarian: *megyei jogú városok,* i.e. county seats plus towns with more than 50,000 inhabitants, of which there are currently 23). As the name indicates, these towns/cities have the authority to perform county functions within their own borders, and they are effectively 'removed' from the territorial fabric of their surroundings.

An expanding territorial public administration system has been 'compensating' for the confusion and the weakening of the regional self-government system. Dozens of decentralised bodies were created on both county and regional levels, which perform tasks that would have undoubtedly required elected representative control and democratic institutionalisation instead. It is not surprising that the past decades were characterised by reform initiatives on regional administration and recurring debates related to these. The so-called 'rescaling debate' dates back to the middle of the 1990s, and led to the programme of political regionalisation. The idea of regions substituting the counties was not due to the recognition of the country's needs but to simply following foreign models. The decisive factor was having access to EU Structural Funds, which in turn led to the creation of seven short-lived NUTS-2 regions. Legislation on regional development was adopted by the Parliament in 1996, setting up the system of regional development councils. These mainly partnership-type institutions operated both in the counties and meso-regions, effectively preventing either level from becoming strong enough. During preparation for EU accession, Brussels continuously criticised the lack of administrative capacities, despite the fact that the model of the law itself and the institutional system of regional development had an almost exemplary role in the region (Pálné Kovács, 2001).

Although the government announced programmes for reorganising regional self-governments both in 2002 and 2006, no reforms have taken place. Following EU accession, the regionalised model of managing Structural Funds has not been introduced in Hungary either. Though every region had its own Operational Programme between 2007 and 2013, decision-making rights were reserved to the National Agency for Regional Development, which was responsible to the central government.

The large change in regional governance happened in 2010 with the conservatives coming to power. A new constitution and self-government law were adopted in 2011, both of which directly – and severely – impaired the position, competences and autonomy of the elected local governments. The institutions of regional development were removed from both counties and NUTS-2 regions. Absorption of cohesion sources now happens through a strongly centralised management. County governments and cities with county rank are authorised for planning and (partly) allocating funds in the framework of the unified Territorial Operational Programme, but final decision-making is reserved to selected managing authorities affiliated with the central government apparatus.

The experiences of the 25 years that have passed since the regime change have shown that the model invented for the local exercise of power did not live up to the challenges of modernisation. The Hungarian state remains centralised, and the past few years even witnessed a substantial set-back in this field. The hectic shaping of the organisational, structural and spatial conditions of governance left a negative impact on the performance of both regional self-governments, and the Hungarian state as a whole.

The roots of the three-tier administration in *Poland* reach back to the Middle Ages (Józsa, 2005). The shaping of the Polish territorial structure was subject to constant debate after the regime change concerning both counties/districts (Polish: *powiaty*) and provinces (Polish: *województwa*). In return, the approximately 2,800 local-level municipalities (Polish: *gminy*), consisting of settlements that belong to joint administration units, proved to be relatively stable. There was a short-lived 'grace period' during transition, when a consensus was reached regarding local self-governance, but the *powiaty* and *województwa* were only created years later. An administrative division similar to the current one in numbers and size had existed before the Second World War, although only Upper Silesia had a right to self-government. The former 49 *województwa* were not eliminated during the regime change: they were maintained as decentralised units, though the nomination of their leaders was influenced by party politics (Swianiewicz, 2011). Instead of the introduction of real decentralisation, several more decentralised government bodies were created regionally between 1993 and 1997.

A new territorial reform took effect in 1999, creating the new *województwa* and once again lending this historical level the right to self-governance. Poland had well-founded fears regarding regionalisation, as the weakness of the central power was usually interlinked with regional fragmentation in Polish history (Swianiewicz, 2011).

Regionalisation in this country – even if the concept has a strong historical background – has been initiated in a top-down fashion, and was primarily fuelled by modernisation and development policy goals instead of the efforts of regional political elites (Gorzelak, 2009). Deepening regional differences – i.e. the factor of accessing Structural Funds – also contributed to the realisation of the need for regionalisation.

Although Parliament voted in favour of regionalisation, the presidential veto related to the number and seats of the new regions (which was largely motivated by towns losing or not receiving a seat) called for a compromise. Thus the legislation creating 16 *województwa*, 308 *powiaty* and 65 cities with county (*powiat*) rank was adopted in 1998 (Józsa, 2005). Gaining the support of towns has always been one of the characteristic objectives of Polish regional reforms introduced by the government; the reform of the *powiaty* was also interlinked with the aim to strengthen larger towns. The pilot phase was conducted based on the voluntary joining of towns, although several of them decided to step back later, for fear of losing their financial sources and competences (Józsa, 2005). Regions that were smaller than before still complied with EU cohesion policy standards, which made them eligible to accept EU funds.

The *województwa* served as both self-governing and administrative units, which had strong historical traditions, but also foretold the possibility of conflict between public and regional self-governance sectors (Emilewicz and Wolek, 2002). In the beginning the creation of regions had not been paired with the decentralisation of different competences and especially resources. Self-governing regions have gained power in a gradual way, forcing the regional public administration to give up certain spheres of their authority in the years leading up to EU accession. The reform process in Poland received significant local support compared to other CEE countries – regional political and non-governmental networks successfully prevented potential regression (Tatur, 2004).

In spite of the elected regions, the management of regional development funds was set up in a centralised way. At the same time, the system of regional planning contracts helped to decentralise the division of funds and competences between the central government and the regions. The latter were gradually getting stronger and acquired more competences, which was partially due to bringing a certain dynamism into political processes through successful lobbying and pushing public opinion to support the deepening of decentralisation. EU cohesion policy had an especially important role in cementing the regions as self-governing units. After initial mistrust in the first EU cycle – when authority was granted to the government-appointed leaders of the *województwa* – the management of cohesion funds now happens through regional self-governments. The government that came to power in 2007 announced a new reform programme to rearrange the division of power between *województwa* leaders and regional self-governments, reduce state influence and introduce special governance forms concerning major town and city areas.

On the whole, the creation and consequent strengthening of regions largely contributed to the fact that these regions play a very substantial role in the current cohesion policy programming period. According to the partnership agreement with the European Commission, around 60 per cent of EU funds are absorbed in the framework of Regional Operational Programmes.

However, plans concerning major cities have yet to reach a similar understanding between the central government, the regions and the cities themselves (Sagan, 2013). It is also indisputable that regionalisation let the genie out of the bottle – regional parties have formed in certain areas with the aim of increasing regional autonomy and amending the constitution that is now based on the model of a unified state. The 2015 'March of the Regions' in Warsaw was a demonstration advertised for this end (Baranyai, 2013, as well as Chapter 12).

The Polish example shows that Europeanisation, and particularly the goal of accessing EU funds, would not have been sufficient to ensure the success of the reforms on its own. Regionalisation in Poland happened as a relatively consistent political decentralisation process, one that was luckily supported by the motivation of gaining access to cohesion funds.

The ethnic and geographic diversity and fragmentation of *Romania* led to a huge variety in territorial divisions throughout history, and it is no coincidence that the level of settlements claims the most stability. At the same time, certain areas of the country that were shaped within or next to various empires have acquired strong regional identities based on national and linguistic foundations. The united Romanian State formed in 1862 aimed to unify these identities as well as asymmetric territorial power structures. First, a county system was created, which lasted for more than 60 years, then – in order to unite the increased territory of the country after the First World War – a lopsided regionalisation was introduced, which in turn effectively served centralisation (Benedek and Bajtalan, 2015). In 1950 the Soviet-model council system was set up, which managed to integrate the previously fragmented spatial system, resulting in 4,000 municipalities, 144 districts and 28 counties. Territorial subdivisions were modified several times during communism, showing the top-down nature of decision-making in regional organisation, and the centralistic and nationalistic goals that were behind it. Despite the above, the county system (39) managed to stabilise itself.

Former centralistic governance traditions were hard to break, even with the regime change and EU accession. Although a new administration law was introduced after the transition, it changed neither the structure, nor the content substantially. Rural communities remained strongly integrated: at the present time, approximately 13,000 villages are governed through 2,686 communes. The administration is divided into 42 counties. Local self-governments are extremely weak, and power gravitates toward the counties – but primarily it is the prefects that have the most authority compared to county self-governments. The entire system is quite questionable when it comes to its legitimacy as self-government, especially regarding the role of prefects and the substantial guarantees of autonomy. The new constitution adopted in 2003 declared the autonomy of regional self-governments and the fact that they were not subjected to prefects (Dobre, 2011).

Similar to other countries, the process of Romanian regionalisation was started with the preparation for EU accession. The fact that the movement of the Hungarian minority toward regional autonomy (as they mostly live in a geographical enclave) already strongly manifested itself after the regime change is a unique feature (Chapter 12). Legislative reform took place between 1998 and 1999, in the framework of preparation for the absorption of cohesion sources.

Similar to Hungary, traditional county administration units were used to create the eight development (NUTS-2) regions. The result was extensively criticised because the division used units built on historical–economical cohesion. Debates were sparked concerning the location of the seats, underlining the fact that borders and seats also broadcast power positions in the eyes of the affected population (Horváth and Veress, 2003). As the eight development regions had no administrations on their own, a new institutional system for development policy was created, setting up development councils and agencies, as well as county government partnerships. A new law regulating regional policy was adopted in 2004, but it is still the 2001 legislation on regional planning – also containing the development system of the town network – that is viewed as a turning point for development policy (Csák, 2013). Even after EU accession, the eight NUTS-2 units did not become the geographical or institutional framework for regional cohesion or regional equalisation, and regions have not had the chance to manage their own Operational Programmes. The Ministry of Regional Development was created in 2007. Currently, this institution claims the right of managing cohesion policy funds (Dobre, 2011). Regional differences have grown further; there are only 10 counties out of the 41 that have GDPs that exceed 100 per cent of the national average (Csák, 2013).

Movements aiming for political regionalisation have gained new momentum since 2011, after the Romanian president announced the programme of reorganising regional administration, basically under the scope of preparing for the next EU programming period. The genie was let out of the bottle, and official, party, professional and civil initiatives have been flooding the political market ever since (Benedek and Bajtalan, 2015). These very different drafts were created along very different interests, but to this point even the NUTS-2 classification remains unchanged. Intentions for decentralisation and regionalisation cancel each other out, pushing any feasible reform of the modernisation of spatial governance into the distant future. Among other reasons, this is due to weak civil society participation and the low potential for advocacy in Romania (Dobre, 2011). Cultural and ethnic tensions within the country and between political parties also do not allow for a grassroots approach to regional decentralisation.

In *Serbia*, the new constitution adopted in 1990 removed any ideological (constitutional) obstacle from the way of creating a Western-type self-government system. However, the disintegration of the federal state of Yugoslavia induced an especially forceful centralisation in Serbia, leading to the nationalisation of the property of municipalities, and setting up a strong level of central (state) control in 1992. In 1997, the opposition won certain positions in major cities and a significant general resistance started to arise against the Milošević system, which also facilitated the start of decentralisation (Meekel, 2008). The 1999 legislation regulating regional self-governance was still considered by the parliamentary opposition as the law of tutelage over local self-governments (Simić, 2001). Finally, in 2002, the Serbian Parliament adopted a new self-government Act that complied with the European Charter of Local Self-Government (Balázs, 2005). In 2003 the Serbian government even announced an action plan regarding preparations for EU accession, including the implementation of the self-government system.

Rural municipalities (Serbian: *opštine*), towns and cities (Serbian: *gradovi*), and the capital city of Belgrade form the basic level of local self-government in Serbia. The country's territory was first divided into 29, then 25 districts (Serbian: *okruzi*). These districts have no constitutional status, and only perform administrative functions, including the supervision of rural communities. The government introduced both the spatial category and the institutional system with a single Act in 1992 – neither of which had existed before in any form (Sević, 2001). The new self-government law has stopped viewing rural villages as social and communal formations: instead, it treats them as territorial units of the state that have the right to self-government. According to the legislation regulating the administrative division of Serbia, there were 169 rural municipalities and 5 cities with municipality status (Serbian: *gradske opštine*). Even on a European scale, the Serbian model is a strongly integrated local government system, comprising almost 5,000 settlements, according to statistics. The average population of a community is 35,000 people (Meekel, 2008). Academic language addresses certain cooperation and partnerships between rural municipalities as regions and regional communities; these do not act as levels of territorial administration, but as partnerships based on the law of self-government.

The quasi member-state status of the so-called provinces that used to have strong autonomy has been changed substantially; they lost their constitutional rights, their separate constitutional courts, etc. The special legal status of these provinces is apparent in the fact that the Serbian constitution lists them by their names, meaning that new provinces can only be formed through amending the constitution and announcing a referendum. Other than Vojvodina, there is only one province (declared autonomous by the constitution): Kosovo and Metohija, over which Belgrade has had no authority since 1999, when the UN took over its supervision.

Rights of the Autonomous Province of Vojvodina were first protected by the so-called Omnibus Law of 2002, then by the new constitution adopted in 2006 – despite the fact that the constitution only contains relatively few provisions concerning administration, most of which serve as a framework only. Although it has received substantial authority, Vojvodina remains dually structured: bodies of the central government are still present and pose a source of conflict regarding different spheres of authority (Korhecz, 2008). All of the above boils down to the fact that generally no medium-level self-government has been created in Serbia, it is present only in provinces claiming a special status due to their ethnic population.

The idea of setting up regions has emerged repeatedly, parallel to the need to give more autonomy to towns and cities, but proposals aiming for reform have not been successful in the Parliament so far. European regional policy meant an obvious challenge in Serbia, as the country committed to the introduction of the NUTS system in the Stabilisation and Association Agreement with the EU. The draft on regional policy divides the country into seven NUTS-2 regions. Development partnerships and agencies will be created within this territorial articulation, but the law specifically declares that administrative or self-government bodies will not function either within these units or in the NUTS-3 regions. Hence the general professional opinion that Serbian regionalisation is mainly a tool of centralisation (Takács, 2009).

Administrative reforms were proceeding slowly and with ample controversy in the newly independent *Slovakia*, which was aspiring to set up both its institutional system and identity of a nation-state. Settlements were granted self-governing autonomy in 1990, but the territorial level of administration underwent significant structural reorganisation. Former districts were eliminated and substituted with 121 township offices. In 1996, the system was reformed again: Slovakia was divided into 2,866 rural municipalities (Slovakian: *obcí*), 8 large administrative regions (Slovakian: *kraje*) and 79 districts (Slovakian: *okresy*), giving self-governing status to the settlements only. A new process to reform public administration started in 1998, but a decision was only reached in 2001 through a constitutional amendment and after a long period of debate marked by ethnic tension. It became possible to elect local governments on the *kraj* level, substituting the former administrative units (Ficza, 2005). The consolidation of the *kraje* happened very slowly: the decentralised administration was obviously reluctant to give up its positions. The year 2004 saw yet another territorial reform abolishing the 79 district offices and dividing their powers between the 8 *kraje* offices, 50 circuit offices and 221 special administrative offices (Mezei and Hardi, 2003), and, more importantly, starting financial decentralisation on the *kraj* level. The *kraje* were getting stronger step by step, systematically gaining ground vis-à-vis the public administration (Buček, 2011). Their role was further strengthened by entering into partnerships with each other, benefiting their capacity for advocacy in the bargaining process with the government.

Four NUTS-2 regions were formed from the *kraje*, while the NUTS-3 level consists of the eight *kraje* themselves. Following EU accession, substantial decision-making mostly took place on the latter level, despite the fact that the regional operational and monitoring committees and their secretariats were set up in the NUTS-2 regions. In the field of development policy, the number of regional development agencies does not follow regional divisions: the network is primarily based on cities and towns. There were approximately 50 agencies operating in 2010 (Buček, 2011). Since Slovakia joined the EU, absorption of cohesion funds has been taking place through a strongly centralised management model. In the 2007–2013 period the *kraje* only participated as contributing authorities in the management of the singular, still integrated ROP, and their cooperation within the NUTS-2 framework is not sufficiently strong yet (Buček, 2011).

Slovenia is still considered a centralised country, which is interlinked with its small size and the difficulty of introducing medium-level governance (Savin, 2008). Another important characteristic of the model is that the regulation of self-governments is strongly saturated with ethnic, linguistic–cultural minority politics, aiming to ensure special rights also on a territorial basis (Dimitrovska Andrews and Ploštajner, 2001). The Slovenian constitution grants self-governing autonomy to rural communities (Slovenian: *občine*), but binds that status to the existence of certain conditions and services, thus containing the danger of possible fragmentation at the local level. Slovenia's 6,000 geographic settlement units were directed by approximately 60 integrated rural communities before the regime change (Ploštajner, 2008). The new comprehensive regulation of self-governments created 147 local self-governments, but – based on the law – only settlements with more than 5,000 inhabitants received the right to self-governance (Brejc and Vlaj, 2001). The local level fragmented further. Currently there are approximately 210 municipalities, and experts consider the Slovenian system fragmented in its nature, which was an argument for centralisation (Setnikar-Cankar et al., 2000). Similar to other countries, the basic problem of the system lies within the underdevelopment of fiscal decentralisation (Savin, 2008): the municipal sector receives only 8 per cent of the GDP and 18 per cent of the budget (Setnikar-Cankar, 2011).

To perform medium-level governance 58 administrative units were set up, but these substantially consist of ministerial departments and proved to be unable to ensure a coordinated model of central control (Leben, 2002).

Adaptation to the EU is naturally an important factor in Slovenia too. The country is striving to strengthen its medium level in order to access Structural Funds. According to previous provisions of the constitution, settlements have the right to congregate into regions. Another possible solution that has arisen before is to set up self-governing regions endowed with legal personhood (Brjec and Vlaj, 2001). Joining the EU eventually sped up the process. The 1999 law regulating regional policy first and foremost aimed to set up an institutional system for regional development in the country. Local self-governments decided to create 22 agencies, but the institutional and funding system of development policy remained rather centralised (Tüske, 2002). The administrative reform was late, and the absorption of EU funds for regional development was set up in a centralised system. The legislation on regional development was amended in 2005, and the NUTS classification was also modified, separating two NUTS-2 units. However, the institutional system of EU cohesion policy stayed characteristically centralised, and no Operational Programme was drafted for the two regions. The 2006 amendment of the constitution opened the way to the creation of self-governing regions, but specific legislation is yet to be adopted. It shows how regionalisation does not primarily belong to the dimension of the spatial division of power; in the hands of the government it would act as a tool of modernisation and, at the same time, centralisation (Dimitrovska Andrews and Ploštajner, 2008; Setnikar-Cankar, 2011).

Summary: diverging development paths?

Local-level self-governments are not stable pillars. If we wish to predict future trends, we need to take into account existing capacities and contemplate the chances of further reforms. Throughout the period of regime change, attention was directed to the strengthening of local self-governments, which in turn made efficiency-oriented modernisation and the future strengthening or rebuilding of medium-level governments more difficult.

Consolidation of the local level of administration and rendering small rural communities capable of decentralisation failed to take place in multiple countries. Certain experts (Swianiewicz 2010; Illner 2013) are questioning if fragmentation is really the root of the problem and whether it truly explains the weakening of self-governance systems. Undoubtedly, the merging of rural communities did not happen in every post-communist country: some preferred forms of partnership instead. Local self-governments created at the time of the regime changes were viewed as the 'restoration' of local democracy. Even though it became apparent that autonomy given to the smallest communities would not solve the problems accumulated in rural areas, any modification of the mainly fragmented structure would have been a very unpopular policy in the following few years (Maurel, 1994). Today it is widely accepted knowledge that economies of scale are not the only factor in local governance (Marcou and Verebélyi, 1993).

The governments of CEE countries have been under-financing their self-governance systems from the start (Pickvance, 1997). This phenomenon cannot be solely explained by the undoubtedly worse economic conditions of the transition period. Local governments continuously found themselves in retreating positions within community budgets compared to other government offices, even in times of prosperity or, at least, during more consolidated periods. It is widely held that fiscal decentralisation in the new democracies of CEE fell far behind the decentralisation of competences (Owens, 2002; Šević, 2008), which made problems more difficult to manage and induced further centralisation during the economic crisis (Blöchliger, 2014). Consolidating the system of local-level self-governments is evidently an important task of the countries discussed. This problem does not only (or maybe does not primarily) require structural reforms, but the rationalisation of funding and public services.

The constant rebuilding of the medium level did not lead to strong territorial/regional self-governance. When it comes to the modernisation of medium-level governance structure, we can say that every one of the discussed countries took certain actions on a smaller or larger scale after the regime changes, but the only truly successful reform in region-building was implemented in Poland (Jordan, 2011). Regional governments are gradually gaining strength in Czechia and Slovakia, although their sizes and capacities to manage cohesion sources proved to be meagre. This is part of the reason for the parallel set-up of the cohesion policy institutional systems in most countries.

Despite the need to introduce the NUTS system, which had an effect on territorial reforms, regional compliance has not yet occurred everywhere. The size of development of 'administrative' regions newly created in smaller countries does not satisfy scale requirements of the NUTS-2 units. These facts warn us that regional administrative reforms need a more comprehensive approach than the imperative implementation of NUTS-2 regions. At the same time, experience shows that without the presence of medium-level governments, ambitions worded during the regime change to create strong self-governance will fail, and the trend of regression will become typical instead.

EU cohesion policy management remains mainly centralised. As medium-level governance reforms mostly failed to occur, the institutional framework of development policy was often set up as a parallel system. Development regions, regardless of whether there are elected regional governments or not, have their own institutions (monitoring committees and agencies), similar to the practice of Western European countries. However, as discussed, Structural Funds are managed through special bodies that were assigned to the central governments after EU accession. Parallel to the centralising administration, centralisation also dominates the field of regional development, to which the only exception is, again, Poland.

Internal motivation is needed to keep up decentralisation that now came to a halt. Decentralisation had no traditional background in the post-communist CEE countries, and reform plans were strongly saturated with the conservative and anti-decentralisation views of the past (Nunberg, 2003; de Vries and Nemec, 2012). When analysing reform processes in the CEE post-communist states, we cannot avoid the question of heritage. However, this heritage has stopped being the only explanation of the failure of reforms – the ambitions of the affected actors, the institutional framework and reform capacities have a strong effect in their own right on the realisation of such reform goals (Haveri, 2012). As discussed, neither so-called Europeanisation, nor EU cohesion policy (both external motivators) can inflict lasting changes in the local and regional governance systems, despite the fact that national governments launched dozens of structural reforms, especially during the preparation phase of EU accession. If we take into account that – as the Sixth Cohesion Report verifies – there is a strong connection between the quality of governance and the absorption capacity of the cohesion countries, then it is the vital interest of every country discussed in this chapter to improve its government performance, as they are all lagging behind according to several statistics that compare government performance (World Bank, EU, OECD). The political enthusiasm characterising the 1990s is no longer perceptible, and political elites need to admit that decentralised governance is important not because of external expectations or because of the pressure of adaptation, but because of their potential to improve their own government performance.

Although only a few of the currently available academic analyses provide us with unambiguous proof of the connection between the quality of governance and decentralisation (cf. Charron et al., 2012; Neudorfer, 2014), it is safe to say that regions are able to influence their own government performance. In other words, besides the successful implementation of decentralisation reforms, the continuous learning and conditioning of regional governments that were given this opportunity is also essential. When analysing background conditions of regional performance, after researching each country's regulation, funding, general structure and capacities regarding its governance systems, we also need to scrutinise characteristics of concrete regional governance. Studying a region's urban governments is also an important lead for future research, as there is a significant possibility that larger cities will become more dynamic actors in areas with weak regional governments.

Acknowledgement

The publication of this research has been supported by project #104985 of the Hungarian National Research, Development and Innovation Office.

Notes

1 Translator's note: In some countries cities and towns are not differentiated by definition. Croatian in particular uses a single word for these two terms.
2 Translator's note: The Hungarian language also uses a single word for towns and cities.

References

Bachtler, J., Mendez, C. and Kah, S., 2013. Reflections on the Performance of Cohesion Policy. In: I. Pálné Kovács, J. Scott and Z. Gál, eds., *Territorial Cohesion in Europe*. Pécs: CERS HAS, pp. 46–58.

Bajo, A. and Jakir-Bajo, I., 2008. Local Government Finance in Croatia. In: Ž. Šević, ed., *Local Public Finance in Central and Eastern Europe*. Cheltenham: Edward Elgar, pp. 117–140.

Balázs, I., 2005. Változások és reformok az európai országok közigazgatásában: Magyarország és szomszédai [Changes and reforms in the public administrations of European countries: Hungary and its neighbours]. *Magyar Közigazgatás*, 55(12), pp. 763–767.

Bankova, I., 2014. Administrative Conflicts in Bulgarian Regional Development Policy: Current Situation, Trends, Possible Solutions. In: I. Pálne Kovács and C.M. Profiroiu, eds., *Regionalisation and Regional Policy in Central and Eastern Europe*. Bratislava: NISPAcee Press, pp. 305–316.

Baranyai, N., 2013. Regionalism in Upper Silesia: The Concept of Autonomous Regions in Poland. In: I. Pálné Kovács, and L. Kákai, eds., *Ten Public Policy Studies*. Pécs: University of Pécs Department of Political Studies, pp. 9–28.

Baun, M. and Marek, D., 2006. Regional Policy and Decentralisation in the Czech Republic. *Regional and Federal Studies*, 16(4), pp. 409–428.

Beluszky, P. and Győri, R., 2006. Ez a falu város! (Avagy a városi rang adományozásának gyakorlata s következményei 1990 után) [This village is a city! (The practice of granting of city rank and its consequences after 1990)]. *Tér és Társadalom*, 20(2), pp. 65–81.

Benedek, J. and Bajtalan, H., 2015. Recent Regionalization Discourses and Projects in Romania with Special Focus on the Székelyland. *Transylvanian Review of Administrative Sciences*, 18(1), pp. 23–41.

Bennett, R.J., ed. 1994. *Local Government and Market Decentralization: Experiences in Industrialized, Developing, and Former Eastern Bloc Countries*. Tokyo: United Nations University Press.

Blažek, J., 2001. Regional Development and Regional Policy in the Czech Republic: An Outline of the EU Enlargement Impacts. *Informationen zur Raumentwicklung*, 11(12), pp. 757–767.

Blöchliger, H., ed., 2014. *Fiscal Federalism. Making Decentralisation Work*. Paris: OECD Publishing.

Bouckaert, G., 2009. Public Sector Reform in Central and Eastern Europe. *Halduskultuur*, 10, pp. 94–104.

Bouckaert, G., Nakrošis, V. and Nemec J., 2011. Public Administration and Management Reforms in CEE: Main Trajectories and Results. *NISPAcee Journal of Public Administration and Policy*, 4(1), pp. 9–29.

Brejc, M. and Vlaj, S., 2001. Local Government in Slovenia (With Special Regard to Regionalisation). In: Z. Gál, ed., *Role of the Regions in Enlarging European Union*. Pécs: Centre for Regional Studies, pp. 101–112. (Discussion Papers/Special).

Brusis, M., 2010. Accommodating European Union Membership: The Regional Level in Bulgaria. In: R. Scully and R.W. Jones, eds., *Europe, Regions and European Regionalism*. Houndmills: Palgrave Macmillan, pp. 221–238.

Buček, J., 2011. Building of Regional Self-government in Slovakia: The First Decade. *Geografický Časopis/Geographical Journal*, 63(1), pp. 3–27.

CEMR (Council of European Municipalities and Regions), 2013. *Decentralisation at a Crossroads. Territorial Reforms in Europe in Times of Crisis*. Brussels.

Charron, N., Lapuente, V. and Dijkstra, L., 2012. Regional Governance Matters. A Study on Regional Variation in Quality of Government within the EU. EC. Working Papers. DG Regional Policy, Brussels.

COR. European Union Committee of the Regions, 2001. *Regional and Local Government in the European Union*. Luxembourg.

Csák, L., 2013. Cohesion Policy in Romania: Intensified Failure. In: I. Pálné Kovács, J. Scott and Z. Gál, eds., *Territorial Cohesion in Europe*. Pécs: CERS HAS, pp. 467–481.

de Vries, M.S. and Nemec, J., 2012. Introduction to Public Sector Dynamics in CEE Countries. In: J. Nemec and M.S. de Vries, eds., *Public Sector Dynamics in Central and Eastern Europe*. Bratislava: NISPAcee Press, pp. 7–18.

Dimitrova, S., 2007. A közigazgatás reformja Bulgáriában – a rendszerváltás kezdetétől az EU-tagságig [The reform in the public administration in Bulgaria – from the beginning of the transition to the membership in the EU]. *Nemzetközi Közlöny*, 1(2), pp. 91–94.

Dimitrovska Andrews, K. and Ploštajner, Z., 2008. Local Government in Slovenia. In: Z. Zlokapa, ed., *Block by Block It's Good to Build Well. Models of Organisation of Local Self-governance*. Banja Luka: EDA.

Dobre, A.M., 2011. Romania: From Historical Regions to Local Decentralization via the Unitary State. In: J. Loughlin, F. Hendriks and A. Lidström, eds., *The Oxford Handbook of Local and Regional Democracy in Europe*. Oxford: Oxford University Press, pp. 685–712.

Drumeva, E., 2001. Local Government in Bulgaria. In: E. Kandeva, ed., *Stabilization of Local Governments*. Budapest: OSI/LG, pp. 141–177.
Ellison, B.A., 2007. Public Administration Reform in Eastern Europe. A Research Note and a Look at Bulgaria. *Administration and Society*, 39(2), pp. 221–232.
Emilewicz, J. and Wolek, J., 2002. *Reformers and Politicians*. Bialystok–Warsawa: SEAP-ELISPA.
Ferry, M. and McMaster, I., 2013. Cohesion Policy and the Evolution of Regional Policy in Central and Eastern Europe. *Europe-Asia Studies*, 65(8), pp. 1502–1528.
Ficza, Cs., 2005. Közigazgatási reform Szlovákiában [Public administration reform in Slovakia]. In: I. Pálné Kovács, ed., *Regionális reformok Európában* [Regional reforms in Europe]. Budapest: TÖOSZ, pp. 230–246.
Gaydardzhieva, V., 2013. EU Regional Policy and the Opportunities for Socio-economic Development of Municipalities in Bulgaria. *Review of Applied Socio-Economic Research*, 5(1), pp. 69–75.
Gorzelak, G., 2009. Regional Development and 'New' Regionalism in Poland. In: J.W. Scott, ed., *De-Coding New Regionalism. Shifting Socio-Political Contexts in Central Europe and Latin America*. London: Routledge, pp. 177–198.
Gorzelak, G. and Goh, Ch., eds., 2010. *Financial Crisis in Central and Eastern Europe. From Similarity to Diversity*. Warsawa: Wydawnictwo Naukowe Scholar.
Göymen, K. and Sazak, O., 2014. *Centralization Decentralization Debate Revisited*. Istanbul: Istanbul Policy Center.
Halász I., 2000. Az új regionális önkormányzatok alkotmányos helye és szerkezete az egyes kelet-közép-európai országokban [The constitutional place and structure of new regional governments in various East Central European countries]. *Állam és Jogtudomány*, 41(1–2), pp. 97–142.
Haveri, A., 2012. Introduction to Twenty Years of Capacity Building in Local Government. In: J. Nemec and M.S. de Vries, eds., *Public Sector Dynamics in Central and Eastern Europe*. Bratislava: NISPAcee Press, pp. 95–97.
Henderson, K., Pettai, V. and Wenninger, A., eds., 2012. *Central and Eastern Europe beyond Transition: Convergence and Divergence in Europe*. Strasbourg: ESF.
Hendriks, F., Loughlin, J. and Lidström, A., 2011. Comparative Reflections and Conclusions. In: J. Loughlin, F. Hendriks and A. Lidström, eds., *The Oxford Handbook of Local and Regional Democracy in Europe*. Oxford: Oxford University Press, pp. 715–743.
Horváth, R. and Veress, E., 2003. Regionális politika és területfejlesztés Romániában [Regional policy and regional development in Romania]. *Magyar Kisebbség*, 8(1) pp. 3–31.
Hughes, J., Sasse, G. and Gordon, C., 2004. *Europeanization and Regionalization in the EU's Enlargement to Central and Eastern Europe*. Houndmills: Palgrave Macmillan.
Illner, M., 2011. The Czech Republic: Local Government in the Years after the Reform. In: J. Loughlin, F. Hendriks and A. Lidström, eds., *The Oxford Handbook of Local and Regional Democracy in Europe*. Oxford: Oxford University Press, pp. 505–527.
Illner, M., 2013. Fragmented Structure of Municipalities in the Czech Republic – An Advantage or a Problem? In: I. Kovács Pálné, J. Scot and Z. Gál, eds., *Territorial Cohesion in Europe*. Pécs: CERS HAS, pp. 444–454.
Ivanišević, S., Koprić, I., Omejec, J. and Šimović, J., 2001. Local Government in Croatia. In: E. Kandeva, ed., *Stabilization of Local Governments*. Budapest: OSI/LG, pp. 179–240.
Jordan, P., 2011. Progress in Administrative Decentralisation in Transformation Countries – A Comparative Survey. *Hravatski Geografsky Glasnik*, 73(1), pp. 71–85.
Józsa, Z., 2005. A lengyel regionális reform [The Polish regional reform]. In: I. Pálné Kovács, ed., *Regionális reformok Európában*. Budapest: Belügyminisztérium IDEA program, pp. 199–230.
Kemkers, R., 2010. Ten Countries – Twenty Years. Administrative Reforms in Ten New EU-Members Two Decades after Revolution. *European Spatial Research and Policy*, 17(1), pp. 117–142.
Kopajtich-Škrles, N., 2008. Bit by Bit – but Where To? Local Self-governance in Croatia. In: Z. Zlokapa, ed., *Block by Block – It's Good to Build Well: Models of Organisation of Local Self-governance*. Banja Luka: Enterprise Development Agency – EDA, pp. 78–123.
Koprić, I., ed., 2003. *Modernisation of the Croatian Public Administration*. Zagreb, University of Zagreb, Konrad Adenauer Stiftung.
Koprić, I., 2007. Regionalism and Regional Development Policy in Croatia. In: I. Kovács Pálné, ed., *Regional Decentralization in Central and Eastern Europe*. Pécs: University of Pécs Department of Political Studies, pp. 87–111. (Political Studies of Pécs, 4).
Korhecz, T., 2008. A Vajdaság Autonóm Tartomány közigazgatásának helyzete és funkciója a Szerb Köztársaság közigazgatási szervezetén belül – aktuális kihívások és kérdések az új alkotmány fényében

[The position and function of the autonomous Province of the Voivodina within the administrative structure of the Republic of Serbia – actual challenges and questions in the light of the new constitution]. *Nemzetközi Közlöny*, 2(2), pp. 22–31.

Leben, A., 2002. Information of Administrative Districts in the Republic of Slovenia. In: G. Wright and J. Nemec, eds., *Public Management in the Central and Eastern European Transition: Concepts and Cases*. Bratislava: NISPAcee Press, pp. 418–440.

Marcou, G. and Verebélyi, I., eds., 1993. *New Trends in Local Government in Western and Eastern Europe*. Brussels: International Institute of Administrative Sciences.

Maslyukivska, O., 2006. Review of the National Policy, Legislative and Institutional Environment Necessary for the Establishment of Municipal Public Private Partnerships (PPPs) for Public Service Delivery and Local Development in the Europe and CIS Region. Western Balkan. UNDP Capacity 2015/PPPUE – Public Private Partnership programme.

Maurel, M., 1994. Local Government Reforms and the Viability of Rural Communities in Eastern Central Europe. In: R. Bennett, ed., *Local Government and Market Decentralization*. Tokyo: UN University Press, pp. 92–112.

Meekel, T., 2008. Local Government Finance in Serbia. In: Ž. Šević, ed., *Local Public Finance in Central and Eastern Europe*. Cheltenham: Edward Elgar, pp. 364–390.

Mezei, I. and Hardi, T., 2003. A szlovák közigazgatás és területfejlesztés aszimmetriái [Asymmetries in Slovak public administration and territorial development]. *Tér és Társadalom*, 17(4), pp. 127–155.

Moxon-Browne, E., 2005. Administrative Capacity in the European Union: How High Can We Jump? www.cfer.org.uk/content.lib_soundings.pdf

Neudorfer, B., 2014. Subnational Authority and the Quality of Government in European Regions. www.falw.vu/~mlg/papers/Neudorfer_decentralization%20and%20qog.pdf

Nikolova, P., 2011. Bulgaria: The Dawn of a New Era of Inclusive Subnational Democracy? In: J. Loughlin, F. Hendriks and A. Lidström, eds., *The Oxford Handbook of Local and Regional Democracy in Europe*. Oxford: Oxford University Press, pp. 664–684.

Nunberg, B., 2003. The Politics of Administrative Reform in Post-Socialist Hungary. In: B.C. Schneider and B. Heredia, eds., *Reinventing Leviathan: The Politics of Administrative Reform in Developing Countries*. Miami, FL: North-South Center Press, pp. 59–88.

Owens, J., 2002. Fiscal Design across Levels of Government: EU Applicant States and EU Member States / Workshop on 'Decentralisation: Trends, Perspectives and Issues at the Threshold of EU Enlargement', Copenhagen, October.

Pálné Kovács, I. 2001. *Regional Development and Governance in Hungary*. Pécs: Centre for Regional Studies. (Discussion Papers, 35).

Perko-Šeparović, I., 2001. The Reform of Self-government in Croatia. In: R. Pintar, ed., *Local Self Government and Decentralization in South-East Europe*. Zagreb: Friedrich Ebert Stiftung, pp. 75–104.

Pickvance, C.G., 1997. Decentralization and Democracy in Eastern Europe: A Sceptical Approach. *Environment and Planning C. Government and Policy*, 15(2), pp. 129–142.

Ploštajner, Z., 2008. Small and Smaller: What Is the Smallest? Local Self-Governance in Slovenia. In: Z. Zlokapa, ed., *Block by Block It's Good to Build Well. Models of Organisation of Local Self-governance*. Banja Luka: EDA, pp. 36–76.

Rácz, Sz., 2014. Regional Development in Croatia from the Turn of Millennium to the Accession. *Regional Statistics*, 4(2), pp. 87–105.

Sagan, I., 2013. Challenges of Regional and Metropolitan Policy in Poland. In: I. Kovács Pálné, J. Scott and Z. Gál, eds., *Territorial Cohesion in Europe*. Pécs: CERS HAS, pp. 402–411.

Savin, D., 2008. Financing Local Communities in Slovenia. In: Ž. Šević, ed., *Local Public Finance in Central and Eastern Europe*. Cheltenham: Edward Elgar, pp. 414–434.

Schneider, B.C. and Heredia, B., eds., 2003. *Reinventing Leviathan: The Politics of Administrative Reform in Developing Countries*. Miami, FL: North-South Center Press.

Scott, J.W., ed., 2009. *De-coding New Regionalism*. Farnham: Ashgate.

Setnikar-Cankar, S., 2011. Slovenia in Transition: Decentralization as a Goal. In: J. Loughlin, F. Hendriks and A. Lidström, eds., *The Oxford Handbook of Local and Regional Democracy in Europe*. Oxford: Oxford University Press, pp. 642–663.

Setnikar-Cankar, S., Vlaj, S. and Klun, M., 2000. Local Government in Slovenia. In: T.M. Horváth, ed., *Decentralization. Experiments and Reforms*. Budapest. OSI/LGI, pp. 388–421.

Šević, Ž., 2001. Local Government in Yugoslavia. In: E. Kandeva, ed., *Stabilization of Local Governments*. Budapest: OSI/LG, pp. 417–469.

Šević, Ž., ed., 2008. *Local Public Finance in Central and Eastern Europe*. Cheltenham: Edward Elgar.
Simić, A., 2001. Local Governments in Serbia. In: R. Pintar, ed., *Local Self Government and Decentralization in South-East Europe*. Workshop Proceedings. Zagreb, Friedrich Ebert Stiftung, pp. 128–159.
Stead, D. and Nadin, V., 2011. Shifts in Territorial Governance and the Europeanisation of Spatial Planning in Central and Eastern Europe. In: N. Adams and R. Nunes, eds., *Territorial Development, Cohesion and Spatial Planning*. London: Routledge, pp. 154–178.
Swianiewicz, P., 2010. Territorial Fragmentation as a Problem, Consolidation as a Solution? In: P. Swianiewicz, ed., *Territorial Consolidation Reforms in Europe*. Budapest: OSI/LGI, pp. 1–23.
Swianiewicz, P., 2011. Poland: Europeanization of Subnational Governments. In: J. Loughlin, F. Hendriks and A. Lidström, eds., *The Oxford Handbook of Local and Regional Democracy in Europe*. Oxford: Oxford University Press, pp. 480–504.
Takács, Z., 2009. Szerbia a regionalizmus útján. Magyar térségfejlesztés a magyar politikai önérdek útján [Serbia on its way to regionalism [Hungarian regional development in line with Hungarian political self-intersts]. In: S. Somogyi and I. Gábrity Molnár, eds., *Évkönyv 2008*. Szabadka: Regionális Tudományi Társaság, pp. 47–70.
Tatur, M., ed., 2004. *The Making of Regions in Post-Socialist Europe – The Impact of Culture, Economic Structure and Institutions*. Wiesbaden: VS Verlag für Sozialwissenschaften.
Tüske, T., 2002. A szlovén regionális politika törvényi háttere [The legal background of Slovenian regional policy]. In: C. Mezei, ed., *Évkönyv 2003*. Pécs: Pécsi Tudományegyetem Közgazdaság-tudományi Kara Regionális Politika és Gazdaságtan Doktori Iskola, pp. 126–143.
Velikova, M. and Mihaleva, D., 2014. New Approaches and Instruments for Specific Territories' Management in Bulgaria: Key Points of the Regional Policy under Reformation. In: I. Pálné Kovács and C.M. Profiroiu, eds., *Regionalisation and Regional Policy in Central and Eastern Europe*. Bratislava: NISPAcee Press, pp. 289–304.
Yanakiev, A., 2010. The Europeanization of Bulgarian Regional Policy: A Case of Strengthened Centralization. *Southeast European and Black Sea Studies*, 10(1), pp. 45–57.

10
Managing regional disparities

László Faragó and Cecília Mezei

Introduction

International academic literature dedicates little attention (space) to the self-evident fact that unified EU regional/cohesion policy and regulations – and changes happening within – will have different effects and readings in developed member states (centre) and transition countries (periphery). Central and Eastern European (CEE) countries are falling behind in the discursive race regarding the future of the European Union, and academic debates concerning European regional policy and planning are also dominated by Western and Northern European actors (Cotella et al., 2012; Lux, 2012). However, in a 'unified Europe' peripheries too have to have their share in reconstructing EUrope.

Further development of European Union governance, deepening integration and the peripheries' catching-up are closely connected questions. Strengthening integration is not possible without regional cohesion. Although an unwelcome change, not only from a CEE point of view but also from a policy angle, the new cohesion policy – despite its declared pursuit of convergence – today primarily serves the *EU's* general political objectives and the interests of those countries determining these, instead of directly supporting regional policy. This might slow down the catching-up process of cohesion countries and less developed regions. Cohesion countries have a stronger interest in renewing the classic regional policy that is based on solidarity rather than continuing with the current changing trends.

EUrope divided into centre and periphery

EU enlargement is not a philanthropic act of the central countries – it is the extension of their influence (existing networks of businesses and institutions) to territories that were previously detached politically, economically, administratively, etc. The EU accession of the CEE countries suited the rearrangement process of the power relationships in the area, fitted into Euro-Atlantic integration and complied with the economic interests of the centre.

Gradual 'integration' of the CEE countries into the western market economy, then into the European Community, took place through one-sided, external control. The building of the capitalist market economy and the transformation of the institutional system had both been initiated before accession – even before the 1990 regime change. Due to a lack of national capital, international companies occupied the foreground during privatisation, and internationalisation happened on a larger scale than in several old EU member states. FDI flows into the periphery came at a significant cost: the meagre growth of less competitive companies with national/local ownership. In 2002, foreign ownership within the banking sector was 90.7 per cent in Hungary, 85.8 per cent in Czechia, 95.6 per cent in Slovakia and 70.9 per cent in Poland. In 2007 FDI stock reached 51.8 per cent of the GDP in Hungary, 48 per cent in Czechia and 31.5 per cent in Slovakia. Poland held a slightly better position with 'only' 24.9 per cent (Nölke and Vliegenthart, 2009, pp. 681, 683).

Pre-accession processes were already characterised by an asymmetric relationship, which limited the transition countries' sovereignty. Countries wishing to join hardly profited from the benefits of their future accession (such as PHARE supports), yet they had to comply with every EU requirement in order to gain membership. The comprehensive *acquis communautaires* and other institutional expectations were a strong tool in the hands of the Commission (Ferry and McMaster, 2013). Staying out of the EU was not a real alternative for any of the countries, which weakened their negotiating positions. The accession process happened exclusively according to the EU's scripts.

The 10 states joined in the middle of the budgetary cycle, which prevented the new members from exercising any influence on the shaping of the budget. When wording the frameworks for the Programming Period 2007–2013, developed member states viewed the 'Eastern bloc' as something to be Europeanised: an area into which developed values and democratic institutions would be channelled. Net contributor countries continue to grant their solidarity subject to the condition that less advantaged countries spend the received support funds on realising *'common'* EU goals (appointed by the former group).

The 'process of European unification' does not automatically elevate all areas and social groups to a higher cultural and economic level. Instead, it reorganises the economy and cultural fields in a way that maintains the centre–periphery relationship. The international financial market and the EU both work based on democratic deficits (Garcia-Arias et al., 2013). 'Europeanisation processes take place in the shadow of globalisation, neoliberal decentralisation and, most recently, global economic crises' (Ferry and McMaster, 2013, p. 1506). Firms and institutions of the newly acceded member states are thrown into a space dominated by stronger actors. Their regional capital is re-evaluated depending on the possibilities of integration into the global economy. The prerogative of following a different path is only given to the strongest EU member states (e.g. the United Kingdom pre-Brexit). After the collapse of Communism, CEE governments were unable to perform a gradualist reform process that was directed from within, and which considered national interests. First, international financial organisations, then the EU, forced these countries to fully open their markets, which resulted in the significant deterioration of their national economies, national industries and agriculture. Domestic products were largely substituted by Western imports. Contrary to previous expectations – with the exception of Slovenia – the proportion of the CEE countries' export directed to the EU27 did not rise after accession; it even fell back due to the crisis (Kengyel, 2014).

In CEE it was primarily the already developed capital regions that were able to benefit from the new possibilities offered by the EU free market, which resulted in a further deepening of regional differences within the countries. Centripetal ('backwash') and technological effects have simply blocked out the impact of the centrifugal ('spread') effect. Marginal productivity of the productive resources in the centre has not been decreasing to an extent that investments in the periphery would have resulted in rapid convergence. The production factors' possible mobility has not led to the automatic 'spreading' of economic activities or to the improving of regional alignment. In the centre, higher wages and higher profit due to the higher concentration allow for the compensation of the negative effects of concentration (higher real estate prices, overcrowding, etc.). Thanks to centrifugal forces (negative externalities), mobile production factors are not flowing to all of the peripheries, but are mainly directed toward new potential centres (e.g. Metropolitan European Growth Areas (MEGA3s) and – because of their proximity – their regions). Facilitating mobility and ensuring better access (Trans European Networks, deregulation, liberalisation) are advantageous to the centres, and instead of helping them can even hinder the economy of the peripheries (Krugman and Venables, 1990). 'Regional equilibrium' is more prominently embodied in the centre–periphery relationship than in actual regional convergence. The innovation-oriented strategy based on endogenous characteristics that is now preferred in the EU can be realised mainly in central regions and in large international corporations.

Nölke and Vliegenhart (2009) demonstrated that neo-liberal capitalism applied in the post-communist CEE countries (the Visegrad 4) is not able to function as a classic 'liberal market economy' (LME) as seen in the Anglo-Saxon countries, or a 'coordinated market economy' (CME) typified by Germany; instead, it creates a peculiar 'dependent market economy' (DME) model which becomes a distinct variety of capitalism. The dependency of CEE markets is also complemented by the fact that developments, investments and decisions of the public sector in these countries all rely on the support system adopted by the EU. This means that *during their post-communist transition, CEE countries slid into a state of dual dependency*. Their dual market economy depends on Western multinational corporations (FDI) and loans, while public sector developments depend on EU funds (European Structural and Investment Funds (ESIFs), agricultural supports) and are thus subjected to their changes. 'Dependent market economies' are competitive in assembling durable goods (e.g. automotive industry), which explains why these industries attract more working capital (FDI). Their comparative advantage lies in their cheap, moderately educated but efficient labour force, and in the in-firm technological transfer happening on an internationally average quality level. The drawback of this CEE model is the vulnerability of these economies against the hierarchic international decision-making system of multinational corporations. The majority of the most significant producing companies and banks are under foreign ownership, and decisions are made at their headquarters located in central areas, while strategic planning is also carried out there. The larger part of national SMEs, through their suppliers' networks, depends substantially on the decisions of international companies. R&D and innovation take place at principal sites, and due to tax optimisation the realisation of profit also occurs mostly abroad. The countries' small amounts of proper development funds are not used 'freely' (according their own preferences) either; instead these are channelled into the EU tender system as a contribution of own funds, subordinating these sums to common EU goals.

Summarising the above, we find that everything declared by Featherstone and Kazamias (2001) describing peripheries proves to be true concerning the CEE region: the area is economically less developed, the economic development of the region has happened in a historically different way, there is high dependency on European supports and it suffers from weaker bargaining positions.

Efforts taken towards strengthening EU governance have been further restricting the discretion of the CEE region's national governments. Cohesion policy and the toolkit it adopted are *increasingly serving general political and governmental integration objectives instead of facilitating traditional/ classic regional policy*. The new regulation of the planning and support systems and the daily practices they apply also fortify central decision-making in Europe, and sanction member states diverging from the joint policy. In the case of less developed countries, part of the decision-making concerning development – through controlling the use of support funds – has been shifted to the Community level, while daily governance tasks have been transferred to the Commission.

The freedom of planning regarding cohesion policy funds is continuously mitigated in cohesion countries. Objectives are now worded in an even more unequivocal way, and an increasing number of methodological regulations have been drawn up (intervention logic, use of indicators, impact assessments, etc.). Neither the target system of Operational Programmes and partnership agreements drafted by member states, nor the appointment of strategic directions happen based on the specifics of the states themselves; they are not carried out according to the priorities of the domestic national institutions or professional and civic organisations, instead they are considered as given facts bestowed upon the countries by authorities above/outside. Member states are not pursuing strategic planning, but 'EU objective driven' fund-oriented planning, meaning that *plans are constructed solely for the purpose of accessing potentially receivable community funds*. Parties that pay can order! Plans are practically being 'customised' on the go, as EU directives need to be adapted to the situation at hand. The inner contradiction of this practice lies in emphasising regional and 'place-based' views, while the possibility of taking macro-regional (e.g. CEE), national and regional specifics into account continues to decrease.

Within the framework of European Economic Governance, the Commission examines whether the fulfilment of the partnership agreement and the Operational Programmes comply with joint strategic objectives and recommendations and, if not, it may initiate the reprogramming of resources. Moreover, granting funds can also be suspended for lack of efficient measures. This current cycle is the first in which the Commission has the discretion of starting such a 'reprogramming' process. If a member state fails to apply certain measures that are deemed effective and accommodate the requirements of the Commission, the latter can advise the Council to entirely or partially stop funding the programmes or priorities in question.

The European Commission – taking into account the 'specifics' of every member state – provides custom recommendations to each country, which contain advice on particular measures to be taken by the country's government in the following 18 months. The evaluation and the recommended measures generally do not concern regional objectives; instead they deal with government deficits, reforms to large benefits systems, job creation, fighting unemployment, enabling innovation and improving the efficiency of the public administration system in every member state. Consequently, it is the degrees to which these macro-regions diverge from EU expectations that are meant by 'national specifics'. Not discussed is the case of a country which, due to its historical background, requires special treatment, or what advantages or disadvantages may result from the country's regional capital or geopolitical position and what possibilities of development they offer or need – these political recommendations are formed solely according to general, 'spatially blind' political and economic policy expectations. These evaluations and recommendations strive neither to remedy specific regional problems (such as the integration of the Roma population in CEE countries, or the rehabilitation of the 'Communist tower blocks'), nor to facilitate 'special' attention and separate treatment for such issues during the allocation of the ESIFs.

Platforms of integration

The strengthening of integration basically happens on three different platforms: within the EU institutions, through the restructuring of member states' inner operations and on a global platform. With European governance opening up and occupying more and more transnational platforms, inner governance possibilities (platforms) are narrowing for member states. The building of European governance is a denationalisation process eliminating state boundaries in the eyes of the member countries, during which roles previously performed by national structures and institutions are partially taken over by supranational agencies (Kohler-Koch, 2005). According to Kohler-Koch (2005), political restructuring decreases loyalty and solidarity subject to state borders and de-nationalises political platforms. Traditional connectivity between space and community is eliminated, turning space-monogamy into space-polygamy (Beck, 2005). A growing portion of business organisations, citizens and especially politicians do not pursue politics based on territorial borders; instead they have moved to a European level.[1] Brussels is continuously opening gateways for lobbying against national governments and regulations, and thus Europeanises political debates.

International organisations (e.g. the International Monetary Fund (IMF)) help reorganise national economies and governance through global recommendations. Provided that the system of 'investor-state dispute settlement' (ISDS) is included, the free-trade agreement between the European Union and the United States can transfer part of the economic debates to a platform fully outside of Europe. Bypassing national jurisdiction, investors can turn to an elected international 'private court' which can award them damages upon finding that their market actions were anyhow restricted (e.g. introducing a special tax). Signing the USA–EU free-trade agreement can further weaken the governments' tools to regulate and restructure their national economies – all within the environment of strengthening competitiveness and striving to 'catch up' with the USA.

These integration (globalisation) steps can disintegrate national governance ability and national institutional systems to a degree when they may even endanger the democratic operation of the EU.

The Lisbon process that aimed to strengthen large European corporations' competitiveness in the global market, and the crisis starting in 2008, have also altered the targets of regional policy. The principle of regional solidarity and the *traditional* logic of the alignment of underdeveloped regions were pushed to the background by the idea of economic growth based on existing facilities and the strengthening of global competitiveness *in the entire European Union*. The support system previously targeted peripheries battling market disadvantages, in order to offer counterweights to existing advantages of the centre. Structural and Cohesion Funds used to be concentrated on selective domains, helping underdeveloped regions with special difficulties. These funds used to create the boundary conditions necessary for cohesion and growth (infrastructure, education, institutional system, etc.), thus compensating the advantages enjoyed by central conurbations. With the emerging discourse of competitiveness – which, in its current form, goes against the core objectives (Treaty of Rome, Article 158) – European regional policy, based on the slogan of concentrating on tangible regional resources, has been aiming considerable attention and spending funds on more developed regions, while mitigating regional differences within the countries themselves remains mostly the task of national governments. The role of metropolitan regions equipped with suitable potential (regional capital) has automatically become more important through the above-mentioned preference for competitiveness factors. The larger part of these regions lies within the EU's 'Pentagon' area, while in CEE it is mostly the capital regions only that count as potential metropolitan growth areas (MEGA3).

Regional differences and their management in CEE

Europeanisation pressure has transformed the institutional system of national regional planning and development. This can be witnessed observing the types of completed plans, the current view of space, the categories of space, and the regional political institutional system and toolkit created to manage regional differences. The alignment of the institutional system and toolkit of regional development to Brussels standards can be viewed as evident, since national resources necessary to receive Structural and/or Cohesion Funds (additionality) basically exhaust development assets in CEE countries. This also means that in spite of these countries having special national or regional problems, there is no room for their customised management, which already excludes the diffusion of genuinely place-based views.

In CEE countries the issue of regional development is important not only on account of the conditions for receiving EU Structural and Cohesion Funds, but also because of the vast regional differences that usually characterise the so-called transition economies or countries. European Structural and Cohesion Funds used to 'reward' regional underdevelopment and mainly served to help the country catch up, instead of managing the differences within the country itself. Before the regime changes of 1990, all the CEE countries had lower inner differences than the spatial disparities measured today.

Despite the convergence of several nation-states, *significant regional (NUTS-2, NUTS-3) differences continue to be visible* even more than 10 years after accession (Figure 10.1). The poorest regions of the EU – with the exception of Mayotte in France – are all located in the 'Eastern bloc'. In 2003 GDP calculated based on purchasing power parity (PPP) was 12 times higher in the most developed regions than in the poorest ones. (The rate was 'only' 11 times higher in 2011.) If we examine disparities on the NUTS-3 level (Eurostat data from 2011), the result becomes even more extreme, close to 30 times higher: Inner London – West reaches 612 per cent, while Vaslui in Romania is at 21 per cent. In 28 NUTS-3 regions the GDP per capita is at least twice as high as the EU average, most of which are found in Germany. In 24 NUTS-3 regions GDP is lower than 30 per cent. The Hungarian county of Nógrád is among these: the rest of the regions are located in Bulgaria (15) and Romania (8) (Eurostat, 2015). GDP has not reached half the EU average in 20 regions. All of these – except for a remote island – are in CEE: 5 in Romania, 5 in Bulgaria, 5 in Poland and 4 in Hungary. When it comes to multi-region countries, GDP per capita surpasses the EU average in every region of Sweden, while all of Bulgaria's regions – even the capital – fall behind.

Although the dominance of capital cities (a macro-regional feature – see Chapter 8), primarily with regards to population density, employment and the ratio of the service sector, can be observed anywhere today, there were certain exceptions to be found before the regime change. The Warsaw region, for example, due to the polycentric nature of the country, the transportation situation and the underdevelopment of the region surrounding the capital, did not have

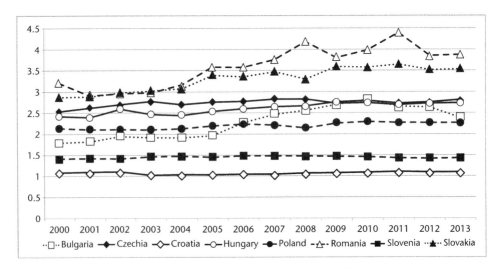

Figure 10.1 Proportion of regional disparities between 2000 and 2013
(GDP per capita: proportion of the wealthiest and poorest NUTS-2 regions)

Source: Author's calculations and construction based on Eurostat data.

outstanding indicators (Mezei and Schmidt, 2013; Horváth, 2004). Similarly, for a long time Sofia fitted into the general backwardness of Bulgaria (Horváth, 2008).

In 2004, with the exception of Prague and Bratislava, every NUTS-2 region within the accession countries was filed under the first category and ranked among the least developed regions, rendering every CEE country a cohesion target region. The picture evolved further with the accession of Romania and Bulgaria in 2007 (Figure 10.2). While CEE countries still unanimously remained cohesion countries, the Budapest-led region of Central Hungary was 'elevated' into the phasing-in (transitional) category, and Bratislava and Prague were promoted to competitiveness regions, leaving the remaining 50 regions as the least developed convergence regions. However, by the 2014–2020 phase, all of the capital regions were elevated to the level of more developed regions except in Slovenia, Croatia and Bulgaria.

While statistics may hide several differences, the 2013 GDP per capita data primarily indicate the deepening of discrepancies instead of the strengthening of regional convergence (Table 10.1). Regional differences are lowest in Croatia and Slovenia, due to each of these countries being divided into two large regions, which are 'assimilating' the capital cities. The spectacular improvement of Bratislava's competitiveness (Annoni and Dijkstra, 2013) has resulted in a 3.5 times difference, whereas Romania shows an even larger variance with the GDP per capita being 4 times higher in some areas than in others.

These serious discrepancies deserve treatment both on the EU and national levels. Mitigating regional differences has proven to be more successful within the group of 10 countries that joined in 2014, while Romania, Bulgaria and Croatia are much further behind (Table 10.2). For a long time Hungary held a leading position within the EU10 branch due to its already EU-compatible legislation on regional development adopted in 1996. The institutional system of regional development was created in accordance with the above law, and a separate funding system earmarked for this objective was in operation. A national regional development plan was adopted in 1998. By today, Hungary's advantage has been completely eroded, as every CEE country has created its own EU-conforming institutional system and toolkit on regional development. By 2001, every country except Croatia had adopted legislation on regional development, and by the middle of the decade all of these articulated a national strategy regarding the issue.

The formulation of the institutional system of regional development happened parallel to public administration reforms in CEE countries. In certain countries, regional development institutions easily fitted into the system of transitioning, reformed public administration, while elsewhere administrative reforms were only imitated, and regional development institutions were set up outside of them (Mezei, 2016). In Poland, a coherent reform of regional administration was carried out in a way that gave the 16 new *voivodeships* an increasingly significant role in regional development. Meanwhile in Hungary regions were created on formal grounds: they have not been integrated into public administration, and the institutional system of regional development has been changed every governance cycle. Czech public administration reform also set up regional municipality units, but European funds are not managed through these *kraje*, but are based on the eight statistical regions (Mezei, 2015).

Local regional problems have been receiving special treatment in many cases, generally resulting in the delimitation of heavily supported areas. Special Economic Zones created in Poland in 1994 are a good example of this: these were set up in order to help industrial regions battling the structural crisis (Baranyai and Lux, 2014). Prekmurje, the least developed NUTS-3 region of Slovenia, has fallen under different regulation since 2009 for the same reason. Croatia separated regions requiring particular treatment directly following the Yugoslav Wars (areas affected by the war or fighting structural problems) in order to facilitate their access to certain state aids (Mezei and Pámer, 2013). The same peculiar tool, universally considered an instrument

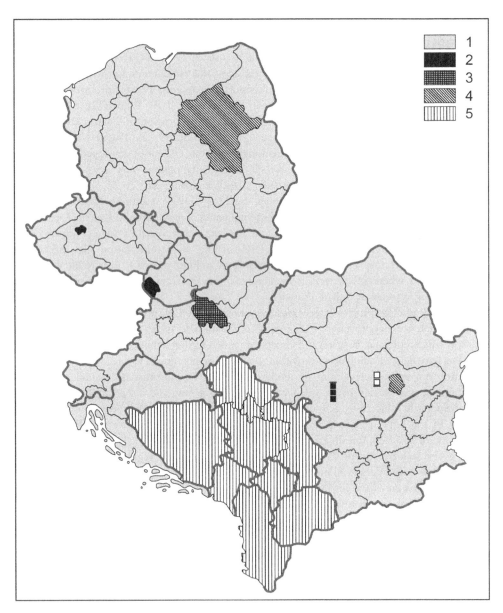

Figure 10.2 Cohesion policy target regions within the macro-region

Source: Authors' construction.
Legend: 1 – Convergence region since accession
2 – (Developed) competitiveness region since accession
3 – Phasing-in region since 2007, then competitiveness region since 2014
4 – Competitiveness region since 2014
5 – IPA region

Table 10.1 Regional disparities in Central and Eastern Europe based on GDP per capita, 2013

Country	Number of NUTS-2 regions	Poorest region	Wealthiest region	Inequality ratio
Bulgaria	6	Severozapaden	Yugozapaden	2.40
Czechia	8	Severozápad	Prague	2.79
Croatia	2	Jadranska Hrvatska (Adriatic Croatia)	Kontinentalna Hrvatska (Continental Croatia)	2.16
Hungary	7	Northern Hungary	Central Hungary	2.72
Poland	16	Lublin Province	Mazovia Province	2.26
Romania	8	Nors-Est	Bucureşti-Ilfov	3.88
Slovakia	4	Východné Slovensko (Eastern Slovakia)	Bratislavský kraj (Bratislava region)	3.55
Slovenia	2	Vzhodna Slovenija (Eastern Slovenia)	Nahodna Slovenija (Western Slovenia)	1.43

Source: Author's calculations and construction based on Eurostat data.

of problem-solving, has characterised Hungary since 2007, when the least advantaged subregions were isolated and special tenders, incentives and forms of aid were introduced to enhance the development of the 33 poorest subregions (LAU-1). The East Poland macro-region was delimited along the same principles in 2007. Partly due to EU pressure, an integrated management system was set up in the five least developed *voivodeships* (Lubelskie, Podlaskie, Podkarpackie, Świętokrzyskie, Warmińsko-mazurskie) to divide the sources granted by European funds.

Special instruments were also applied to solve regional problems in the CEE countries, among which we find national regional development strategies aimed at the convergence of struggling, special areas or for the balanced development of the whole country. Apart from particular regional plans, other instruments were also used in the region. Examined from a CEE perspective, Polish planning contracts are unique in this regard. These were introduced in 2001 based on a French model, and laid down the framework for contracts between the state and the *voivodeships* concerning the allocation of national and EU support funds. Separate monetary funds targeting regional development and convergence objectives operated for shorter or longer periods in Croatia and Hungary as well. Local tax exemptions primarily aimed at attracting FDI, entrepreneurship incentives (fee discounts) or the ensuring of infrastructure also surfaced in multiple countries (e.g. Slovakia, Poland, Hungary) as special regional development tools. After their EU accession, the countries in question needed to create an EU-conforming framework for the above incentives. These discounts were in many cases tied to special zones, industrial parks, business/economic districts – the installation, regulation and funding of which are also viewed as special regional development tools (Mezei, 2006).

The situation of regional development is perfectly mirrored by the setup and operation of its institutional system. In countries that have stable ministries for regional development (Poland, Czechia), the position of regional development is much stronger, resulting in better cross-sectoral coordination and the prevailing of regional interests, as opposed to countries having no such ministries where the task of regional development is carried out sporadically, attached to ministries from various sectors. In Slovenia an independent minister without portfolio is responsible for this area, and Croatia also decided to create a separate ministry for this purpose. In Slovakia, a ministry responsible for regional policy operated between 1999 and 2010 (Ministry of Construction and Regional Development), whereas Hungary had an independent minister without portfolio between 2008 and 2010 (Mezei, 2013).

Table 10.2 Regional policy toolkit and institutional systems of the Central and Eastern European countries

Country	First legislation on regional development	First national strategy on regional development	Minister/ministry responsible for regional development	Regional Operational Programmes 2004–2006	2007–2013	2014–2020[1]
Hungary	1996	1998	Distributed between various ministries	Integrated	Separated	Integrated (4) and one for Budapest (9)
Slovakia	2001	2001	Distributed between various ministries	No	Separated	Integrated (6)
Bulgaria	1999	1999	Ministry of Regional Development and Public Works	–	Separated	Integrated (5)
Romania	2001[2]	2002[3]	Ministry of Regional Development and Public Administration	–	Integrated	Integrated (9)
Slovenia	1999	2004	Minister without portfolio responsible for Development, Strategic Projects and Cohesion	No	Integrated	Integrated (11)
Croatia	2009	2010	Ministry of Regional Development and EU Funds	–	Integrated[4]	Integrated (9)
Czechia	2000	2006	Ministry of Regional Development	Integrated and one for Prague	Separated	Integrated (8) and one for Prague (5)
Poland	2000	2000	Ministry of Economic Development[5]	Integrated	Separated	Separated (10)

Source: Author's construction based on EC, 2014; Mezei and Pámer, 2013; Kolev, 2002; Horváth, 2008; Lőrinczné Bence, 2014; Nagy, 2004.

Notes: 1 – The number of thematic objectives integrated into the Operational Programme is shown in parentheses.
2 – The institutional system of regional development was created earlier with legislation adopted in 1998.
3 – The 2002–2005 National Development Plan already contained a chapter on regional development, and the Strategic Conception on Regional Development: 'Romania 2030' was adopted in 2008, and was included in the 2007–2013 Plan.
4 – The IPA programme prepared for the programming period contained a ROP, which was included in the convergence programme after EU accession in 2013.
5 – Between 2005 and 2013, the Ministry of Regional Development was responsible for regional policy.

The way Regional Operational Programmes are prepared and managed also shows how deeply the regional view and the bottom-up principle are accepted in a certain country's practices. While in the 2004–2006 period the EU did not support drawing up separate ROPs, in the following seven years independent development plans were created for NUTS-2 regions almost everywhere (with the exception of small countries, and Croatia and Romania, with both joining at the end of the cycle). However, the presence of ROPs is not everything: in Hungary for example, programmes were created through centralised, top-down planning and were based on unified models, focusing on priorities that were omitted from sectoral programmes. They were also carried out through almost fully centralised management (Mezei, 2016). In Poland, however, the bottom-up approach and decentralised (*voivodeship*-based) management were effectively realised. It is no coincidence that it was Poland where the system of separate ROPs remained in operation even after 2014.

Competitiveness and/or regional cohesion

The issue of spending European Structural and Investment Funds in a way that improves the competitiveness of the entire Community and acts as an incentive for economic growth – but does not deepen regional differences to an extent which will hinder further development and integration – was always on the agenda of the European Union.

Recently, it is the classic economic approach that has been receiving a greater emphasis: if EU funds are spent in developed central regions possessing the necessary capacities (highly educated labour, functional institutional system, agglomeration advantages, critical mass, etc.), the overall performance of the EU improves more than with the support going to moderately developed peripheries and to the least advantaged regions. If developed countries keep financing non-sustainable investments in peripheries in the long term, it will have a significant distorting effect and can even slow down the pace of development in the central countries (Fernandez, 2011, pp. 169–170, describes the debate). Traditional strategies generally prefer faster and higher returns and opt for investing in developed regions, especially during crisis periods (Sapir, 2003; World Bank, 2009). This view is mirrored in the Lisbon process, which places the main emphasis on improving competitiveness – and thus undermines regional cohesion. Leftist economist Fabrizio Barca (2009) advises that economic development should be based on agglomeration advantages, and then the produced surplus can be spread out to avoid social tension.

According to Camagni and Capello (2015), modern regional development policy can maximise the output of community investments only if it activates unused resources and *if it utilises regional capital in the most efficient way everywhere*. This is how aggregated impact will be largest on the European level. The new cohesion policy conceived with a place-based view in mind and targeting almost all 'places' will satisfy both efficiency and fairness objectives, and will produce growth everywhere. However, we believe that the above policy does not enhance growth and mitigate regional differences at the same time. Developed areas will have more significant regional capital than peripheries, and *basing development on local resources of various size and quality will result in the further deepening of these differences.*

As Mendez (2013, pp. 645–646) summarised, the opposition between the efficiency objectives of the Lisbon strategy and the fairness objectives of cohesion policy (dualism, trade-off) is based on misunderstandings. First, development policy can remain growth-oriented even if funds target poorer regions (efficiency criteria), which will improve the opportunities of the local population. Second, efficiency and fairness objectives can mutually reinforce each other; negative trade-off is not the general rule. Third, the further growth of developed regions boasting larger economic concentration cannot remain an infinite process either.

According to the ESPON ET2050 project[2] calculations (Camagni and Capello, 2015, p. 38), focusing development on second- and third-level cities ('urban scenario') instead of concentrating on potential metropolitan growth areas (MEGAs) holds numerous advantages. The expected regional growth rate will be the highest in this case, both regarding the older 15 and the newer 13 member states. Regional capital will be used most efficiently, and we can avoid negative impacts due to excessive concentration, while also widely facilitating the spillover effect of urban poles. Gál and Lux (2014, p. 12), who participated in the project, drew attention to the fact that this strategy can actually prove risky in the CEE area, since the region lacks a dense network of second- and third-level cities, i.e. the ones that exist are too weak and their development should be a priority. According to the project's model calculations, when it is exclusively rural and peripheral regions that are being developed ('regional scenario'), both older and newer member states will experience a slower growth rate. This may be true in this form – we accept that on the EU level it is the 'urban scenario' that promises a greater growth rate. However, if developed countries/regions also receive their share of development funds, cohesion countries will receive a relatively lower sum than the full budget of the ESIFs. This will result in the cohesion member states experiencing slower growth compared to previous times when they had been granted the full amount of support funds. For example, growth rates will be the highest in the CEE countries if they receive the largest possible share of ESIFs, and regional differences will decrease if these sources are spent on developing urban areas outside of the capital regions.

If we accept that – due to agglomeration effects, external economies and returns to scale – return on capital is higher in developed central areas than in peripheries, it follows that development is already happening in these regions on a market basis (see the arguments in Chapter 2). The existence of regional differences not only foretells that less advantaged regions will have worse prospects, etc., it also means that production factors are not fully utilised, and their better capitalisation would benefit the entire Community. The 'place-based' paradigm could be applied, and the growth rate as well as the strengthening of inner cohesion could be maximised if cohesion supports facilitated the best possible utilisation of resources in peripheries and in the least advantaged regions, rendering these areas more competitive.

Cohesion supports equal less than half a percentage of the EU's GDP (0.36). This can be considered an acceptable *social price to pay for mitigating regional differences*. It does not equal the reallocation of earnings or the 'overtaxation' of the most developed states. *The centre will also profit from development investments effectuated in less developed regions*: more equal regional development ensures growth opportunities to companies operating in the centre, and it can mitigate migration pressure in central areas. Moreover, adequately targeted investments in less advantaged regions (resolving bottlenecks, connecting infrastructure into networks) can induce even larger growth than in the already developed regions. Investments and consumption will both increase due to the supports received, which will improve tax revenue and overall facilitate the sustainability of the entire process. Naturally, due to lower population density and base development level, this will have a smaller effect on average EU growth, but development induced in the centre will compensate for this.

In core regions, a considerable part of growth (or at least averting recession) relies on economic relationships maintained with the peripheries. Net contributor countries 'buy' growth potential for a relatively low sum, while a substantial part of the funding flows back into the central regions. This and the expansion of the market (creation of demand) compensates for the contributions. Investments supported by cohesion policy are mainly realised by Western European companies (or their CEE subsidiaries), granting them significant revenues through the creation of demand and the repatriation of profit. Companies in the 'old' member states receive a considerable share of every euro spent in the cohesion countries. More than half of the support granted to the Visegrad 4 countries appears as – direct or indirect – profit in the books of the EU15, primarily German, French, Austrian, Spanish, etc., firms. In the case of Germany, Ireland and Luxembourg, revenues gained from increased exports due to the creation of trade have exceeded the sum of contributions these countries had provided (c.f. Kengyel, 2014).

Conclusion and discussion points

In the European Union, integration reached through the strengthening of supranational governance does not automatically lead to the social, economic and regional cohesion of the less developed countries of the East and the South. If the centre and peripheries are drawn under unified governance (from multiple points of view), and countries with different levels of development need to compete under the same market conditions, regional discrepancies will not be mitigated automatically. Moreover, the enlargement that brought the number of EU members up to 28 states has actually increased regional differences within the EU – European governance now needs to be much more flexible and interactive in alleviating these disparities (Jachtenfuchs and Kohler-Koch, 2004). The fact that differentiated development strategies are needed due to the member states' internal differences (economic development, governance structure, member state preferences) has to be accepted and managed on the EU level too. Two sovereignties (EU and nation-state) sharing the governance of the same group of citizens and territories is only possible with *multilayered governance based on the principle of subsidiarity*. It is not enough for the EU's political and administrative centres to aim for the strengthening of supranational governance while neglecting the fact that – according to the requirement of subsidiarity – in many areas, national governance needs reinforcement.

It is evident that conserving the EU's position on the world market and improving its competitiveness is an interest shared by every member state, but this can be reached in more than one way. Further economic boosting of the already highly developed central areas can promise quick returns, but it will lead to the further deepening of regional differences, which in turn will come with a higher cost: negative externalities that can eliminate short-term gains.

The Mediterranean and CEE regions have an interest in maintaining solidarity and cohesion support, stressing that the latter should be allocated to their regions. The overall economic performance of the EU could also be improved if cohesion countries, and less developed regions holding significant growth potential received exclusive supports in the future in order to enhance their economic development and competitiveness. This would result in the considerable decrease of development discrepancies between the countries: the economic potential of these regions would be better utilised, while central areas would draw significant profit from developments realised in the peripheries. The fact also needs to be taken into account that developed countries (particularly core regions) that are severely affected by the crisis and currently display a modest growth rate – but still exercise decisive influence in EU politics – are, according to their domestic interests, primarily interested in reviving their own economies.

The CEE region needs to set up its own European core area which would contribute to creating a polycentric spatial structure in Europe. However, if EU regional policy shaped by the current centre (the most developed member states) does not reach the same conclusion, countries of the periphery need to initiate this process themselves:

- Cohesion countries with similar assets and experiencing similar international impacts have to cooperate in order to ensure that cohesion policy continues to serve exclusively the realignment of less developed countries.
- High dependency on FDI can be eased through the strengthening of domestic funding options. Improving the economic financing capacity of national bank systems is a good

example, as well as increasing the proportion of national ownership in a way that avoids monopolisation.
- Besides job creation, emigration from the peripheries can also be mitigated through increasing wages. In order to keep competitiveness and continue to attract investors, higher wages need to be compensated with other benefits (e.g. lower taxes and dues, strengthened predictability, skilled labour, higher innovation). Mobility may also be 'limited' via market-friendly instruments (e.g. localisation can be facilitated through the widening of supplier networks). Improving external economic viability can attract and tie companies to a place in the long run (see Chapter 3).
- Member states have to extend development to metropolitan areas outside their capital regions *through concentrated decentralisation*. Geographic–economic concentrations need to be created within the least advantaged regions. Once a critical mass is achieved, it will facilitate agglomeration effects and improve external economic viability.
- *Decentralisation at the member state level:* the prevalence of regional, 'place-based' and 'bottom-up' views, and the subsidiarity principle would demand that various groups of countries determine the main priorities, objectives and actual tasks of the following cycle themselves. It would be beneficial if these groups of countries – besides complying with joint objectives – had the possibility to word their specific goals for one part of the multi-annual financial framework. An opportunity should be granted to member states to use part of their indicative member state financial support funds freely, according to their own preferences. This would differ from the pre-1989 system, as these developments would need to be indicated in partnership agreements and Operational Programmes, and the utilisation of these funds would take place in the framework of the EU institutional system. Each member state and region has to find the most suitable policy for their own development, 'in the light of its particular economic, social, environmental, cultural and institutional conditions' (EC, 2005, p. 8).
- Supports should not be extended to developed regions; instead, *priority funding* should be given to regions where the GDP per capita does not reach 50 per cent of the EU average. (In practice this equals the 20 least developed regions of the EU.)

Essentially, our recommendations refer to the growth-oriented reduction of the social and economic backwardness in the cohesion countries' least developed regions. Taking into account the regional concentration trend of economic activities, this can be realised through certain regionally concentrated measures (mitigation of the shortage of capital, polycentric development, fulfilling concentration requirements), which can utilise the less advantaged areas' regional capital with great efficiency. The further deepening of the EU's joint governance is only possible after curbing significant current regional welfare differences.

Acknowledgement

The publication of this research has been supported by project #104985 of the Hungarian National Research, Development and Innovation Office.

Notes

1 Hungary provides us with a good example. The opposition and certain NGOs do not pursue politics within the national framework (or within the country's borders), and they do not attempt to realise their goals in Hungary anymore. Instead they look to the European community and to the assistance of the EU institutional system with their plans. Kohler-Koch (2005, p. 7) provides other examples of how

the EU secures 'escape routes' regarding taxes imposed on capital revenues, genetically engineered agricultural products and university tuition fees.
2 www.espon.eu/main/Menu_Projects/Menu_AppliedResearch/ET2050.html

References

Annoni, P. and Dijkstra, L., 2013. EU Regional Competitiveness Index RCI 2013. Luxembourg: European Commission Joint Research Centre Institute for Security and Protection of the Citizens. http://ec.europa.eu/regional_policy/sources/docgener/studies/pdf/6th_report/rci_2013_report_final.pdf [Accessed 14 March 2015].
Baranyai, N. and Lux, G., 2014. Upper Silesia: The Revival of a Traditional Industrial Region in Poland. *Regional Statistics*, 4(2), pp. 126–144.
Barca, F., 2009. *An Agenda for a Reformed Cohesion Policy. A Place-based Approach to Meeting European Union Challenges and Expectations. Independent Report Prepared at the Request of Danuta Hübner*. Brussels: Commissioner of Regional Policy.
Beck, U., 2005. *Mi a globalizáció?* [What is globalisation?]. Szeged: Belvedere.
Camagni, R. and Capello, R., 2015. Rationale and Design of EU Cohesion Policies in a Period of Crisis. *Regional Science Policy & Practice*, 7(1), pp. 25–49.
Christiansen, T., Jørgensen, K.E. and Wiener, A., eds., 2000. *The Social Construction of Europe*. London: Sage.
Cotella, G., Adams, N. and Nunes, R.J., 2012. Engaging in European Spatial Planning: A Central and Eastern European Perspective on the Territorial Cohesion Debate. *European Planning Studies*, 20(7), pp. 1197–1220.
EC, 2005. *Cohesion Policy in Support of Growth and Jobs: Community Strategy Guidelines, 2007–2013.* COM(2005) 299 final. Brussels: Commission of the European Communities.
EC, 2014. Country Progress Reports (Hungary, Slovakia, Poland, Czech Republic). http://ec.europa.eu/index_en.htm [Accessed 15 September 2014].
Eurostat 2015. Gross Domestic Product (GDP) at Current Market Prices by NUTS-2 Regions. [nama_10r_2gdp]. Available from: http://appsso.eurostat.ec.europa.eu/nui/show.do [Accessed 26 November 2015].
Fernandez, J., 2011. Why Location Matters: The Terms of a Debate. In: OECD, ed., *Regional Outlook. Building Resilient Regions for Stronger Economies*. Paris: OECD, pp. 167–174.
Ferry, M. and McMaster, I., 2013. Cohesion Policy and the Evolution of Regional Policy in Central and Eastern Europe. *Europe–Asia Studies*, (65)8, pp. 1502–1528.
Featherstone, K. and Kazamias, G., 2001. Introduction: Southern Europe and the Process of 'Europeanization'. In: K. Featherstone and G. Kazamias, eds., *Europeanization and the Southern Periphery*. London: Frank Cass, pp. 1–22.
Garcia-Arias, J., Fernandez-Huerga, E. and Salvado, A., 2013. European Periphery Crises, International Financial Markets, and Democracy. *American Journal of Economics and Sociology*, 72(4), pp. 826–850.
Gál, Z. and Lux, G., 2014. Territorial Scenarios and Visions for Central and Eastern Europe. *ET2050 Territorial Scenarios and Visions for Europe, Final Report*, Vol. 8, 30 June.
Horváth, Gy., 2004. Regionális egyenlőtlenségek Európában [Regional inequalities in Europe]. *Magyar Tudomány*, 49(9), pp. 962–977.
Horváth, Gy. ed., 2008. Az új közép-európai uniós tagországok és a szomszédos országok regionális intézményrendszere és azok szerepe a kohéziós politika végrehajtásában c. kutatási zárótanulmány [The regional institution system of new Central European Union member states and their neighbours and the role of such in the execution of cohesion policy, closing report]. Pécs: MTA Regionális Kutatások Központja.
Jachtenfuchs, M. and Kohler-Koch, B., 2004. Multi-level Governance. In A. Wiener and T. Diez, eds., *European Integration Theory*. Oxford: Oxford University Press.
Kengyel, Á., 2014. Az európai uniós tagság mint modernizációs hajtóerő [European Union membership as a driving force for modernisation]. *Közgazdasági Szemle*, 61(4), pp. 493–508.
Kohler-Koch, B., 2005. European Governance and System Integration. *European Governance Papers*. (EUROGOV) No. C–050.
Kolev, B., 2002. Legal Framework of Regional Development in Bulgaria. In: Gy. Horváth, ed., *Regional Challenges of the Transition in Bulgaria and Hungary*. Pécs: Centre for Regional Studies, pp. 78–85. (Discussion Papers; Special Issue).

Krugman, P. and Venables, A., 1990. Integration and the Competitiveness of Peripheral Industry. In: C.J. Blis and J. Braga de Macedo, eds., *Unity with Diversity in the European Community*. Cambridge: Cambridge University Press.

Lőrinczné Bencze, E., 2014. A horvát területi politika jogszabályi hátterének változásai [Changes in the background of legal regulations for spatial policy in Croatia]. In: M. Lukovics and B. Zuti, eds., A területi fejlődés dilemmái. SZTE Gazdaságtudományi Kar, Szeged, pp. 93–103.

Lux, G., 2012. Saját tereikbe zárva? Közép- és Kelet-Európa a regionális tudomány európai integrációjában [Trapped in their own spaces? Central and Eastern Europe in the European integration of regional studies]. In: J. Rechnitzer and Sz. Rácz, eds., *Dialógus a regionális tudományról*. Győr: Széchenyi István Egyetem, Magyar Regionális Tudományi Társaság, pp. 151–170.

Mendez, C., 2013. The Post-2013 Reform of EU Cohesion Policy and the Place-based Narrative. *Journal of European Public Policy*, 20(5), pp. 639–659.

Mezei, C., 2006. Helyi gazdaságfejlesztés Közép-Kelet-Európában [Local economic development in East Central Europe]. *Tér és Társadalom*, 20(3), pp. 95–108.

Mezei, C., 2013. Case Study Report Building Structural Fund Management Systems: Learning by Doing or Imitating? In: ESPON TANGO – Territorial Approaches for New Governance: Annex A: Case Studies. Luxembourg: ESPON, Final version. pp. 91–110. www.espon.eu/export/sites/default/Documents/Projects/AppliedResearch/TANGO/ESPON_TANGO_Case_Study_Annex_A_final.pdf [Accessed 25 February 2016].

Mezei, C., 2015. Building Institutions for the Structural Funds in the Visegrad Countries. In: Á. Bodor and Z. Grünhut, eds., *Cohesion and Development Policy in Europe*. Pécs: Institute for Regional Studies, Centre for Economic and Regional Studies, Hungarian Academy of Sciences, pp. 60–71.

Mezei, C., 2016. Limited Involvement: The Role of Local and Regional Actors in the Hungarian Structural Fund Management. In: P. Schmitt and L.V. Well, eds., *Territorial Governance across Europe: Pathways, Practices and Prospects*. London: Routledge, pp. 157–170.

Mezei, C. and Pámer, Z., 2013. Területfejlesztés és intézményrendszere [Spatial development and its institutional system]. In: Z. Hajdú Z. and I. Nagy, eds., *Dél-Pannónia* [South Pannonia]. Pécs: MTA KRTK Regionális Kutatások Intézete, pp. 430–455.

Mezei, C. and Schmidt, A., 2013. A lengyel regionális politika és intézményei [Polish regional policy and its institutions]. *Tér és Társadalom*, 27(3), pp. 109–126.

Nagy, B., 2004. A területiség megjelenítése a romániai Nemzeti Fejlesztési Tervben [The presentation of spatiality in the Romanian National Development Plan]. In: *Fiatal regionalisták IV. országos konferenciája*. Győr: Széchenyi István Egyetem Multidiszciplináris Társadalomtudományi Doktori Iskola. (CD-ROM).

Nölke, A. and Vliegenthart, A., 2009. Enlarging the Variety of Capitalism: The Emergence of Dependent Market Economies in East Central Europe. *World Politics*, 61(4), pp. 670–702.

Sapir, A., 2003. *An Agenda for a Growing Europe. Making the EU Economic System Deliver. Report of an Independent High-Level Study Group Established on the Initiative of the President of the European Commission*. Brussels.

World Bank, 2009. *World Development Report*. Washington, DC: World Bank.

11
Rebordering Central and Eastern Europe

Cohesion Policy, cross-border cooperation and 'differential Europeanisation'

James Wesley Scott

Introduction

Central and Eastern Europe's post-World War I history – and destiny – is inextricable from the emergence and transformation of state borders. With the 1993 division of Czechoslovakia into two independent states, and the subsequent disintegration of Yugoslavia, the question of 'rebordering' CEE appears now to have been generally settled. Yet border politics are still very much alive in this part of the world, despite eastern enlargement of the European Union and Schengen Area and more than 25 years' experience in using the tools of local and regional cross-border cooperation. One explanation for this is the idea that nation-building in CEE represents unfinished business, a process that was curtailed by war and its geopolitical aftermath and that now coincides, often uncomfortably, with the simultaneous project of European integration. Consequently, this chapter reflects research on the significance of borders in CEE borders, interrogating, among other things, the impacts of European integration and EU Cohesion Policy. Concretely, the objective is to consider rebordering processes as a post-1989 CEE development context. This also involves studying tensions between normative elements of European policy agendas and their application as Cohesion policies. In a normative, policy-oriented sense this is a question of borders as framing conditions for regional development. In a more critical and analytical sense this involves interrogating the actual use of borders in politically and ideologically framing national development within a wider European context.

Within the EU context, the development of a 'post-national' sense of community has been understood to be an overlying goal. In fact, the project of building a European Union has involved attempts to transcend national divisions and thus create conditions for durable peace, prosperity and more effective interstate cooperation. At the same time, the EU seeks to create a supranational community based on a shared sense of political, social and cultural identity. As a result, processes of 'Europeanisation' – defined in terms of a gradual diffusion of transnational understandings of citizenship, territoriality, identity and governance – are closely related to changing political understandings and uses of Europe's many state borders (Scott and Liikanen, 2011). In addition, the development of the EU has gone hand in hand with the emergence of shared policy agendas and policy instruments that have created much closer links between member states. Since the end of the 1980s, cross-border cooperation has been firmly embedded within EU structural policies, Cohesion Policy in particular.

State borders reflect and thus help us interpret tensions as well as points of connection within intercultural and interstate relations. They can also indicate tensions and contradictions inherent in Europeanisation processes that have taken place since 1989. A major assumption informing this research is that the symbolic meanings, functional impacts and socio-political significance of borders are in fact central to understanding the overall significance of European Union as a project of political community building. Consequently, the global objectives of the chapter are to interrogate the significance of cross-border cooperation in the re-bordering and Europeanisation of CEE. Due to its geographical location and experiences with many CEE border contexts, Hungary will receive specific attention. For historical and geographical reasons, Hungary can be understood to be a laboratory of Europeanisation both in terms of the local adaptation of EU norms and practices (Ágh, 2003), progressive legislation regarding minority rights (Vizi, 2009) and practices of cross-border cooperation (Baranyi, 2008; Hardi, 2010). Having championed a 'spiritualisation' of borders well before the end of the Cold War, the country has advocated greater political and cultural opening but also a politics of outreach to Hungarian ethnic communities in neighbouring countries – a goal not shared by its neighbours.

The chapter begins with a brief discussion of Europeanisation processes understood in terms of the promotion of cross-border cooperation and then proceeds with an overview of factors conditioning the post-1989 re-bordering of CEE. Particular attention will be drawn to the significance of cross-border cooperation within EU Cohesion and regional policies. As EU policies and their local adaptation are a major focus within regional research, I will seek to explore the links between the notion of European Cohesion and cross-border (territorial) cooperation as an indicator of the scope and local impacts of Europeanisation. One result that will be presented here is the confirmation of the hypothesis that while EU–European principles of cross-border cooperation have been partly mainstreamed into regional development policies they have at the same time been superimposed by the domestication of EU policies in the interest of nation-building. This is clearly evident in the case of Hungary: the sizable Hungarian ethnic minorities in neighbouring countries (a legacy of the June 1920 Treaty of Trianon) and the loss of major second-tier cities after World War I partly explain Hungary's interest in more dynamic cross-border relations, but rather at a more formal political level in terms of interstate cooperation.

Europeanisation, cohesion and cross-border cooperation

Since the end of the 1980s, cross-border cooperation has been firmly embedded within EU structural policies, Cohesion Policy in particular. It is also an important element of the new European Neighbourhood Programme operating at the EU's external borders. Consequently, questions of territoriality, regional policy and cross-border cooperation (CBC) have been central to the emergence of the European Union as a political community. According to Manzella and Mendez (2009, p. 9), regional policy has been 'perceived as a crucial instrument for the identity of a European model of society, and for the legitimacy and viability of the whole political process of integration'.

However, in order to signify something more than the sum of national concerns, the EU has needed an *exceptionalist and idealist* narrative that goes beyond state-centred political thinking and that is open-ended – territorially and conceptually. I thus argue that a major element of the EU's political identity lies precisely in reconciling flexible idealism – breaking down borders between societies – with more fixed or what could be called realist territorial perspectives. EU Cohesion

Policy has thus emerged concurrently with paradigms of cross-border cooperation and notions of territory based on spatial relationships, cross-border and transnational networks and supranational geostrategies rather than exclusively on administrative and legalistic frameworks. A central aspect here is the recasting of national spaces as integral elements of an international political community; from this also emerges the attempt to create a common set of discourses in which various political and social issues are negotiated.

Formally introduced by the Single Act of 1985, partly in response to the regional challenges posed by southern enlargement, the notion of 'Economic and Social Cohesion' has become a central unifying idea legitimising an EU role in territorial development. Furthermore, Territorial Cooperation (TC) within the European Union is understood as a form of local and regional promotion of Cohesion that transcends state borders. In academic debate, territorial cooperation is most generally known as 'cross-border cooperation' but generally both terms have equivalent meanings. More pointedly, I define CBC/TC in terms of political projects carried out by private, state and, to an extent, third sector actors with the express goal of extracting benefit from joint initiatives in various economic, social, environmental and political fields. This has been associated with state–society paradigms that suggest that new forms of politically relevant action can (or must) increasingly take place beyond the seemingly inflexible territoriality of the state (see Faludi, 2013). Through new forms of political and economic interaction – both institutional and informal – it has been suggested that greater cost-effectiveness in public investment can be achieved, economic complementarities exploited, the scope for strategic planning widened and environmental problems more directly and effectively addressed.

The concept of CBC in the European context is not new; it began as a number of subnational political projects already in the 1950s between Dutch and German communities. However, it is the context of post-Cold War change that has elevated CBC to the paradigmatic status it has enjoyed in EU policy. As such, CBC has been appropriated by the European Union as a unique social innovation and as part of the EU's new and progressive political identity (Perkmann, 2002, 2007; Scott, 2014). CBC is thus an element of Europeanisation – which can be understood as a diffusion of supranational notions of post-national stateness and a hybrid, multilevel sense of governance, citizenship and identity (Scott and Liikanen, 2011). Through its support of CBC, the European Union has promoted a self-image or role model for intercultural dialogue and local/regional development.[1] With specific regard to Europeanisation and its role in the construction of cross-border cooperation contexts, European policies have been aimed at networking cities and regions within a theoretically borderless European space (but without violating the formal space of administrative regulation). This is evidenced by a proliferation of initiatives aimed at promoting transnational networking, including Research, Training and Development schemes (such as the multibillion EURO framework programmes), the European Spatial Development Perspective (ESDP), Visions and Strategies for the Baltic Sea Region VASAB, INTERREG, and the ESPON (European Spatial Planning Observatory Network) programme.

Cross-border cooperation has been promoted by the EU on the assumption that national and local identities can be complemented and goals of co-development realised within a broader European vision of community. As such, borders have been used as explicit symbols of European integration, political community, shared values and, hence, identity by very different actors (Langenohl, 2015; Lepik, 2009; Perkmann, 2005). Consequently, the Euroregion concept has proved a powerful tool with which to transport European values and objectives to CEE and beyond (see Bojar, 2008). Cross-border associations and territorial cooperations are now a ubiquitous feature within the EU's many border regions. Euroregions, cross-border city partnerships, European Groupings for Territorial Cooperation (EGTCs) and similar cooperation vehicles have also come into being (Medve-Bálint and Svensson, 2013; Lepik, 2012; Popescu, 2008, 2011;

Zhurzhenko, 2006, 2010). Thus, a significant degree of 'de-bordering' through CBC appears to have taken root within the enlarged and wider European Union context.[2] Nevertheless, in order to go beyond a more normative understanding of cross-border cooperation it is essential to interrogate the complex nature of borders and border-related identities. These reflect, for example, 'Europeanising' and 'nationalising' influences upon cross-border interaction as well as opportunity structures providing CBC incentives.

CEE has been attributed a crucial role in promoting Cohesion and CBC within the context of the EU's enlargement process. Given its historic significance, CEE's post-1989 de-bordering was a milestone, indeed a major challenge, in the development of European Union as a political community. The elimination of border defences and barbed-wire between East and West was, for example, highly symbolic in that it portended a reconstitution of a pan-European space and good neighbourhood relations between individual states. For citizens of CEE countries, and in terms of everyday life, de-bordering was perceived as a new liberty to travel and to express oneself as a 'normal' European. The momentum of European enlargement and process of pre-accession, accompanied by large development subsidies, served to de-emphasise the significance of borders as barriers. This was also promoted by the gradual integration of CEE into the logic of European Cohesion Policy, which strategically targeted local and regional cross-border cooperation.

A (re)bordering perspective on Central and Eastern Europe

As the above discussion suggests, the European Union has had an important impact on the nature of cross-border relations in CEE (Hajdú, 1998; Hardi, 2003; Lados, 2005; Scott, 2006; Zhurzhenko, 2006, 2010). In preparing CEE countries for membership, the EU adopted a strategy based on institutionalised cross-border cooperation and aimed at a gradual lessening of the barrier function of national borders. These policies have also been aimed at integrating previously divided border regions in order to build a more cohesive European space. Indeed, the popularity of the Euroregion concept during the 1990s and early 2000s was undeniable, given the generous incentives offered to local and regional actors. Nevertheless, research suggests that this normative political language of Europeanisation (e.g. as a process of de-bordering regional development) has in several ways contrasted with realities at CEE borders – a situation where cross-border cooperation has reflected competing territorial logics at the EU, national, regional and local levels and conflicting attitudes towards more open borders.

It is also clear that instead of an across the board de-bordering, different processes of re-bordering have taken place in CEE after 1989 (Jańczak, 2013a, 2013b). Herrschel (2011) has suggested that the momentum of European integration has contributed to overlapping processes of border transcendence and confirmation, not least because of deeply rooted historical memories that continue to imbue national borders with highly symbolic meaning. Following the *bordering* approach now widely utilised in comparative border studies (van Houtum and Naerssen, 2002; Newman, 2006; Scott, 2012) the making of borders is not only understood as the traditional process of confirming state sovereignty but is something that takes place in everyday situations: it involves the creation of socio-spatial distinctions based on different and often shifting criteria.[3] Furthermore, this approach emphasises the roles of borders in framing social action. For the sake of this discussion it is important to distinguish between *normative de-bordering*, which was vital for the orchestration of a new democratic European order, and political, social and cultural *re-borderings* that resulted from CEE states repositioning themselves within a changing regional and wider European context.

While there is no room here for a detailed discussion of the research state of the art on cross-border cooperation in relation to Cohesion Policy in CEE, it is possible to arrive at a number of important generalisations. As Orlowski (2010) has suggested, the consequences of enlargement have been stark; the clear lack of East–West convergence has in fact cemented divisions within Europe as a whole and with CEE in particular. As several EU Reports on Social and Economic Cohesion document, despite increases in general welfare the imbalances between Europe's core areas and its vast peripheries remain and depopulation of many rural zones continues unabated (see EC 2004, 2007, 2014). Furthermore, regional disparities as well as cultural and political heterogeneity are certain to increase as a long-term result of enlargement. Gorzelak and Smetkowski (2010) have also shown that, in stark contrast to the objectives of Cohesion Policy, a consolidation and 'petrification' of territorial patterns based on core–periphery inequalities is taking hold in CEE states (Gorzelak and Smetkowski, 2007). This process of growing territorial differentiation is based on relative abilities: first, to attract/generate investment, especially into innovative sectors; and, second, relative proximity and accessibility to economically dynamic urban centres. As a result, regional polarisation has been a fact of life since the political transitions from 1989 and large dynamic cities have virtually detached themselves from their regional contexts. Future prospects for CEE regions are thus interpreted in terms of continuing polarisation and a danger of stagnation of internal and external peripheries. In CEE, both EU internal and external borders are with a few notable exceptions characterised by pronounced marginalisation regions at these borders are relatively underdeveloped in both quantitative and qualitative dimensions and continue to lose highly qualified workers to metropolitan cores. As a result, domestic polarisation reinforces structural conditions of West–East dependence, and this, in addition to ethno-political tensions, has tended to fragment the region and thus limit the overall impact of cross-border cooperation.

Another impact of socio-economic and territorial divisions is that, having achieved general de-bordering goals in a functional sense, impetus for greater social and socio-economic interaction across CEE borders has diminished (Jańczak, 2013a). In fact, and as will be elaborated below, it is national consolidation, rather than cross-border cooperation, that has taken overall priority within Cohesion policies. Indeed, as Gorzelak (2006) was able to document in the Polish–German case, the main benefit of CBC according to Polish local actors was the accumulation of experience in working with EU programmes and funding opportunities, rather than the results of cooperation projects. While the functional de-bordering of Polish–German relations was indeed vital, the step towards more substantial partnerships has remained tentative (Balogh, 2014). Moreover, the situation of local and regional cross-border cooperation at other Polish borders has been generally difficult as well: in specific areas of social interaction (education, culture) progress has been documented, but CBC has not come close to satisfying expectations of – or contributing to – a more profound de-bordering of CEE (Böhm, 2015; Domaniewski and Studzińska, 2016; Jańczak, 2013b). In the case of CEE, important shifts in appropriations and practices of cross-border cooperation have taken place since 1989. While avid learners of EU norms and governance practices – and thus of CBC – during the 1990s and early 2000s, CEE countries have developed a much more localised and nationally focused interpretation of CBC and Cohesion.

To sum up, CBC is an area where Europeanisation has exerted considerable adaptive pressure in countries such as Hungary, Poland, Slovakia and Czechia. However, these countries have not merely reacted to EU policies but have duly domesticated and incorporated them into specific national development strategies in ways that have not always conformed to EU expectations. Perhaps the most conspicuous example of national reinterpretations of EU Cohesion Policy is that of Hungary, which has gradually converted Europeanisation into a national development strategy with geostrategic border/crossing implications.

James Wesley Scott

Cross-border cooperation and Europeanisation in the Hungarian case

Hungary's borders in the years directly after 1989 are an excellent example of attempts on the part of local, national and EU actors to appropriate CBC as a multilayered exercise in regional development and historical de-bordering of post-Cold War Europe. Through the use of symbolisms of the border as a bridge between neighbours, Hungary's relationships with its neighbours was recast in a wider European context of overcoming the 'scars of history'.[4] Political cooperation, and most certainly cross-border cooperation, was closely intertwined with rapprochement and a desire to develop a culture of mutual goodwill, underlined by the Hungarian government's decision to resist any form of revisionist political initiative that might result in new border-related tensions (Csaba et al., 2009). Conversely, the 1990s reflected a 'drive for convergence' of CEE countries to European standards and the universal adoption of overall Cohesion Policy goals as a means to secure EU membership.

Europeanisation – and the extent of the EU's influence on Hungary with particular reference to CBC and Cohesion Policy – can be roughly divided into three phases: (1) the period of systemic transition that brought in external political perspectives and new models of domestic governance (1989–1993): (2) the phase during which Hungary prepared for EU accession (1994–2003); and (3) the post-accession period (since 2004). During the first phase, the impact of the EU was rather indirect. Elite views were focused on the 'completely new' situation and on 'forgetting the past'. Only abstract notions of Europe were present at that time (as the national development policy document of 1993 gives evidence of). The period was also one of overcoming the odium of state socialism and of rejecting the socialist past and its legacies.

The second phase was one of preparing Hungary for EU accession and was the most decisive in terms of Europeanisation. The 1993 agreement signed with the EU introduced an intensive period of political, social learning and the impact of pre-accession was crucial (SAPARD, PHARE programmes). Regional policy and CBC were part of this learning process and these policies were developed and executed more or less strictly according to EU guidelines – unlike the situation today. EU experts were directly involved in a period of tutelage in which Hungarian actors carefully followed EU directives and counselling. Furthermore, a process of regionalisation was implemented which created new decision-making structures and programming bodies at the subnational level (Pálné Kovács, 2009a, 2009b). One example of this was the Plan for South Transdanubia and the creation of a Regional Development Agency for the region, both financed out of PHARE funds.

During this phase, openness to EU notion of de-bordering and regional cooperation was at its apogee and within the post-1989 context of European integration Hungarian borders were conceptualised as regional development contexts in close alignment with a wider European reading of Cohesion Policy (see Barta, 2006). For example, in the case of Hungary, 'post-socialist' regional studies since the first studies of Rechnitzer (1990) very much focused on the development of new local economic networks between Hungary and its neighbours and the roles that border regions play in their creation. As development trends in the early 1990s clearly indicated, post-socialist economic transformation and differential border effects contributed to the exacerbation of core–periphery relationships. Distinctions were made in terms of characterising Hungary's borders between successful and dynamic Western border areas (those with Austria and Slovakia) and declining eastern border regions neighbouring Romania, Eastern Slovakia and Ukraine (Baranyi, 2001; Bihari and Kovács, 2005; Szörényiné Kukorelli et al., 2000). Similar to the regional development context, specific Hungarian borders have also been conceptualised within a wider European context of economic integration and Cohesion. The Concept of Border as National Periphery was perhaps most vigorously expressed by Baranyi

(2001) who depicted Hungary's eastern border regions as 'peripheries of the periphery'. Baranyi developed a number of arguments in which border areas and settlements were clearly identified as regional development problems where cross-border cooperation held out prospects of linking in to more general EU development processes.

However, the years following accession in 2004 have seen the impact of the EU considerably weaken – perhaps, paradoxically, Hungary, now an EU member, a 'good learner', began to appropriate Cohesion Policy and CBC more intensively in terms of domestic concerns regarding the situation of ethnic Hungarians in neighbouring countries. By the same token, EU insensitivity to Hungary's needs did represent a major stumbling block to a more balanced 'learning process' between the two partners. One principal source of disaffection with the EU is the lack of support for Hungary's attempts to promote linguistic and nationality rights for ethnic Hungarians in neighbouring countries.[5]

Smaller-scale CBC projects were relatively effective in the pre-accession phase as a learning process in obtaining resources and CBC actors were much more open. After 2004, however, the scale and significance of CBC decreased appreciably. Reasons for this included: less money (and thus less local interest) and the fact that National Operational Programmes (OPs) became much more important in guiding regional development and development strategies of local actors. This shifted the emphasis away from CBC and more towards domestic development. With Romanian accession in 2007 a further element came into play. Romania became much more active in CBC and Hungarian actors lost their privileged status. In addition, CBC declined in importance with regard to Hungarian minorities as more nationally defined and oriented programmes emerged.

Given the above, what can be said about CBC as a force for regional development? The opening of borders has certainly led to greater functional interrelationships within the Carpathian Basin as cross-border labour markets and shopping patterns indicate (Mezei, 2008; Nagy, 2011). Generally speaking, however, CBC at Hungary's borders has not been significant in direct economic and infrastructural terms, but rather in 'soft' areas of development. CBC provided and still provides an important level of institutional learning but the lack of strong subnational governments limits the actual ability of these actors to engage in development projects across borders. The only true working CBC institutions are in the West, on the border with Austria as well as the Istergránum European Grouping of Territorial Cooperation (EGTC) between Hungary and Slovakia (Esztergom–Komárom–Komarno). Most Euroregions appear to have outlived their functions as regional mediators – made redundant by a combination of institutional flux (recentralisation), a lack of institutional support, EU bureaucracy and political uncertainty in CEE. It is remarkable, for example, that Nagy (2011) in his study of cross-border urban networks makes almost no mention of Euroregions and institutionalised forms of CBC. Instead, he focuses on functional networks and relationships as drivers of CB interaction between Hungarian and Romanian towns. Nagy also states that the cross-border centrality of Hungarian cities such as Gyula is not a mere function of ethnic relations but involves Hungarian–ethnic Romanian interaction as well.

At the same time, shifts in European Union priorities and approaches in programming Cohesion and Territorial Cooperation Policy have also played an important role in conditioning Hungary's 'domestication' of European Union policies. There is in fact little doubt that since the historic turn of events of 1989/1991 and the heady days of a 'new European order' there has been a shift in the EU's focus on CBC. EU rhetoric about the benefits of CBC is today a far cry from the prosaic language of the 1990s. Most recently, CBC has been subsumed within the more inclusive notion of Territorial Cooperation (TC) and its main aim remains to 'reduce the negative effects of borders as administrative, legal and physical barriers, tackle common problems and exploit untapped potential'.[6] It is clear from recent debate on European Cohesion that the EU stakes much of its political capital on more traditional instruments of redistribution that are nationally oriented even if subject to supranational guidelines. Indeed, the 2007–2013 budget of €8.7 billion for Territorial Cooperation amounted to a mere 2.5 per cent of the total Cohesion Policy budget. Furthermore, a major overall share of Cohesion funds are targeted to Central and Eastern European countries where there appears to be less enthusiasm for CBC as a regional development resource.

Divergent patterns in the appropriation of CBC as a Cohesion Policy

In the case of Hungary, instrumental understandings of CBC characterise the present situation and there appears to be a lack of understanding of the border as a resource; regional cooperation has also tended to be weak. In addition, the trend of Hungarian recentralisation has made CBC more complex – there is a further reduction in institutional capacity, and political will to engage in cooperation projects has diminished. Nevertheless, one means to counteract the marginalisation of cross-border cooperation and insensitivity to specific regional situations at and around Hungary's borders could very well be the recognition of trans-sovereignty which understands rights to nationality and national identity as independent of and/or even complementary to state-based citizenship. This is a form of national belonging that would uphold the spirit of European integration but that has been frustrated by national/political tensions, new administrative borders, questions of linguistic rights and inaction at the European level with regard to minority rights. Local autonomies based on a trans-sovereign model (see Bakk and Öllös, 2010) would be a major potential means of empowering local CBC but also improving prospects for more general economic and social integration within Central and Eastern Europe.

Hence, as argued above in the case of Hungary, CBC is certainly understood in terms of European Cohesion but is heavily influenced by overlying political goals of 'nation-building'. Hungarian interests are clearly defined by a desire to improve connections between Motherland and Hungarian communities around the country as well as improve the living standards and stabilise the conditions in neighbourhood areas as a means to keep Hungarians thriving there – effectively de-bordering to create new (trans)national spaces. However, this has also engendered the distrust of Hungary's neighbours who at times have interpreted CBC as a means to extend Hungarian extraterritorial sovereignty claims. Indeed, a partial resurgence of national rivalries and historical animosities has taken place between several EU member states and has, for example, affected local cooperation between Hungary, Slovakia and Romania. Different regional interpretations of CBC thus indicate a highly variable appropriation of Europeanisation policies. Concretely, there is a notable East–West divide in the acceptance and adaptation of CBC as a set of regional development practices.

However, since 2004 the situation has changed markedly. Rather than reflect conformism to 'Core Europe', Hungary (as well as Poland, Slovakia, Czechia and other CEE countries) now appears to interpret the construction of Europe in terms of a redoubled focus on national development priorities. Batory (2010) in fact argues that Hungarian political interpretations of European integration (and Europeanisation) have become rather formal and statist whereas Hungary's political identity with regard to its neighbours is seen in affective, i.e. ethnic, cultural and historic rather than European terms. In the case of Hungarian national development strategies of 1998 and 2007 we in fact see a marked change. While in the first document Hungarian border areas and regional development issues related to cooperation with neighbouring states received generous coverage, the 2007 National Development Strategy (National Development Agency, 2007) only gives very brief mention of CBC. In practical terms, CBC remains as a minimalist exercise – national strategic plans generally consider it as an extension of national development.

The minimalist, instrumental Hungarian approach indicates a relative lack of policy mainstreaming as well as a focus on national consolidation. Reasons for difficult CBC contexts in the Hungarian case also include: lack of local capacity to promote cooperation, cumbersome EU regulations and project management rules, frequent interstate tensions regarding ethno-linguistic issues, as well as local orientations to national centres and European core regions rather than to neighbouring states (see, for example, Baranyi, 2008; Hajdú et al., 2009; Hardi, 2010; Mezei, 2008). Furthermore, as my research has suggested, EU-inspired strategies of institutionalised CBC – an area of complex social, economic and political diversity – have tended to be 'co-opted' in Hungary by specific nationally defined interests: Euroregions have generally been 'top-down' creations, inhibiting processes of region-building through local initiative. At the same time, institutional legacies, such as strong central control, have contributed to variegated Europeanisation processes (Pálné Kovács, 2009a, 2009b). Hence, CBC in the Carpathian Basin (which involves Hungary, Romania and Slovakia) is not a self-evident phenomenon and appears to have lost momentum since the days of PHARE–INTERREG.

These observations confirm that political and social de-bordering in CEE has been a patchy, selective process and that the transcendence of 'mental', historically contingent borders remains challenging. Resentment, a lack of engagement and in some cases fear have all played a major role in reducing the role of de-bordering in the reconstitution of CEE as a cooperation space (e.g. the Carpathian Basin). At the same time, the focus in the region has been on nation-building, on repositioning CEE states within a wider European context. Highly symbolic cooperation vehicles, such as the Visegrad 4 and the Euroregion Carpathia have proved too weak to actually form a basis for concrete networking on interstate projects and initiatives. As has been suggested above, de-bordering has also brought with it a new East–West orientation rather than a reconstitution of neighbourhood relations. This process has also been influenced by core–periphery relations and market competition which weaken a sense of common purpose between CEE countries.

In stark contrast to these developments in CEE, CBC in Western Europe is no longer as dependent on external funding as it once was. Here, we see a routinisation of local and regional cross-border cooperation that is generally embedded within multilevel governance structures. It is therefore no coincidence, as the ESPON–TERCO project has shown, that EGTCs are concentrated in western areas of the EU. In the most 'successful' – that is, the most well-organised – border regions (e.g. the Dutch–German Euroregions), public-sector and NGO cooperation has been productive in many areas, especially in questions of environmental protection, local services and cultural activities. Additionally, successful cases (e.g. German–Dutch) seem to involve a process of pragmatic incrementalism, with 'learning-by-doing' procedures and a gradual process of institutionalisation. As working relationships have solidified, experience in joint project development has accumulated and expertise in promoting regional interests increased, as has the capacity of regional actors to take on large-scale problems and projects.

Concluding thoughts on cohesion and European borders

The transcendence of borders as barriers to interaction and cooperation remains an inherently EU–European idea, recent crisis situations notwithstanding. However, in terms of de-bordering CEE this idealistic European message is domesticated in specific ways that give evidence of a reluctance to fully engage in cross-border cooperation. Admittedly, there has been a general shift away from prioritising CBC and to refocusing on core 'national' goals of European Cohesion Policy. However, while CBC has become routinised and even independent of structural fund support in western European contexts, it is seen more in instrumental, even opportunistic, terms in Central and European countries. Arguably therefore, we find a notable East–West divide in the acceptance of CBC as a set of regional development practices.

Viewed from a contemporary perspective it appears that the normative language of (European) integration has contrasted with local realities. At the local level cross-border cooperation reflects competing territorial logics at the EU, national, regional and local levels and conflicting attitudes towards more open borders.[7] As a result, cross-border cooperation within CEE has been a difficult undertaking. A resurgence of nationalism (e.g. in Slovakia and Romania) and a new focus on national development has also tended to de-emphasise local forms of cooperation in the Hungarian case (as well as in Poland and Slovakia).[8] Conflicts between 'Europeanising' and 're-nationalising' conceptions of borders can in fact be identified, interpreted in terms of identity politics serving specific groups within border regions. The research background reveals that CBC was never a prioritised area of political endeavour or a privileged element of transformation.

This reflection on CBC thus takes into consideration the *longue durée* nature of creating cross-border political practices at the local and regional level. Indeed, CBC has rarely produced rapid results in terms of economic growth and regional development. Furthermore, local and regional actors develop cooperation mechanisms situationally and in ways that reflect both political opportunities and social and structural constraints. Despite all the shortcomings of the EU model of institutionalised CBC, institutional change elicited by EU policies and funding mechanisms has led to a degree of 'Europeanisation' of cooperation contexts and thus of spatial planning and development dialogue. This is evident in the discourses, agendas and practices of cross-border actors; they very often legitimise their activities by referring to the wider political, economic and spatial contexts within which their own region must develop. Nevertheless, actual patterns of CBC practices indicate a rather disjointed and complex reality. The European Union itself cannot provide a central template for de-bordering Europe. This will rather depend on how a post-national Europe is interpreted, negotiated and constructed 'at the margins'.

Acknowledgement

The publication of this research has been supported by project #104985 of the Hungarian National Research, Development and Innovation Office.

Notes

1. This is reflected, for example, in the EU's Green Paper on Territorial Cohesion (subtitled 'Turning Territorial Diversity into Strength') in which the need to develop strong cross-border linkages as well as more robust forms of regional and local cooperation with neighbouring states is emphasised (see EC, 2008).
2. See, for example, Scott and Liikanen (2011); the special issue of *Regional Studies*, Vol. 33, No. 7, 1999, edited by James Anderson and Liam O'Dowd; Scott (2006); as well as *European Research in Regional Science*, Vol. 10, 2000.
3. This perspective is based on a notion of conceptual change that involves shifts from largely functional to cognitive and symbolic perspectives on borders. Bordering can take place, for example, as an everyday construction of borders through ideology, media representations, policy agendas and political institutions, popular attitudes, everyday forms of border-crossing, etc.
4. Robert Schuman's pronouncement that national borders in Europe represented scars of history ('Les cicatrices de l'histoire') has become an evocative political discourse in the processes of European integration and enlargement.
5. See for example, the article titled 'Elfelejtett magyar kisebbség' in *Magyar Nemzet*, 17 May 2006: in conjunction with the EU accession of Romania the Hungarian government voiced complaints that the interests of the Hungarian minority were not being appropriately taken into consideration. As Kinga Gál exclaimed: 'how did one-and-a-half million ethnic Hungarians disappear in this (EU Commission's)

report?' (*Magyar Nemzet*, 2006). Only the Roma population were named in the EU Commission report that was released in May 2006. Both government and opposition expressed their dismay at the EU Report's omission of Hungarian minority rights.
6 http://ec.europa.eu/regional_policy/en/policy/cooperation/european-territorial/cross-border/2007-2013/
7 See Popescu (2008) who comments on the case of Romania.
8 See Bürkner (2006). In its edition of 20 October 2009, the Hungarian daily *Népszabadság* ('Nem jött létre a 'régiók Európája', reporter: István Tanács) lamented a lack of true cross-border cooperation with neighbouring states, citing national particularisms and limited European vision.

Bibliography

Ágh, A., 2003. *Anticipatory and Adaptive Europeanization in Hungary*. Budapest: Hungarian Centre for Democracy Studies.
Bakk, M. and Öllös, L., 2010. Politikai közösség és kulturális identitás Magyarország és a szomszédos országok magyar kisebbségeinek viszonyban. In: B. Bitskey, ed. *Határon túli magyarság a 21 szazadban*, Budapest: Köztársasági Elnöki Hivatal, pp. 41–68.
Balogh, P., 2014. *Perpetual Borders: German–Polish Cross-Border Contacts in the Szczecin Area*. Stockholm: Stockholm University.
Baranyi, B., 2001. A periféria perifériájá – egy kérdőives vizsgálat eredményei és tanulságai az Északkelet-Alföldi határ menti területein. In: B. Baranyi, ed. *A Határmentiség kérdőjelei az Északkelet-Alföldön*. Pécs: MTA Regionális Kutatások Központja, pp. 55–86.
Baranyi, B. ed., 2008. *Magyar–ukrán határrégió*. Debrecen: MTA Regionális Kutatások Központja.
Barta, Gy., 2006. *Hungary – The New Border of the European Union*. Pécs: Centre for Regional Studies. (Discussion Papers, 54).
Batory, A., 2010. Kin–State Identity in the European Context: Citizenship, Nationalism and Constitutionalism in Hungary. *Nations and Nationalism*, 16(1), pp. 31–48.
Bihari, Z. and Kovács, K., 2005. Slopes and Slides: Spatial Inequalities in Employment Opportunities. In: Gy. Barta, É.G. Fekete, I. Kukorelli Szörényiné and J. Timár, eds. *Hungarian Spaces and Places: Patterns of Transition*. Pécs: Centre for Regional Studies of the Hungarian Academy of Sciences, pp. 360–378.
Böhm, H., 2015. Czech–Polish Borders: Comparison of the EU Funds for Cross-border Cooperation of Schools in Selected Euroregions. In: M. Pete, ed. *Cross-Border Review. Yearbook 2015*. Budapest: European Institute, CESCI, pp. 59–74.
Bojar, E., 2008. Euroregions in Poland. *Tijdschrift voor Economische en Sociale Geografie*, 87(5), pp. 442–447.
Bufon, M. and Markelj, V., 2010. Regional Policies and Cross-border Cooperation: New Challenges and New Development Models in Central Europe. *Revista Romana de Geografia Politica*, 12(1), pp. 18–28.
Bürkner, H-J., 2006. Regional Development in Times of Economic Crisis and Population Loss: The Case of Germany's Eastern Border Regionalism. In: J.W. Scott, ed. *EU Enlargement, Region Building and Shifting Borders of Inclusion and Exclusion*. Aldershot: Ashgate, pp. 207–215.
Csaba, L., Jeszenszky, G. and Martonyi, J., 2009. *Helyünk a világban. A magyar külpolitika útja a 21. században*. Budapest: Éghajlat Könyvkiadó.
Domaniewski, S. and Studzińska, D., 2016. The Small Border Traffic Zone between Poland and Kaliningrad Region (Russia): The Impact of a Local Visa-Free Border Regime. *Geopolitics*, 21(2), pp. 243–262.
EC, 2004. *A New Partnership for Cohesion, Convergence, Competitiveness, Cooperation. Third Report on Economic and Social Cohesion*. Luxembourg: Office for Official Publications of the European Communities.
EC, 2007. *Growing Regions, Growing Europe. Fourth Report on Economic and Social Cohesion*. Luxembourg: Office for Official Publications of the European Communities.
EC, 2008. *Green Paper on Territorial Cohesion. Turning Diversity into Strength*. COM(2008) 616 final. Brussels: Commission of the European Communities.
EC, 2014. *Investment for Growth and Jobs. Promoting Development and Good Governance in EU Cities and Regions. Sixth Report on Social, Economic and Territorial Cohesion*. Brussels: European Commission.
Faludi, A., 2013. Territorial Cohesion and Subsidiarity under the European Union Treaties: A Critique of the 'Territorialism' Underlying. *Regional Studies*, 47(9), pp. 1594–1606.
Fritsch, M., 2009. European Territorialization and the Eastern Neighbourhood: Spatial Development Co-operation between the EU and Russia. *European Journal of Spatial Development*, 35, pp. 1–27.
Für, L., 2012. *Magyar sors a Kárpát Medencében. Népdeséünk évszázadai 896–2000*. Budapest: Kairosz Kiadó.

Gorzelak, G., 2006. Normalizing Polish–German Relations: Cross-border Cooperation in Regional Development. In: J.W. Scott, ed. *EU Enlargement, Region Building and Shifting Borders of Inclusion and Exclusion.* Aldershot: Ashgate, pp. 195–205.

Gorzelak, G. and Smetkowski, M., 2007. Regional Dynamics in Central and Eastern Europe. Paper presented at the International Conference, Regional Development in Central and Eastern Europe, 20–22 September, University of Warsaw, Poland.

Gorzelak, G. and Smetkowski M., 2010. Regional Development Dynamics in Central and Eastern European Countries. In: G. Gorzelak, J. Bachtler and M. Smetkowski, eds. *Regional Development in Central and Eastern Europe.* London: Routledge, pp. 34–58.

Hajdú, Z., 1998. *Changes in the Politico-geographical Position of Hungary in the 20th Century.* Pécs: Centre for Regional Studies. (Discussion Papers, 22).

Hajdú, Z., 2008. A Kárpát-medence államosodási folyamatainak változásai és történeti földrajzi elemzésük. *Korall,* 9(31), pp. 75–100.

Hajdú, Z., Lux, G., Pálné Kovács, I. and Somlyódyné Pfeil, E., 2009. *Local Dimensions of a Wider European Neighbourhood: Cross-border Relations and Civil Society in the Hungarian–Ukrainian Border Area.* Pécs: Centre for Regional Studies. (Discussion Papers, 71).

Hardi, T., 2003. Az EU-csatlakozás hatása a határ menti térségek fejlődésére a Nyugat-Dunántúlon. In: M. Kis, L. Gulyás and E. Erdélyi, eds. *Európai kihívások 2. Tudományos konferencia.* Szeged: Szegedi Tudományegyetem Szegedi Élelmiszeripari Főiskolai Kar, pp. 158–162.

Hardi, T., 2010. Határ menti térségek, határon átnyúló kapcsolatok. In: V. Szirmai, ed. *Közép-Dunántúl,* Pécs, Budapest: MTA Regionális Kutatások Központja, Dialóg Campus Kiadó.

Herrschel, T., 2011. *Borders in Post-Socialist Europe: Territory, Scale, Society.* Farnham: Ashgate.

Horváth, Gy., 2010. *Territorial Cohesion in the Carpathian Basin: Trends and Tasks.* Pécs: Centre for Regional Studies. (Discussion Papers, 81).

Jańczak, J., 2013a. Revised Boundaries and Re-frontierization. Border Twin Towns in Central Europe. *Revue d'études comparatives Est-Ouest,* 44(4), pp. 53–92.

Jańczak, J., 2013b. *Border Twin Towns in Europe: Cross-Border Cooperation at a Local Level.* Berlin: LogosVerlag.

Keményfi, R., 2006. Az 'ezeréves határok' etnikai területek legitimációs eszközei a 20. század első felének földrajzban. In: B. Bakó, R. Papp and L. Szarka, eds. *Mindennapi előítéletek. Társadalmi távolságok és etnikai sztereotípiák.* Budapest: Balassi Kiadó, pp. 325–344.

Lados, M., 2005. A határmentiségtől az integrált határrégióig. *Tér és Társadalom,* 19(2), pp. 1–5.

Langenohl, A., 2015. *Town Twinning, Transnational Connections and Trans-local Citizenship Practices in Europe.* Houndmills: Palgrave Macmillan.

Lepik, K., 2009. Euroregions as Mechanisms for Strengthening of Cross-border Cooperation in the Baltic Sea Region. *TRAMES,* 13(3), pp. 265–284.

Lepik, K., 2012. *Cross-border Cooperation Institutional Organization.* Saarbrücken: Lampert Publishing.

Magyar Nemzet, 2006. Elfelejtett magyar kisebbség [The Forgotten Hungarian Minority]. *Magyar Nemzet,* 17 May.

Manzella, G.P. and Mendez, C., 2009. *The Turning Point of EU Cohesion Policy. Report Working Paper, EPRC.* Strathclyde: University of Strathclyde.

Medve-Bálint, G. and Svensson, S., 2013. Diversity and Development: Policy Entrepreneurship of Euroregional Initiatives in Central and Eastern Europe. *Journal of Borderlands Studies,* 28(1), pp. 15–31.

Mezei, I., 2008. *Városok Szlovákiában és a magyar határ mentén.* Somorja, Pécs: Fórum Kisebbségkutató Intézet, MTA Regionális Kutatások Központja.

Nagy, G., 2011. Gyula – város a határon. A központi funkciók határon átnyúló hatása. *Tér és Társadalom,* 25(4), pp. 127–147.

National Development Agency, 2007. *New Hungary Development Plan. National Strategic Reference Framework 2000–2013.* Budapest: NDA.

Newman, D., 2006. Borders and Bordering: Towards an Interdisciplinary Dialogue. *European Journal of Social Theory,* 9(2), pp. 171–186.

Orlowski, W., 2010. The Puzzles of Convergence. In: G. Gorzelak, J. Bachtler and M. Smetkowski, eds. *Regional Development in Central and Eastern Europe.* London: Routledge, pp. 7–18.

Pálné Kovács, I., 2009a. Europeanization of Territorial Governance in Three Eastern/Central European Countries. *Halduskultuur,* 10, pp. 40–57.

Pálné Kovács, I., 2009b. Regionalization in Hungary: Options and Scenarios on the Road to Europe. In: J.W. Scott, ed. *De-Coding New Regionalism. Shifting Socio-political Contexts in Central Europe and Latin America.* Aldershot: Ashgate, pp. 199–214.

Perkmann, M., 2002. *The Rise of the Euroregion. A Bird's Eye Perspective on European Cross-border Co-operation.* Department of Sociology, Lancaster University.

Perkmann, M., 2005. Cross-border Co-operation as Policy Entrepreneurship: Explaining the Variable Success of European Cross-border Regions. CSGR Working Paper, 166/05, University of Warwick.

Perkmann, M., 2007. Policy Entrepreneurship and Multilevel Governance: A Comparative Study of European Cross-border Regions. *Environment and Planning C*, 25(6), pp. 861–879.

Popescu, G., 2008. The Conflicting Logics of Cross-border Reterritorialization: Geopolitics of Euroregions in Eastern Europe. *Political Geography*, 27(4), pp. 418–438.

Popescu, G., 2011. *Bordering and Ordering the Twenty-first Century: Understanding Borders.* Lanham, MD: Rowman and Littlefield.

Radaelli, C.M. and Pasquier, R., 2006. Conceptual Issues. In: P. Graziano and V. Maarten, eds. *Europeanization: New Research Agendas*, Basingstoke: Palgrave Macmillan, pp. 35–45.

Rechnitzer, J. ed., 1990. *A nyitott határ.* Győr: MTA Regionális Kutatások Központja.

Scott, J.W. ed., 2006. *EU Enlargement, Region-building and Shifting Borders of Inclusion and Exclusion.* Aldershot: Ashgate.

Scott, J.W., 2011. Reflections on EU Geopolitics: Consolidation, Neighbourhood and Civil Society in the Reordering of European Space. *Geopolitics*, 16(1), pp. 146–175.

Scott, J.W., 2012. European Politics of Borders, Border Symbolism and Cross-Border Cooperation. In: T. Wilson and H. Donnan, eds. *A Companion to Border Studies.* Hoboken, NJ: Wiley-Blackwell, pp. 83–99.

Scott, J.W., 2014. From Europhoria to Crisis. Cross-Border Cooperation, Euroregions and Cohesion. In: L. Dominguez and I. Pires, eds. *Cross-Border Cooperation Structures in Europe. Learning from the Past, Looking to the Future.* Basel: Peter Lang, pp. 81–94.

Scott, J.W. and Liikanen, I., eds., 2011. *European Neighbourhood through Civil Society Networks.* Abingdon: Taylor and Francis.

Szörényiné Kukorelli, I., Dancs, L., Hajdú, Z., Kugler, J. and Nagy, I., 2000. Hungary's Seven Border Regions. *Journal of Borderlands Studies*, 15(1), pp. 221–254.

van Houtum, H. and van Naerssen, T., 2002. Bordering, Ordering and Othering. *Tijdschrift voor Economische en Sociale Geografie*, 93(2), pp. 125–136.

Vizi, B., 2009. Hungary: A Model with Lasting Problems. In: R. Bernd, ed. *Minority Rights in Central and Eastern Europe.* London: Routledge, pp. 119–134.

Zhurzhenko, T., 2006. Regional Cooperation in the Ukrainian–Russian Borderlands: Wider Europe or Post-Soviet Integration? In: J.W. Scott, ed. *EU Enlargement, Region Building and Shifting Borders of Inclusion and Exclusion.* Aldershot: Ashgate, pp. 95–111.

Zhurzhenko, T., 2010. *Borderlands into Bordered Lands. Geopolitics of Identity in Post-Soviet Ukraine.* Stuttgart: Ibidem Verlag.

12

(Ethno-)regional endeavours in Central and Eastern Europe

Nóra Baranyai

Introduction

Reviving (ethno-)regional movements in the 1960s and 1970s played a significant role in the territorial and public administration reforms of Western Europe. In forming the structures at the time, attention was paid to historical, cultural and ethnic boundaries, and as a result of decentralisation, territorial units with great independence came about in Belgium, Italy, Spain and the United Kingdom, among others. Following the change of regimes in Central and Eastern Europe (CEE), (ethno-)regional movements aiming to shape bottom-up territorial–public administrative units appeared in this macro-region, too. Having inherited centralised state socialist structures, the CEE countries did not take into account the requests of these movements, apart from some exceptional cases, when they carried out their necessary public administration reforms and, where it was, regionalisation. Thus (ethno-)regionalism still aims at reshaping the present territorial structures and deepening decentralisation in this part of Europe, aiming sometimes to change the whole state structure, while at others to establish the autonomy of historical or ethnic-multicultural regions.

Our study attempts to synthesise the CEE characteristics of (ethno-)regionalism by giving an overview of the history and objectives of these movements as well as of the territorial–public administrative and ethnic minority policies of the given states. We are focusing on the regional endeavours of populous minority groups (living in South Slovakia and in Székely Land) and on historical and multiethnic regions having ethnic or regional identity (Upper Silesia, Istria, Moravia, Subcarpathia, Vojvodina). The study is based on relevant legal regulations, census data, and on the draft and final programmes of movements.

The state socialist legacy

Following the Second World War the CEE countries came under Soviet political influence. Ukraine belonged to the Soviet Union since 1922, whereas after democratic transition the other countries in this area saw the scenario of communist parties seizing power one after the other. There was only one exception which was able to escape Soviet influence, namely communist Yugoslavia. In the other countries the Soviet pattern was adopted in every area of the state structure, public administration included, entailing several reforms and new legal regulations.

Council systems based on the principle of democratic centralism were established and provisions were made that often resulted in structural transformation. These led to unnatural territorial units having nothing to do with traditions. Additional reforms were only made (e.g. in Poland, Romania) to crush the power of political opponents (Illés, 2002). Internationalist ideology made disputes about individual and territorial autonomy superfluous, and regions having had self-governing status were fitted into the general system of public administration. In the state socialist era only two areas had special status: Székely Land within the (Mureș–) Hungarian Autonomous Region[1] and Vojvodina in Yugoslavia which had independent foreign policy.

In the transition period following the regime changes in the 1990s the key task was to transform basic institutions, introduce democracy and market economy, and change the structure of the economy. In public administration this meant eliminating the overcentralised council system and creating a traditional European system in its place (Illés, 2002). First the local self-governments were created and given a wide scope of authority; while meso-level formations were divested of their power, their functions became void (Pálné Kovács, 2000; Illés, 2002; Horváth, 2006). Decentralisation, that is, the establishment of territorial self-governments, started much later in this area than in the West, actually by the pressure of adapting to the European Union. There was insufficient social–political support for setting up new really functioning meso-level institutions, and this slowed down the process of decentralisation which was finally accomplished by way of superficial public administration reforms (Pálné Kovács, 2009). Aversion to decentralisation originated from historical tradition and/or fear in some states of their ethnic minorities, while in countries lacking regional traditions, political controversy hindered the public administration reforms (Illés, 2002; Gulyás, 2005).

(Ethno-)regional endeavours: CEE examples

In addition to great structural transformations, the regime changes initiated other processes as well. By the last years of the state socialist era there were significant political and social movements that started a new epoch of self-organisation in the CEE region. Having been forced to 'dissolve' in internationalism or specific nationalisms (like Czechoslovakianism, Yugoslavianism, Soviet identity), national, ethnic and regional identities[2] started to revive in the course of transition, which became evident by the growing number of their organisations and the assertion of their goals. There were political movements based on national identity endeavouring to become independent and to redraw the borders within the region, and some of them were successful in creating new nation states[3] (disintegration of socialist federations), while others attempted to achieve personal and/or territorial autonomy within the given country. Personal autonomy is now provided (in principle) in several CEE countries. This might be intended to prevent larger demands like territorial autonomy, or to send a message to or put pressure on the neighbouring countries (e.g. the Hungarian law on ethnic minorities;[4] Dobos, 2015; and Chapter 11). Some of the ethnic minority groups (e.g. the Turkish minority in Bulgaria) had to renounce their territorial autonomy claims for political reasons (Curtis, 1992), whereas others (e.g. the Albanians in Macedonia) practically possess special status, albeit not in a constitutional sense (Pap and Végh, 2009).

There are several types of active regionalism in the CEE region (Table 12.1). Following the disintegration of large empires and as a consequence of peace treaties, several ethnic groups became minorities in the new territories of small states. These groups started to formulate their claim for self-government after the change of regime. Several multiethnic regions had been formed as a result of changing state borders and the resulting migration over the centuries. These regions would prefer territorial autonomy to show that they are different and multifarious both culturally and linguistically, irrespective of whether they are in a more or less advantageous

Table 12.1 (Ethno-)regional organisations and endeavours in the individual countries

Country	Organisation	Reason for autonomy	Objective (state structure / autonomous unit)
Croatia	Istarski Demokratski Sabor/Istrian Democratic Assembly (1990)	historical multicultural region	regionalised unitary state Istria Autonomous County
Czechia	Moravané/Moravians (2005)	historical (ethnic) region	federal state Moravian–Silesian Region
Poland	Ruch Autonomii Śląska/Silesian Autonomy Movement (1990)	historical multicultural (ethnic) region	regionalised unitary state Upper Silesian Voivodeship
Romania	Székely Nemzeti Tanács/Székely National Council (2003)	historical ethnic region	unitary state Székely Land Autonomous Region
	(Romániai Magyar Demokratikus Szövetség/Democratic Alliance of Hungarians in Romania, Magyar Polgári Párt/Hungarian Civic Party)		
	Erdélyi Magyar Néppárt/ Hungarian People's Party of Transylvania (2011)	historical ethnic region	regionalised unitary state Székely Land special status region
Serbia	Liga Socijaldemokrata Vojvodine/League of Social Democrats of Vojvodina (1990)	historical multicultural region	regionalised unitary or federal state Autonomous Province of Vojvodina
	Vojvođanska partija/ Vojvodina's Party (2005)	historical multicultural region	federal state Republic of Vojvodina
Slovakia	Magyar Közösség Pártja/Party of the Hungarian Community (1998)	ethnic region	unitary state regional self-governance with special status
	Nomos Regionalizmus (2007)	ethnic region	unitary state South Upland Autonomous Area
Ukraine	Soym pídkarpats'kikh rusinív/Soim of Subcarpathian Rusyns (2000) Respublika Podkarpatskaya Rus'/ Subcarpathian Rusyn Republic (2008)	historical multicultural (ethnic) region	federal (independent?) state Subcarpathian special self-governing territory

Source: Author's compilation based on political programmes.

situation than the rest of the country. Usually there is a dominant minority in these multiethnic regions that leads the movements, but frequently a kind of territorial/regional identity develops and strengthens, and this becomes a new identity for the members of these groups. Regional identity might also appear in historical regions without a dominant minority. Due to the great variety of regionalisms, the aims of the individual movements and political parties differ; they endeavour to achieve their self-governing status on different grounds and in different state organisational structures. This also holds true where there are several organisations in the same area.

Next we are going to briefly review the main characteristics of the individual regions and their societies as well as the organisations active in these regions, their activities and their results achieved so far.

Istria, in the westernmost part of Croatia, is a multiethnic region having experienced several border changes. At the time of the Austro-Hungarian Monarchy it was a province subordinated directly to Vienna, then following the First World War it became part of Italy, and after the Second World War it was annexed to Yugoslavia (Urošević, 2011). Its final division between Italy and Yugoslavia was accomplished in 1954 (Ballinger, 1999). After gaining independence (1991), Croatia started to establish a centralised nation state (Beovic, 2013) as it wanted to ensure its independence through national unity (Cocco, 2010). The regional movement organised in opposition to Franjo Tuđman's centralisation and nation-building policy could build on Istrian regional identity which did not develop as a 'counter-identity', neither did it prevail over national identities (Urošević, 2011; Beovic, 2013); instead, it expressed strong bonds to the region (Cocco, 2010). Decentralisation announced by the new government in 2000 mitigated regionalism and regional identity temporarily, but then they again gained in strength due to insufficient reforms. Struggling for independent Istrian identity and establishing an autonomous Istrian region since its inception, the Istrian Democratic Assembly (*Istarski demokratski sabor*, IDS, 1990) is an important political factor in the county (*županija*); it is a permanent member of the county assembly despite having lost much of its initial support.[5] County assemblies have a wide range of functions and make their own statutes.[6] However, IDS claims autonomy for this region with an even wider sphere of authority and economic independence to be realised in a regionalised structure – either symmetric or asymmetric (Beovic, 2013).

Advocates of Moravian regionalism intend to constitute the autonomy of historic *Moravia* (and Silesia) in the eastern part of Czechia, reforming the Republic as a federal state consisting of three federal provinces, namely, Czechia, Moravia–Silesia and Prague. Abolished by the communists in 1949, the Moravian–Silesian Province was to revive promptly after the change of regime (Bureš and Just, 2010). This endeavour has remained dominant ever since the independent Czech Republic was established. Having had significant political support initially, the Moravian–Silesian parties were unable to enforce their claims for autonomy, because based on the experiences obtained from the disintegration of the former federal state, the government instituted a unitary state (Illner, 2002). This failure, complementing the 2000 public administration reform which disregarded the borders of provinces, led to a sharp drop in the voter base of these parties[7] and also to declining Moravian and Silesian identity which was strong at the time of the regime change. Later a new party, The Moravians, was formed (Moravané, 2005) which revived the old endeavours, and political support for this movement started to grow again. Due to political mobilisation, identity became stronger, but it lags far behind the peak at the time of transition. Despite having strengthened, the party has not become a significant political factor, so it is doubtful whether its programme, which intends to restore the system of provinces (*Obnova zemského zřízení*), and aims to fundamentally change the state structure and public administration, balancing the Moravian–Silesian and Czech territories, and providing sufficient resources for the province, will ever become a fundamental document for discussion in parliament. Moravians have kept the restoration of provinces on the agenda by initiating a referendum to be held in 2016, although they cannot expect sufficient support even in the territory of historical Moravia.

Upper Silesia is located in the southern part of Poland bordering Czechia. After the final division of Poland, Silesia became a border area of three large empires. Its largest part belonged to Prussia, later to the German Empire. After the First World War, the eastern part of Silesia, i.e. Upper Silesia, was annexed to the new Polish state and had autonomy between 1920 and 1939. Since the shift of borders following the Second World War almost the whole of Silesia has belonged to Poland. Established at the time of the regime change, the Silesian Autonomy Movement (*Ruch Autonomii Śląska*, RAŚ, 1990) aimed to restore the autonomy that existed between the two world wars, but the public administration reform executed in 1998, instead of unification, divided Upper Silesia between two voivodeships. RAŚ abandoned its earlier strategy to fight for special status and prepared a programme in 2010 which aimed at setting up an asymmetric regionalised state (Poland of Regions, *Polska Regionów*). The 2020 deadline it set turns out to be impossible to keep as it has not succeeded in gaining enough support for its aims in other areas of the country. RAŚ can use as its political base the group with Silesian identity, concentrated in Silesian Voivodeship, that has revived since the change of regime. The group of people intending to achieve recognition as an ethnic minority is much bigger than the number of those who vote for RAŚ, which has twice succeeded in winning mandates in the regional self-government and has thus become a regional political factor. It follows from the above that autonomy is not sufficiently supported in this voivodeship either.

Székely Land, situated at about the middle of Romania, has a majority Hungarian population. It became part of Romania as a consequence of the peace treaties closing the First World War. Its area is about the same as the total area of the Harghita, Covasna and Mureş counties together. The Hungarian Autonomous Region established in this area in 1952 on Stalin's explicit command naturally had only limited independence due to the main characteristics of the state socialist era. In 1960 the borders of this autonomous region were changed so that the ratio of those speaking Hungarian decreased from 77.35 per cent to 61 per cent within the population of the new Mureş-Hungarian Autonomous Region (Bottoni, 2008; Gulyás, 2010). Its self-governing status ceased to exist as a result of the 1968 public administration reform which restored the system of counties. Following the change of regime there were many ideas about how to create autonomy. Some of them related only to Székely Land, others to all of the Hungarians living in Romania.[8] Despite several documents, draft statutes and other draft laws made at different times, the cause of territorial autonomy did not advance. This was only partly due to the unchanged attitude of the Romanian state – the introduced statute was rejected without discussion by both the parliament and the senate. It should be remembered, too, that there is no Minority Act in Romania, and no public administration reform (regionalisation) has been carried out. In addition, Hungarians are politically divided,[9] which may threaten their parliamentary representation,[10] and this makes it very difficult to shape a common standpoint, as there are different attitudes, approaches and ideas. To make it worse, looking to the future, although the Hungarian/Székely population forms the majority in Székely Land taken as a whole (more precisely in the three counties), its ratio is decreasing, which in the long run might question one of the basic pillars of claiming autonomy.

Located in the northern part of Serbia, *Vojvodina* had belonged to several states before the First World War. After the war it became part of Yugoslavia and following the south Slavic wars it has belonged to Serbia. Due to historical reasons, it is multiethnic with a Serbian majority. Since 1945 it has had autonomy of various degrees. In 1974 Vojvodina and Kosovo were granted independence, with a sphere of authority very similar to that of the member republics. However, Milošević's nationalist-centralising movement put an end to this and, subsequent to the regional *coup d'état* in 1988, autonomy was reduced to a formality by the new constitution (Korhecz, 2009). In democratising Serbia, the so-called Omnibus Law in 2002 and the new constitution in 2006 laid the foundations for asymmetric autonomy of a certain degree, but the sphere of authority granted is far from the earlier one. Therefore, political forces in Vojvodina aim to restore the earlier level of independence. Vojvodinian endeavours were laid down in the new statutes on autonomy adopted by the Parliament of Vojvodina in 2009. However, they could not come into force up until now, as the situation has been greatly affected by Kosovo gaining independence in 2008. The League of Social Democrats of Vojvodina (*Liga Socijaldemokrata Vojvodine*, LSV, 1990), which regularly gained mandates (mostly in coalition) at provincial elections, aims to further deepen asymmetric decentralisation, while Vojvodina's Party (*Vojvođanska Partija*, VP, 2005), which unites several political movements, calls for changing the status of the province into a republic and establishing a federal state. Deepening autonomy is supported by the majority of all the ethnic groups in Vojvodina (even by the Serbs). This regional commitment, which prevails over national-ethnic identity, has strengthened Vojvodinian identity.

The territory of present-day Slovakia belonged to the Hungarian Kingdom (within the larger territorial unit called Upper Hungary). It became part of Czechoslovakia as a consequence of the post-First World War peace treaties and the international agreement after the Second World War which confirmed the 1920 boundaries of Hungary. In the southern part of this area, often called *Upland* (the former territory of one-time Upper Hungary), there lived and still live a significant number of Hungarians[11] who, however, do not form a concentrated ethnic majority. The change of regime and Slovakian independence activated the movements of Hungarians fighting for autonomy, and several political organisations were formed at that time. Related to the public administration reform planned for 1993, claims for considering the ethnicity principle were proclaimed and in the so-called Komarno draft (1994) proposals were made for delimiting one large or three smaller public administration units with Hungarian majority (Gulyás, 2008a, 2009). However, the 1996 public administration reform did not take into account the county delimitations proposed by the Hungarian parties and thus the areas inhabited by Hungarians were divided among several regions (*Krajs*) (Gulyás, 2008a). With this the doors closed for the organisations fighting for the recognition of 'associated nation' status (Duray, 1999), especially when the 2001 public administration reform did not change the regional borders (Gulyás, 2008a). When three parties merged[12] to form the Party of the Hungarian Coalition (MKP), which participated in both parliament and government, not even this could further the cause of autonomy. The political organisation MOST–HÍD (Bridge) (2009), formed by dissidents from the coalition, has abandoned the struggle for autonomy and has supported the idea of a multiethnic, multicultural, multilingual state (Bochsler and Szőcsik, 2013). Hungarians in Slovakia seem to prefer the ideas of MOST–HÍD, because since it was established MKP has lost its representation in parliament. Territorial autonomy is supported by the movement Nomos Regionalizmus – which was to be registered in 2007 – led by János Bósza. The draft of South Upland Autonomous Area – rejected even by MKP – was worked out on the basis of the Hodža–Bartha line drawn in 1918, albeit the ethnic borders have changed since and there is a Slovak majority in the delimited area (Gulyás, 2008b). It is doubtful whether the issue of autonomy in Slovakia will again enter the centre of public life for the following reasons: the possibility of gaining autonomy has been weakening in social consciousness; MKP, which has been fighting for autonomy, has been devalued; and the Hungarian population in Slovakia has been decreasing and is politically divided.

Subcarpathia lies in the westernmost oblast of Ukraine. Its historical development is different from that of the rest of the country, having been a periphery and having had a specific social and ethnic structure. The Rusyn movement here was formed at the time of gaining independence from the Soviet Union and aimed to achieve autonomous status within independent Ukraine. At the regional referendum, held simultaneously with the referendum about the independence of the country, 78 per cent of the people living in this area voted for Subcarpathia's self-governing administrative territory status[13] (Sasse, 2001; Kuzio, 2005). The government, however, disregarded this and included the territory in the unitary state.[14] Following unsuccessful attempts to gain autonomy, in 1996 Kiev passed a document entitled 'Action plan to resolve the Ukrainian–Rusyn problem'[15] (Požun, 2000) in which the Rusyns were stigmatised as a group threatening the sovereignty and territorial unity of the Ukrainian state. Dealing with the Rusyn problem with such a negative attitude broke down the Rusyn movements and significantly decreased the proportion of the population acknowledging their Rusyn identity. As a consequence of the minority policies of the Soviet Union and Ukraine, as well as of the Rusyns having been proclaimed Ukrainians, the Rusyns have lost their earlier majority, although their number is estimated to be between 650,000 and 800,000 (Sasse, 2001; Požun, 2000). Separatist and irredentist endeavours in this area were expressed in the following documents and ideas: the 2008 declaration of restoring Carpathian Ukraine abolished by its annexation in 1945 (Hajdú, 2009), the official establishment of the virtually existing Subcarpathian Rusyn Republic,[16] and the 2014 draft legislation on instituting an asymmetric federation. In recent years, the Rusyn movement expected the help of Russia in achieving wider autonomy similar to the one in the Minsk Agreement valid for the areas inhabited by Russians.

Main characteristics of the CEE (ethno-)regional movements

As seen above, the endeavours of the (ethno-)regional movements are wide-ranging, their chances of achieving their aims are different, as are their spheres of manoeuvring. Below we shall summarise the regional endeavours in CEE, paying special attention to both external factors, i.e. legislative obstacles in specific countries, and internal ones, namely those within the social–political base of the individual movements and parties.

Following the change of regimes, the CEE countries aimed primarily to establish and strengthen their independent states, for which they employed unitary state structures with an emphasis on the nation state. Despite this, there are significant differences among the concepts of nation as they occur in the constitutions. In the Constitution of Romania, for example, its nation state character is declared by stating: 'Romania is a sovereign, independent, unitary and indivisible National State' (Constituția 2003, Article 1), which is based on 'the unity of the Romanian people' (Article 4). In the Slovakian constitution we can find sovereignty practised together with the national minorities and ethnic groups, and there is similar phrasing in the Croatian, Hungarian, Polish, Serbian and Ukrainian constitutions.[17] We can also find examples of mentioning a state of citizens (instead of nation or peoples): 'We, the citizens of the Czech Republic in Bohemia, Moravia and Silesia' (Ústava 1992, Preambulum). As can be seen, with the exception of Romania, these states consider all of their citizens to be subjects of sovereignty; however, in their constitutions very little is said about the fundamental rights of national, ethnic minorities and regional groups, and only in general terms. Constitutions usually state their basic attitude towards minorities in highlighted places (e.g. Preambulum, general principles, fundamental rights). In some countries (e.g. Poland, Slovakia) it is placed in later parts. National, ethnic minorities are rarely enumerated in the introductory part of constitutions. The Croatian constitution is an exception, providing a long list of minorities in the introductory part. The language used by the majority is defined as the official language (except in Czechia), the use of minority languages are guaranteed in the constitutions, with the exception of Poland where only the development of languages is mentioned. Minority rights should be covered in detail by laws on minorities, but these were not considered to be among the most important tasks in the majority of the CEE countries. This is clearly expressed by their signing the Charter of Regional or Minority Languages rather late. Having significant Hungarian minority populations, Slovakia and Romania still have not passed laws on minorities, thereby generating several conflicts both with the minority organisations in their own country and with the mother country. Where such laws do not exist, minorities which have representatives in the Slovakian Council of the Government for National Minorities and Ethnic Groups[18] and the Romanian National Minority Council,[19] respectively, can be regarded as recognised. In the other CEE countries we can find laws on minorities, but there are significant differences among definitions, and their lists of recognised groups. The Ukrainian law emphasises the subjective element, namely, the declaration of identity. Accordingly, the law on minorities does not enlist the recognised groups, as the list of regional languages can be found in the law on languages.[20] The Serbian law also stresses minority rights and the free expression of identity, and lists the minorities in the context of minority language use.[21] The Czech law gives a detailed definition of minorities, but as proposed by the Government Council for National Minorities, the list of minorities is given in a government decree reviewed from time to time.[22] The Polish Act on National and Ethnic Minorities and on the Regional Language differentiates between national/ethnic minorities and regional languages and specifies which minority or regional groups are recognised in a given category.[23] The Croatian minority law neither defines minority, nor does it enlist minorities, as it is based on the Croatian constitution that names them.

Among the examples above those frameworks in which regionalism is based – at least partly – on national/ethnical identity can be considered relevant for recognising minorities and safeguarding their individual or collective rights. As seen in the review above, states with a significant Hungarian population have the most negative attitude towards minorities; minority rights in these countries are not formulated in unified legislation. This attitude has hardly changed in the last 25 years, which is due partly to the majority population still fearing 'medium-sized' minorities, and partly because minorities have remained very weak in asserting their interests politically (Halász, 2007). Groups with (also) an ethnic component to their identity (like Moravians, Rusyns, Silesians) are not in a better position either, as they have been mostly dealing with making their identity recognised, which has not brought any results.[24] In the regionalism of multiethnic areas (Istria, Vojvodina)[25] minority issues are less important. Regional identity here does not have a minority character: instead, it expresses multiculturalism and identification with the region.

From the aspect of how successful regionalism can be, the character and the structure of the state as laid down in the constitution are also important. Most of the CEE countries declare their state as unitary and indivisible in order to safeguard their territorial integrity. In spite of this it may happen that special regulations relate to the whole country (like in Spain) or part of it (e.g. Vojvodina, or earlier Kosovo). Based on Western European experiences, regime changes and reforms to the state structure, as well as ensuing public administration reforms and regionalisation itself, may create circumstances favourable for satisfying the claims of (ethno-)regional movements. As mentioned earlier, and referred to in the mini case studies, public administration reforms in the CEE countries are rather problematic, as the territorial structures are rarely fitted to the ethnic or historical boundaries either inherited (like Székely Land) or created (e.g. South Upland, Upper Silesia, Moravia). At places where public administrative and historical structures (almost) coincide (Istria,[26] Subcarpathia,[27] Vojvodina[28]), the degree of decentralisation is insufficient. In the first case, the territorial–public administrative structure should also have been changed to help realise regional endeavours. In this respect the level of legislation regulating the territorial units is important. Czechia has the most rigid system, since the upper medium-level units (*krajs*) are protected by the constitution, as are the regions by enlisting their names and giving their delimited area. The sphere of manoeuvring is somewhat bigger in Romania where although the levels of public administration units are given in the constitution, the counties (*judeţs*) constituting the medium level are enlisted in a 1968 law and its amendments. The Polish and the Slovakian territorial–public administrative systems are the easiest to change in principle, because the medium-level units are not even mentioned in their constitutions. As stipulated in these constitutions,[29] both the lower and the upper medium levels (*powiat* and *województwo*, and *okres* and *kraj*, respectively) are defined and delimited in statutes.

As can be seen from this short review, the prospects for regionalism in the CEE countries are different from a legal aspect. However, looking at the governments' intentions and the political atmosphere, we can state that regionalism has little hope in these countries. Having the status of autonomy, Vojvodina is undoubtedly in the best situation. However, Serbia is not likely to widen independence which might threaten its integrity with a further shrinking of the territory of the country. It is no accident that a significant part of the (ethno-)regional movements are fighting to reform the whole state structure instead of establishing territorial autonomy. Placing the public administration system of a classical or decentralised unitary state on a regionalised or federal basis greatly reduces fears of autonomy and this might serve as a viable solution for modernisation. The proposed state structures are of course different – just like in Western Europe – depending on traditions and possibilities. A federal state is aimed at in Czechia, Serbia and Ukraine, while there are endeavours to create symmetric or asymmetric regionalised states in Croatia, Poland and in Romania (e.g. the proposal of the Hungarian People's Party of Transylvania). Although it is easier to make territorial autonomy accepted by both government and society if it is realised alongside a comprehensive reform, not even such proposals have been able to gain mass support. The reformists still endeavour to achieve their classical aims (in Romania: see proposals by the Democratic Alliance of Hungarians in Romania, the Hungarian Civic Party and the Székely National Council; and there are similar proposals in Slovakia and Ukraine). The only aim of these draft proposals is to grant these areas special status. In terms of spheres of authority transferred to the self-governing territorial units, there are no big differences among the various ideas. Spheres of authority can be restricted not only by the exclusive rights of the state (these usually relate to tasks indispensable for preserving the integrity of the state), but also by power wielded simultaneously or jointly by the region and the state. Irrespective of the type of their ideas, these organisations emphasise that (ethno-)regionalism does not intend to change the existing state structure. However, separatism can be found in the rhetoric of several movements from time to time, sometimes in drafts made public (e.g. in Vojvodina earlier) or documents (like in Subcarpathia in 2008), and at other times in declarations in a covert form. Either way, these are a source of tension (as was the case earlier in Upper Silesia).

Regional movements support their ideas of creating autonomy with various motives and reasons. Referring to ethnic majority based on census data holds only for Székely Land; this reason is mentioned most of the time in relation to introducing territorial autonomy. Similar reasoning is attempted by Rusyns, not recognised as a minority, whose population according to census is not significant even at regional level; their clout to enforce their interests is rather restricted. Emphasising the legal continuity of the autonomy[30] they had earlier, mostly between the two world wars, is another reason employed. Several organisations have demanded the restoration of independence on this basis (in Moravia, Subcarpathia, Upper Silesia). Vojvodina also belongs here, as it aims to retrieve the spheres of authority it had under the state socialist era. The experiences of Western European movements show that the economic situation of a region and its development or its underdevelopment might as well serve as a basis for making claims. Such economic reasoning is also used in the CEE region; organisations often criticise the redistribution by the state both in the relatively rich (Istria, Upper Silesia, Vojvodina) and in the relatively poor (Moravia) regions. The aim is the same: the region wants to have a greater part of the locally generated taxes at its disposal. Only the reasoning is different: the rich regions do not want to play too large a role in mitigating territorial inequalities, while the poor regions think that their taxes contribute to the further growth of the more developed parts of the country.

Regional identities based on historical–economic–social specificities and regional commitment might also serve as bases for regionalism. Actually, the principal basis of these organisations consists of groups with national, ethnic or regional identity. Since the regime changes censuses have again paid attention to measuring identity; in several countries it is possible to declare double bonds. In the course of transition when new structures were being shaped, the number of those declaring ethnic and regional identity was significant (Moravians and Istrians). However, due to the lack of expected results and the failure of regional reforms, their number decreased considerably by the next census. In the last decade there was a renaissance of ethnic/regional identity in CEE, significantly increasing the number and the ratio of these groups within their regions (Table 12.2). In spite of this, the 'critical mass' has not been reached. At the same time, national/ethnic identity has been declining in the two areas inhabited by Hungarians. This might be the result of demographic processes (population decrease, migration) as well as assimilation and the 'absence of identity'.[31]

Nationwide or regional election results (Table 12.3) show a more precise picture of the real support for these movements than does the size of groups with national, ethnic or regional identity. Organisations based on ethnic or regional identity usually wield their political weight at the local and/or regional elections where most of them have been successful recently. The Istrian Democratic Assembly (IDS) has been the strongest group in the county assembly since the beginning, irrespective of having lost votes despite having taken part in the last two elections in coalition. The League of Social Democrats of Vojvodina (LSV), which was also in coalition at earlier elections, participated independently in the 2012 elections and it achieved better results, gaining 10 mandates in the Assembly of the Autonomous Province of Vojvodina. The Silesian Autonomy Movement (RAŚ) had its first success in the 2010 regional self-government electionS.

Table 12.2 Size of population having regional or ethnic identity per areas

Identity and area	Former census (2001 or 2002)		Last census (2011)	
	population (No.)	population (%)	population (No.)	population (%)
Istrian (Istria)	8,865	4.3	25,203	12.1
Hungarian (South Upland)	454,186	28.8	401,948	25.3
Moravian (Moravia and Silesia)	371,345	8.0	512,248	11.2
Rusyn (Subcarpathia)	10,069	0.8	–	–
Székely/Hungarian (Székely Land)	668,471	59.2	609,033	56.8
Silesian (Upper Silesia)	172,743	3.0	828,518	14.7
Vojvodinian (Vojvodina)	10,154	0.5	28,567	1.48

Source: Author's compilation based on census data.

Note: The size of the groups is given according to the territorial units stated by the regional organisations, but in some cases such data were not available. In the case of Istria, Subcarpathia and Vojvodina we worked with the existing public administration units. South Upland here contains the districts mentioned in the draft on autonomy, also those (e.g. Nitra) the greater part of which is not considered to belong to the region. Moravia and Silesia here mean the whole of six districts (South Morva, Zlín, Morva-Silesian, Olomouc, Pardubice, Vysočina), whereas Székely Land means the whole of three counties (Harghita, Covasna, Mureş), finally Upper Silesia means here the Opole and the Silesian voivodeships.

Table 12.3 Regional support of (ethno-)regional organisations at the last elections*

(Ethno-)regional organisation	Number of votes	Proportion of votes (%)
Moravians (2012)* (Moravia and Silesia)	14,772	0.56
Ilstrian Democratic Assembly (2013)* (Istrian)	35,969	43.93
Silesian Autonomy Movement (2014)* (Upper Silesia)	97,131	7.20
League of Social Democrats of Vojvodina (2012)* (Vojvodina)	111,397	11.00
Democratic Alliance of Hungarians in Romania (2012)** (Székely Land)	189,613	49.90
Hungarian People's Party of Transylvania (2012)** (Székely Land)	33,491	8.80
Party of the Hungarian Coalition (2012)** (South Upland)	98,800	14.21
MOST–HÍD (Bridge) (2012)** (South Upland)	112,145	16.13

Source: Author's compilation based on election results.

Notes: The size of the groups is given according to the territorial units stated by the regional organisations, but in some cases such data were not available. In the case of Istria, Subcarpathia and Vojvodina we worked with the existing public administration units. South Upland here contains the districts mentioned in the draft on autonomy, also those (e.g. Nitra) the greater part of which is not considered to belong to the region. Moravia and Silesia here mean the whole of six districts (South Morva, Zlín, Morva-Silesian, Olomouc, Pardubice, Vysočina), whereas Székely Land means the whole of three counties (Harghita, Covasna, Mureș), finally Upper Silesia means here the Opole and the Silesian voivodeships.
The table contains the results of the first round in the case of two-round elections and the results of the lower-house elections in the case of parliaments with two chambers.
* Regional elections.
** Parliamentary (lower-house) election.

In 2014 it won four mandates in the Silesian Voivodeship Assembly (*sejmik*) despite getting fewer votes than earlier. In 2014 it also succeeded in putting up a candidate again in neighbouring Opole Voivodeship after 2006. Although it takes part in the competition at European, national and regional levels alike, Moravané has only been able to succeed at the local level. In the 2014 local self-government elections[32] a total of 72 of their candidates won mandates in the territory of Czechia supported by Moravians and sometimes also by some other social organisations.

We found it more important to analyse the national elections than those at local, county or district levels in the case of national/ethnic political organisations, because it is parliamentary representation that is of primary importance for these communities. However, the internal conflicts of the mass parties like the Party of the Hungarian Coalition (MKP) or the Democratic Alliance of Hungarians in Romania (RMDSZ) and the resulting fractures, as well as the creation of an opposition to the parties representing Hungarians threaten the effective assertion of interests in both Romania and Slovakia. Székely Land exemplifies how the Hungarian People's Party of Transylvania drew a significant number of votes from RMDSZ which had 59.8 per cent of the votes in the 2008 national election. Rivalry between these two parties for Hungarian votes at the national level led to a situation whereby it became doubtful for a while whether RMDSZ could reach the 5 per cent threshold to get directly into parliament. Compared to earlier results (6.8, 6.2 and 6.17 per cent), the votes won by RMDSZ at the last election (5.13 per cent) are very telling, especially when we look at the votes won by the Hungarian People's Party of Transylvania (0.64 per cent) which was worthless from the point of view of parliamentary mandates. The Slovakian Party of the Hungarian Coalition has already had the bitter experience of losing parliamentary representation. The earlier support it had (11.2 per cent and 11.7 per cent) dropped to less than 5 per cent (4.3 per cent in both 2010 and 2012) after MOST–HÍD (Bridge) entered the political arena, and it is clear that it was the organisation having got into parliament that took the Hungarian votes away from the coalition. The Party of the Hungarian Coalition has also seen its popularity decline in South Slovakia, where instead of the earlier 34.5 per cent, only 14.2 per cent voted for the coalition at the last elections, and it is very probable that MOST–HÍD won many of these votes.

Summary

CEE organisations could draw lessons from the practices of Western European (ethno-)regional movements. Territorial autonomies have proven to be an effective means to settle conflicts between majority and minority groups and to strengthen the territorial integrity of states. True, a long and troublesome path led to these autonomies. There has been no uniform pattern to follow to succeed, but some circumstances offering opportunities for (ethno-)regionalism can be identified. One of them was the regime change entailing deep and necessary multifarious changes which created favourable conditions for establishing new territorial–power structures that also involved (ethno-)regional claims (cf. the case of Spain). Another one consists of EU principles, like decentralisation, subsidiarity and, as a consequence, regionalisation, which also offered opportunities for regionalisms. Public administration reforms brought about the frameworks that could be utilised to gain real independence as claimed by the movements. We could mention Belgium here with its regions and communities, but South Tyrol, which was granted real autonomy only as a result of Italian regionalisation, also belongs here.

CEE has undergone a completely different development process,[33] and even in the course of transition it had problems and challenges significantly different from those in Spain or Portugal. The disintegration of federal states (Czechoslovakia, Soviet Union, Yugoslavia) into nation states – not always in a peaceful way – gave rise to fears in these countries, which determined the attitude to the establishment of territorial units with wide-scale independence. The almost traditional distrust among some nations also worked against regionalisation, which was clearly discernible in policies relating to minorities with significant populations. The intention to create an independent nation state went against the endeavours of the minority and (ethno-)regional movements, which became clear from both the stipulations of the important laws and the slow process of decentralising state power. The necessity of adapting to the European Union has finally led to public administration reforms (Czechia, Poland, Slovakia); however, the delimitation of self-governing units corresponding to NUTS-2 regions was only carried out in Poland. Together with all its problems in practice, this step taken by Poland made it the leading country of CEE. The delimitation of statistical regions in Croatia, Romania and Serbia, and the proposal for structural change in Ukraine based on the Polish pattern should also be mentioned here. These changes raised the issue of forming an upper medium level of self-governance, but this is still awaiting realisation. With the failure of the 2013 Romanian public administration reform, the period of large structural changes seems to have come to an end. The CEE countries are past the first wave of regionalisation: the planning–statistical units have been delimited and where the state considered it important, it granted spheres of authority to the respective levels. In addition to political intentions, new impulses would also be needed to resume this course, but EU processes do not move in this direction. The Europe of Regions principle, which served as a reference point for (ethno-)regional movements, has been losing ground in the 2000s (Pálné Kovács, 2007) and thus the role of regions has been getting devalued, perhaps irreversibly (this matter and its consequences are further discussed in Chapter 19). Handling the world economic crisis and providing help to the self-governments in trouble entail centralisation and the widening of state functions, which greatly rearranges the competencies and the financing of medium-level institutions (Pálné Kovács, 2014). These events have made the circumstances rather difficult for the CEE (ethno-) regional movements and parties, many of which (Moravané, RAŚ and in observer status the Hungarian People's Party of Transylvania) are members of the European Free Alliance which represents the Europe of Regions movement.

The West European autonomous regions have been considered as benchmarks and have been a reference point for the CEE movements. The crisis of these autonomous regions further decreases the chances of the movements in CEE. Separatist endeavours claiming wider self-governance (Flanders), independence referenda (Scotland) or referendum attempts (Catalonia) might make the CEE states even more mistrustful of territorial autonomy. Even more so, as this mistrust has been strengthened by the process leading to an independent Kosovo (2008), by the Crimean crisis (2014) and the ongoing Eastern Ukrainian conflict.

Irrespective of being based on national, ethnic or regional identity, the CEE (ethno-)regional movements, as they are, have little chance to realise their aims. The minority policies in these countries and their defence of state integrity above all provide no opportunity for starting substantial discussions about autonomy or widening the present limited sovereignty of the respective regions. Some of the organisations are weak and can only play a political role at the local–regional level. They are unable to assert their interests on a national scale. Some others have enough social–political support, but because of their internal conflicts they cannot represent the issue of territorial autonomy strongly and uniformly at the central government level. Hence, the reaction of governments to (ethno-)regional endeavours cannot be other than belittling the claims of these movements or, after a successful action, declaring it perilous, while the majority in these societies mostly ignores the claims of minority or regional groups.

Acknowledgements

The publication of this research has been supported by project #104985 of the Hungarian National Research, Development and Innovation Office. While writing this paper, Nóra Baranyai was supported by the Pál Erdős Research Scholarship for Young Researchers.

Notes

1 This autonomous region was established on Stalin's command as a means of political window dressing. It existed between 1950 and 1960, when its borders were redrawn, its rights curtailed, and it was renamed the Maros–Hungarian Autonomous Region. This region was entirely dismantled in 1968.
2 The emergence and strengthening of regional identities might have been helped in some areas by state socialist public administration units and development policies, too (Illés, 2002).
3 Created as a consequence of the Dayton Peace Agreement, concluded in order to stop armed ethnic conflicts, Bosnia-Herzegovina – with a relative Bosnian majority – can at the same time be characterised by having highly independent ethnic units, a weak administration system and many centrifugal forces able to burst the state (Reményi, 2012).

4 Ethnic minorities living in Hungary are not populous, they are scattered and their linguistic assimilation is advanced (Dobos, 2015). The most populous and politically the most active, Gypsies are extremely divided and the parties representing them have not been able to become important political factors due partly to the electoral laws, and partly to the dominance of the ethnic minority self-governing system.
5 At the 2015 Croatian parliamentary election, IDS (in coalition) was given 19.77 per cent of the votes in Istria, thus gaining three parliamentary mandates.
6 For example, it is stipulated in Istria that the Italian minority must have at least one representative in the assembly.
7 The Moravian and Silesian parties have become insignificant, which was also affected both by their internal conflicts resulting from their failure, and the abandonment of their mass party character.
8 The three-tier autonomy model of the Democratic Alliance of Hungarians in Romania should be mentioned here. This model proposes different solutions for the Hungarians living in different situations (those who are scattered, those who are the majority in that area, and those who live in one territorial block). Without naming it, the model separates territorial autonomy for Székely Land, personal autonomy for all Hungarians living in Transylvania, and special status for certain settlements not clarified in detail.
9 Started as a mass party, the Democratic Alliance of Hungarians in Romania now has several rivals in the Romanian political arena. Having seceded from the Democratic Alliance of Hungarians in Romania, the Hungarian National Council of Transylvania later turned into a political party under the name of Hungarian People's Party of Transylvania. There exists a Hungarian Civic Party, as well. Established specifically to achieve autonomy for Székely Land, the Székely National Council must also be taken into account when dealing with these issues.
10 The electoral threshold is 5 per cent for parties, but the alternative threshold (earlier 6 won mandates in the House of Deputies and 3 in the Senate, according to the new electoral law (2015) 20 per cent of the votes at least in four counties) allows regional parties to enter the Parliament.
11 The Hungarian population here is still significant in spite of its systematic expatriation in the first period of the state socialist era (see, for example, the Beneš decrees).
12 The Coexistence Political Movement (1990), the Hungarian Christian Democratic Movement (1990) and the Hungarian Civic Party (1992).
13 There were several ideas about how to realise autonomy. There was a declaration of restoring the autonomous republic; negotiations started with Czechoslovakia about accessing to the federal state as an autonomous territory; there were endeavours to establish the Subcarpathian Rusyn Republic; steps were taken toward joining the Commonwealth of Independent States (Belitser, n.d.).
14 At the same time the Crimean Autonomous Republic was established and given wide-ranging autonomy. (Now it belongs to Russia.)
15 The document was approved by the State Committee of Ukraine on Nationalities and Migration.
16 The 2nd European Congress of Rusyns issued an ultimatum for the Ukrainian state leadership, but the demanded territorial autonomy for the Rusyns was not granted by the deadline of 1 December 2008, so they 'established' their independent state with its draft constitution, own budget and government.
17 'We, the Slovak Nation ... together with members of national minorities and ethnic groups living in the Slovak Republic ... we, the citizens of the Slovak Republic' (Ústava 1992, Preambulum). 'The Republic of Croatia is hereby established as the nation state of the Croatian nation and the state of the members of its national minorities' (Ustav, 2001, Part 1). 'The national and ethnic minorities living in the Republic of Hungary participate in the sovereign power of the people: they represent a constituent part of the State' (Alkotmány 1989, Art. 68.) or 'We proclaim that the nationalities living with us form part of the Hungarian political community and are constituent parts of the State' (Alaptörvény, 2011, National Avowal). 'We, the Polish Nation – all citizens of the Republic…' (Konstytucja, 1997, Preambulum). The 'Republic of Serbia is a state of Serbian people and all citizens who live in it' (Устав 2006, Art. 1). 'Ukrainian people – citizens of Ukraine of all ethnicities' (Конституція 1996, Preambulum).
18 It is a consulting body for the government (Halász, 2007). Based on this, minorities in Slovakia are as follows: Hungarians, Czechs, Romas, Rusyns, Ukrainians, Croatians, Germans, Jews, Bulgarians, Poles, Moravians and Russians.
19 Being the federation of the national/ethnic minority organisations, this council is an advisory body of the Department for the Protection of Minorities. According to this, minorities in Romania are: Hungarians, Germans, Romas, Serbs, Armenians, Tatars, Turks, Ukrainians, Ruthenes/Rusyns, Poles, Czechs, Slovaks, Bulgarians, Russian Lipovans, Jews, Croats, Macedonians, Albanians, Greeks and Italians (Jánosi, 2008).

20 This law mentions the following regional or minority languages: Russian, Belorussian, Bulgarian, Armenian, Gagauz, Yiddish, Crimean Tatar, Moldavian, German, modern Greek, Polish, Roma, Slovakian, Hungarian, Rusyn, Karaim and Krymchak (Закон 2012 Art. 7.2).
21 In addition to the official language of the country, Croatian, Romanian, Ruthenian, Hungarian, Slovakian and Czech can be used in Vojvodina, while Bosnian, Albanian and Bulgarian in the central part of Serbia (Gojkovic, 2008).
22 The government decree approved on 3 July 2013 mentions Polish, German, Roma, Slovakian, Belarusian, Bulgarian, Greek, Croatian, Hungarian, Rusyn, Russian, Serbian, Ukrainian and Vietnamese minorities.
23 Belarusians, Czechs, Lithuanians, Germans belong to the group of national minorities; Russians, Armenians, Slovaks, Ukrainians and Jews to minorities; and Karaims, Lemkos, Romas and Tatars to ethnic minorities. In addition, there is one regional language, namely, Kashubian.
24 Rusyns have been acknowledged as a native nationality in Subcarpathia since 2008. However, this status was awarded by the Subcarpathian County Council and they have been unable to achieve recognition at the national level ever since.
25 Basically Moravia and Upper Silesia also belong here, albeit in these areas two types of movements – the ethnic–linguistic and the autonomist–regionalist ones – are interconnected at several points.
26 The Croatian constitution merely specifies that territorial self-governments can only be established at the level of counties (Ustav 2001, Art. 134). According to the respective law, there are 20 counties in Croatia, one of which is Istria (*Istarska županija*) (Zakon, 2006, Art. 3).
27 The definition of territorial levels in Ukraine is given in its constitution where the upper medium-level territorial units are delimited and enlisted; one of these is Subcarpathia Oblast (*Zakarpattia oblast*) (Конституція 1996, Art. 133).
28 The Serbian constitution stipulates that 'State power is restricted by the right of citizens to provincial autonomy and local self-government' (Устав 2006, Art. 12). Later it names the Autonomous Province of Vojvodina (*Аутономна Покрајина Војводина*) and the Autonomous Province of Kosovo–Metohija as autonomous territorial communities (Art. 182).
29 The Polish constitution formulates it like this: 'Other units of regional and/or local government shall be specified by statute' (Konstytucja, 1997, Art. 164.2). In Slovakia 'The self-administration of higher territorial units and their bodies will be established by law' (Ustava, 1992, Art. 64.3).
30 From among the special statuses of modern times, we can only speak of 'classical', really independent territorial autonomy in the case of Silesia.
31 At the last census significant masses of people claimed no identity: 2,742,669 in Czechia and 1,236,810 in Romania.
32 Regional and local elections are not held at the same time.
33 According to Illés (2002), a significant role was played by feudalism in the differences in regionalisms as well as in the development and strengthening of regional identities. Feudalism developed much later in CEE than in the western part of the continent, and in the eastern part of Europe the dominance of the central power was characteristic instead of territorial bonds.

References

Ballinger, P., 1999. The Politics of the Past: Redefining Insecurity along the 'World's Most Open Border'. In: J. Weldes, M. Laffey, H. Gusterson and R. Duvall, eds. *Cultures of Insecurity. States, Communities, and the Production of Danger*. Minneapolis: University of Minnesota Press, pp. 63–90.

Belitser, N., n.d. Political and Ethno-Cultural Aspects of the Rusyns' Problem: A Ukrainian Perspective. Kyiv: Pylyp Orlyk Institute for Democracy.

Beovic, D., 2013. Europe's Rising Regionalism and the Quest for Autonomy: The Case of 'Istrian Identity'. Budapest: Central European University. MA Thesis.

Bochsler, D. and Szőcsik, E., 2013. The Forbidden Fruit of Federalism. Evidence from Romania and Slovakia. *West European Politics*, 36(2), pp. 426–446.

Bottoni, S., 2008. *Sztálin a székelyeknél. A Magyar Autonóm Tartomány története (1952–1960)* [Stalin at the Székelys. History of the Hungarian Autonomous Province, 1952–1960). Csíkszereda: Pro-Print Könyvkiadó.

Bureš, J. and Just, P., 2010. The Origin of the Czech and Slovak Pluralist Party System. *Politics in Central Europe*, 6(1), pp. 41–82.

Cocco, E., 2010. Borderland Mimicry: Imperial Legacies, National Stands and Regional Identity in Croatian Istria after the Nineties. *Narodna Umjetnosc*, 47(1), pp. 7–28.

Curtis, G.E., ed., 1992. *Bulgaria: A Country Study*. Washington, DC: GPO for the Library of Congress. http://countrystudies.us/bulgaria/ [Accessed 6 November 2015].

Dobos, B., 2015. *A szakpolitikai célok meghatározásának problémája a magyarországi kisebbségpolitikában* [Problems of determining the objectives of minority policies in Hungary]. Budapest: Magyar Tudományos Akadémia (MTA Law Working Papers, 10).

Duray, M., 1999. *Önrendelkezési kísérleteink* [Our attempts at autonomy]. Somorja: Méry Ratio.

Gojkovic, N., 2008. System of Minorities' Protection in Serbia. www.kas.de/upload/auslandshomepages/serbien/Gojkovic_en.pdf [Accessed 31 August 2015].

Gulyás, L., 2005. *Két régió – a Felvidék és a Vajdaság – sorsa az Osztrák–Magyar Monarchiától napjainkig* [Fate of two regions – Upland and Vojvodina – from the Austro–Hungarian Monarchy up to now]. Pécs: Hazai Térségfejlesztő Rt.

Gulyás, L., 2008a. Regionalizációs törekvések és etnoregionalizmus a poszt-kommunista Szlovákiában 1989–1998 [Regionalisation attempts and Ethno-regionalism in Post-communist Slovak Republic 1989–1998]. *Tér és Társadalom*, 22(4), pp. 189–204.

Gulyás, L., 2008b. A szlovákiai magyar kisebbség újabb etnoregionalista kísérlete: a Bósza-tervezet 2008 [Another ethno-regionalist attempt of the Hungarian minority in Slovakia: the Bósza draft proposal, 2008]. *Közép-európai Közlemények*, 1(1) pp. 110–114.

Gulyás, L., 2009. Regionalizáció, területi reformok és közigazgatási térfelosztás Közép-Európában I. Csehszlovák és szlovák tanulságok [Regionalisation, territorial reforms and public administrational delimitations in Central Europe 1. Czechoslovakian and Slovakian lessons]. *Közép-európai Közlemények*, 2009, 2(1), pp. 86–92.

Gulyás, L., 2010. A tartományi rendszer és a nemzeti kérdés Romániában 1950–1968 [Province-sytem and National-question in Romania 1950–1968]. *Tér és Társadalom*, 24(3), pp. 163–176.

Hajdú, Z., 2009. Kárpátalja területi autonómiája: alapvető államstruktúra- és határváltozások, de a kérdés örök? [Territorial autonomy of Subcarpathia: important changes in state structure and borders, but does the question remain?] In: L. Kupa, ed. *Kisebbségi autonómia-törekvések Közép-Európában a múltban és a jelenben*. Pécs: Pécsi Tudományegyetem, Bookmaster Kft., pp. 208–215.

Halász, I., 2007. Nemzeti és etnikai kisebbségek és a kisebbségi jogi szabályozások Közép-Európában [National and ethnic minorities and legal regulations of minority rights in Central Europe]. *Kisebbségkutatás*, 16(3), pp. 563–585.

Horváth, Gy., 2006. Regionális helyzetkép a Kárpát-medencéről [Regional snapshot of the Carpathian Basin]. In: Sz. Rácz, ed. *Regionális átalakulás a Kárpát-medencében*. Pécs: Magyar Regionális Tudományi Társaság, pp. 9–22.

Illés, I., 2002. *Közép- és Délkelet-Európa az ezredfordulón. Átalakulás, integráció, régiók* [South and Southeast Europe at the turn of the millennium. Transformation, integration, regions]. Budapest–Pécs: Dialóg Campus Kiadó.

Illner, M., 2002. Multilevel Government in Three East Central European Candidate Countries and Its Reforms after 1989. European University Institute. (EU Working Papers, 7).

Jánosi, D., 2008. A romániai nemzeti kisebbségek védelmének intézményi kerete [Institutional framework of defending national minorities in Romania]. *Magyar Kisebbség*, 13(3–4). pp. 167–190.

Korhecz, T., 2009. A vajdasági tartományi autonómia és nemzetiségi önkormányzat [Autonomy and ethnic self-governance of the province of Vojvodina]. In: L. Szarka, B. Vizi, N. Tóth, Z. Kántor and B. Majtényi, eds. *Etnopolitikai modellek a gyakorlatban*. Budapest: Gondolat, pp. 172–190.

Kuzio, T., 2005. The Rusyn Question in Ukraine: Sorting out Fact from Fiction. *Canadian Review of Studies in Nationalism*, 32, pp. 1–15.

Pálné Kovács, I., 2000. Régiók Magyarországa: utópia vagy ultimátum? [Hungary of regions: Utopia or ultimatum?] In: Gy. Horváth and J. Rechnitzer, eds. *Magyarország területi szerkezete és folyamatai az ezredfordulón*. Pécs: MTA Regionális Kutatások Központja, pp. 73–92.

Pálné Kovács, I., 2007. Magyar területi reform és az uniós fejlesztéspolitika [Hungarian territorial reform and EU cohesion policy]. *Magyar Tudomány*, 168(10). pp. 1306–1315.

Pálné Kovács, I., 2009. A területi érdekérvényesítés átalakuló mechanizmusai [Changing mechanism of asserting territorial interests]. In: I. Lengyel and J. Rechnitzer, eds. *A regionális tudomány két évtizede Magyarországon*. Budapest: Akadémiai Kiadó, pp. 149–172.

Pálné Kovács, I., 2014. *Jó kormányzás és decentralizáció* [Good governance and decentralisation]. Budapest: Magyar Tudományos Akadémia. (Székfoglaló előadások a Magyar Tudományos Akadémián).

Pap, N. and Végh, A., 2009. A kelet-macedóniai albánok autonómia-törekvései és azok etnikai-földrajzi háttere [Autonomy endeavours of East Macedonian Albanians and their ethnic-geographic background].

In: L. Kupa, ed. *Kisebbségi autonómia-törekvések Közép-Európában – a múltban és a jelenben*. Pécs: Pécsi Tudományegyetem, Bookmaster Kft., pp. 139–147.

Požun, B.J., 2000. Multi-Ethnic Outpost. *Central European Review*, 2(40). www.ce-review.org/00/40/pozun40.html [Accessed 12 May 2014].

Reményi, P., 2012. The Statehood of Bosnia-Herzegovina according to the Hartshorne Model. *Historia Actual Online*, 2012/4, pp. 129–140. www.historia-actual.org/Publicaciones/index.php/haol/article/viewFile/677/642 [Accessed 11 March 2015].

Sasse, G., 2001. The 'New' Ukraine: A State of Regions. *Regional & Federal Studies*, 11(3), pp. 69–100.

Urošević, N., 2011. The Role of Culture in Creation of Regional and a Common European Cultural Identity – Istrian Case Study. In: M. Frederiksson, ed. *Current Issues in European Cultural Studies*. Norrköping: Linköping University, pp. 147–158.

Croatia

Constitutional Law on the Rights of National Minorities in the Republic of Croatia, 2002 (Ustavni zakon o pravima nacionalnih manjina)
www.vsrh.hr/CustomPages/Static/HRV/Files/Legislation__Constitutional-Law-on-the-Rights-NM.pdf [Accessed 9 June 2015].

Law on counties', cities' and communities' territories in the Republic of Croatia, 2006 (Zakon o područjima županija, gradova i općina u Republici Hrvatskoj)
www.propisi.hr/print.php?id=5006 / [Accessed 9 June 2015].

Law on Local and Regional Self-government, 2001 (Zakon o lokalnoj i područnoj (regionalnoj) samoupravi)
www.legislationline.org/documents/action/popup/id/5864 [Accessed 3 June 2015].

Statutes of the Istrian Region, 2003 (Statut Istarske županije)
www.istra-istria.hr/index.php?id=587 [Accessed 3 June 2015].

Programme declaration of the Istrian Democratic Assembly, 1991 (Programska deklaracija Istarskog Demokratskog Sabora)
www.ids-ddi.com/ids-ddi/dokumenti/programska-deklaracija/ [Accessed 2 June 2015].

Declaration of the autonomous Istria County, 1994 (Deklaraciju o Autonomnoj Županiji Istarskoj)
www.ids-ddi.com/ids-ddi/dokumenti/rovinjske-deklaracije/ [Accessed 16 June 2015].

Ustav 2001 = Constitution of the Republic of Croatia (Ustav Republike Hrvatske)
www.constitution.org/cons/croatia.htm [Accessed 31 August 2015].

Czechia

Ústava 1992 = Constitution of the Czech Republic (Ústava České republiky)
www.ilo.org/dyn/travail/docs/1967/Constitution%20of%20the%20Czech%20Republic.pdf [Accessed 24 March 2014].

Charter of fundamental rights and freedoms, 1993 (Listina základních práv a svobod)
www.usoud.cz/en/charter-of-fundamental-rights-and-freedoms/ [Accessed 16 October 2014].

Constitutional Act on the Creation of Higher Territorial Self-Governing Units and on Amendments to Constitutional Act of the Czech National Council, No. 1/1993 Sb., the Constitution of the Czech Republic, 1997 (Zákon o vytvoření vyšších územních samosprávných celků a o změně ústavního zákona České národní rady č. 1/1993 Sb., Ústava České republiky)
www.usoud.cz/fileadmin/user_upload/ustavni_soud_www/Other%20constitutional%20acts/347_1997.pdf [Accessed 14 April 2014].

Statutes of the National Minority Council, 2001 (Statut Rady vlády pro národnostní menšiny)
www.vlada.cz/assets/ppov/rnm/130703_statut_usneseni_530.pdf [Accessed 24 March 2014].

Statutes of Moravané/The Moravians, 2009 (Stanovy Strany Moravané)
http://moravane.cz/o-nas/dokumenty [Accessed 24 March 2014].

The Programme of Moravané/The Moravians (no date) = Restoration of the system of provinces (Obnova zemského zřízení)
www.moravane.cz/files/download/9517fd0bf8faa65 [Accessed 24 March 2014].

Law on the rights of national minorities and on amendments to Acts, 2001 (Zákon o právech příslušníků národnostních menšina o změně některých zákonů)
http://usefoundation.org/view/951 [Accessed 24 March 2014].

Hungary

Alkotmány 1989 = The Constitution of the Republic of Hungary (A Magyar Köztársaság Alkotmánya) www2.ohchr.org/english/bodies/cescr/docs/E.C.12.HUN.3-Annex2.pdf [Accessed 6 November 2015].
Alaptörvény 2011 = The Fundamental Law of Hungary (Magyarország Alaptörvénye) www.kormany.hu/download/e/02/00000/The%20New%20Fundamental%20Law%20of%20Hungary.pdf [Accessed 6 November 2015].

Poland

Constitution 1997 = The Constitution of the Republic of Poland (Konstytucja Rzeczypospolitej Polskiej) www.sejm.gov.pl/prawo/konst/angielski/kon1.htm [Accessed 23 April 2005].
Law on national and ethnic minorities and on regional languages, 2005 (Ustawa z dnia 6 stycznia 2005 r. o mniejszościach narodowych i etnicznych oraz o języku regionalnym) http://isap.sejm.gov.pl/DetailsServlet?id=WDU20050170141 [Accessed 20 November 2008].
Law on the Polish language, 1999 (Ustawa z dnia 7 października 1999 r. o języku polskim) http://isip.sejm.gov.pl/Download?id=WDU19990900999&type=3 [Accessed 20 November 2008].
Law on the introduction of the country's three-tier territorial delimitation, 1998 (Ustawa z dnia 24 lipca 1998 r. o wprowadzeniu zasadniczego trójstopniowego podziału terytorialnego państwa) http://isap.sejm.gov.pl/Download?id=WDU19980960603&type=3 [Accessed 17 August 2006].
Draft on changing The Constitution of the Republic of Poland, Silesian Autonomy Movement, 2010 (Projekt zmian Konstytucji Rzeczypospolitej Polskiej, Ruch Autonomii Śląska) www.autonomia.pl/n/konstytucja [Accessed 17 May 2011].
Statutes of Silesian Autonomy Movement, 2011 (Statut Ruchu Autonomii Śląska) http://autonomia.pl/n/statut-ruchu-autonomii-slaska [Accessed 15 September 2012].
Organic Statute of the Autonomous Silesian Voivodeship (draft), Silesian Autonomy Movement, 2011 (Statut Organiczny Śląskiego Województwa Autonomicznego, Ruch Autonomii Śląska) www.autonomia.pl/n/statut-organiczny [Accessed 15 September 2012].

Romania

Constituția 2003 = Constitution of Romania (Constituția României) www.cdep.ro/pls/dic/site.page?id=371 [Accessed 24 March 2014].
Law on the territorial–public administrative reorganisation of the Socialist Republic of Romania, 1981 (Legea nr. 2/1968 privind organizarea administrativă a teritoriului Republicii Socialiste România, republicată în 1981) www.legex.ro/Legea-2-1968-384.aspx [Accessed 24 March 2014].
Bill on the Rights of National Minorities in Romania (draft) Democratic Alliance of Hungarians in Romania, 2005 (Törvény a romániai nemzeti kisebbségek jogállásáról. Tervezet. Romániai Magyar Demokratikus Szövetség) Magyar Kisebbség 2005, 1–2, pp. 83–107. www.jakabffy.ro/magyarkisebbseg/pdf/2005_1-2_9.pdf [Accessed 24 March 2014].
Proposals for the Regional Transformation of Romania, Hungarian People's Party of Transylvania, 2013 (Javaslatok Románia regionális átalakítására, Erdélyi Magyar Néppárt) www.neppart.eu/javaslatok-romania-regionalis-atalakitasara.html [Accessed 8 May 2014].
Proposals for Amending the Constitution of Romania, Hungarian People's Party of Transylvania, 2013 (Javaslatok Románia Alkotmányának a módosításához, Erdélyi Magyar Néppárt) http://neppart.eu/javaslatok-romania-alkotmanyanak-a-modositasahoz1.html [Accessed 8 May 2014].
Framework Law on Regions (draft), Hungarian People's Party of Transylvania, 2013 (Kerettörvény a régiókról, Erdélyi Magyar Néppárt) www.neppart.eu/admin/data/file/20140320/kerettOrveny-a-regiokrol_jo.pdf [Accessed 8 May 2014].
Bill on Creating a Region of Special Status for Székely Land (draft), Hungarian People's Party of Transylvania, 2013 (Törvény Székelyföld különleges jogállású régió létrehozásáról, Erdélyi Magyar Néppárt) www.neppart.eu/admin/data/file/20140320/tOrveny-szekelyfOld-kUlOnleges-jogallasu-regio-l.pdf [Accessed 8 May 2014].
The New Opportunity – Political Framework Programme of the Hungarian Civic Party (in Romania), 2008 (Az új lehetőség – a Magyar Polgári Párt politikai keretprogramja)

www.polgaripart.ro/index.php?option=com_rubberdoc&view=category&id=55%3Aprogram&Itemid=892 [Accessed 8 May 2014].

Statutes of the Democratic Alliance of Hungarians in Romania, 2013 (A Romániai Magyar Demokrata Szövetség Alapszabálya)
www.rmdsz.ro/uploads/pages/attachements/RMDSZ%20alapszabalyzat%202013.pdf [Accessed 8 May 2014].

Programme Adopted at the 4th Congress of the Democratic Alliance of Hungarians in Romania, 1995 (A Romániai Magyar Demokrata Szövetség IV. Kongresszusán elfogadott programja) Cluj-Napoca, pp. 47–61.
www.rmdsz.ro/uploads/pages/attachements/RMDSZ%20program.pdf [Accessed 8 May 2014].

Programme of the Democratic Alliance of Hungarians in Romania, 2011 (A Romániai Magyar Demokrata Szövetség Programja) Oradea, pp. 225–276.
www.rmdsz.ro/uploads/pages/attachements/RMDSZ%20program.pdf [Accessed 8 May 2014].

Smaller, More Effective Regions! Congress Documents, Democratic Alliance of Hungarians in Romania 2013. (Kisebb, hatékonyabb régiókat! Kongresszusi dokumentumok, Romániai Magyar Demokrata Szövetség) 11th Congress of the Democratic Alliance of Hungarians in Romania Miercurea Ciuc
www.rmdsz.ro/uploads/fileok/dok/RMDSZ11Kongresszus%20regiok.pdf [Accessed 8 May 2014].

Statutes of the Autonomy of Székely Land in Romania (draft), Democratic Alliance of Hungarians in Romania – Hungarian Civic Party (in Romania), 2014 (A romániai Székelyföld autonómia statútuma, Romániai Magyar Demokrata Szövetség – Magyar Polgári Párt)
http://rmdsz.ro/uploads/fileok/dok/A_romaniai_Szekelyfold_autonomia_statutuma.pdf [Accessed 22 September 2014].

Statutes of the Autonomy of Székely Land (draft), Székely National Council, 2004 (Székelyföld Autonómia Statútuma, Székely Nemzeti Tanács)
www.sznt.ro/fr/index.php?option=com_content&view=article&id=15%3Aszekelyfoeld-autonomia-statutuma&catid=10%3Astatutum&Itemid=14&lang=fr [Accessed 4 December 2013].

Let Székely Land Be an Autonomous Public Administration Region, Székely National Council, 2013 (Legyen Székelyföld önálló közigazgatási régió, Székely Nemzeti Tanács)
http://sznt.sic.hu/hu-sic/index.php?option=com_content&view=article&id=642%3Aszekelyfoeld-oenallo-koezigazgatasi-regio&catid=11%3Ahatarozatok&Itemid=15&lang=fa [Accessed 5 May 2014].

Together We Can Attain Autonomy for Székely Land! Székely National Council, 2013 (Együtt ki tudjuk vívni Székelyföld autonómiáját! Székely Nemzeti Tanács)
http://sznt.sic.hu/hu-sic/index.php?option=com_content&view=article&catid=11%3Ahatarozatok&id=643% 3Aegyuett-ki-tudjuk-vivni-szekelyfoeld-autonomiajat-&Itemid=15&lang=fa [Accessed 8 May 2014].

Serbia

Устав 2006 = Constitution of the Republic of Serbia (Устав Републике Србије)
www.ustavni.sud.rs/page/view/139-100028/ustav-republike-srbije [Accessed 4 June 2015].

Programme of the League of Social Democrats of Vojvodina, 2009 (Program Lige socijaldemokrata Vojvodine)
http://lsv.rs/UserFiles/pdf/Program_LSV.pdf [Accessed 16 June 2015].

Republican Declaration, Vojvodina's Party – Republicans of Vojvodina, 2012 (Republikanska deklaracija, Vojvođanska partija – Republikanci Vojvodine)
www.vojvodjanskapartija.org.rs/index.php/vesti/iv-vojvodanska-konvencija-romski/ [Accessed 16 June 2015].

Law on the protection of the rights and freedoms of national minorities, 2002 (Zakon o zaštiti prava i sloboda nacionalnih manjina)
www.seio.gov.rs/upload/documents/ekspertske%20misije/protection_of_minorities/the_law_on_the-protection_rights_nat_minorities.pdf [Accessed 9 June 2015].

Law on the territorial system of organisation in the Republic of Serbia, 2007 (Zakon o teritorijalnoj organizaciji Republike Srbije)
www.paragraf.rs/propisi/zakon_o_teritorijalnoj_organizaciji_republike_srbije.html [Accessed 9 June 2015].

Slovakia

Ústava (1992) = Constitution of the Slovak Republic (Ústava Slovenskej republiky)
https://www.prezident.sk/upload-files/46422.pdf [Accessed 31 August 2015].
Law on the territorial and public administrative delimitation of the Slovak Republic, 1996 (Zákon o územnom a správnom usporiadaní Slovenskej republiky)
www.zakonypreludi.sk/zz/1996-221 [Accessed 9 June 2015].
Law on the state language of the Slovak Republic, 1995 (Zákon o štátnom jazyku Slovenskej republiky)
www.culture.gov.sk/vdoc/462/an-act-of-parliament-on-the-state-language-of-the-slovak-republic--1ab.html [Accessed 9 June 2015].
Draft on the autonomy of South Upland Autonomous Area, Nomos Regionalizmus, 2007 (Dél-Fölvidéki Autonómia Környék autonómia-tervezete, Nomos Regionalizmus)
http://commora-aula.gportal.hu/gindex.php?pg=23545946&nid=3849084 [Accessed 9 June 2015].
Resolution of the Komarno Congress, 1994 (A komáromi nagygyűlés állásfoglalása)
www.duray.sk/index.php?option=com_content&view=article&id=94:-a-komaromi-nagygyles-allasfoglalasa&catid=1:dm-cikk&Itemid=60 [Accessed 9 June 2015].

Ukraine

Конституція 1996= Constitution of Ukraine (Конституція України)
www.legislationline.org/documents/action/popup/id/16258/preview [Accessed 24 March 2014].
Law of Ukraine on the principles of state language policy, 2012 (Закон України Про засади державної мовної політики)
http://zakon2.rada.gov.ua/laws/show/5029-17 [Accessed 24 March 2014].
Law of Ukraine on national minorities, 1992 (Закон України Про національні меншини в Україні)
http://zakon3.rada.gov.ua/laws/show/2494-12 [Accessed 24 March 2014].
Draft federal agreement on delimiting the spheres of authority between the authorities of Ukraine and the special autonomous area of Subcarpathia, Subcarpathian Rusyn Republic, 2015 (Федеративного договора о разграничении полномочй между органами государственной власти Украиныи и органами государственной власти Срециальной Самоуправляемой Территории, Республика Подкарпатская Русь)
http://getsko-p.livejournal.com/2015/01/19/ [Accessed 22 March 2015].

Homepages of Statistical Offices and Election Commissions

Državno izborno povjerenstvo Republike Hrvatske
www.izbori.hr/ws/index.html
Državni zavod za statistiku – Republika Hrvatska
www.dzs.hr/
Český statistický úřad
www.czso.cz/
Główny Urząd Statystyczny
http://stat.gov.pl/
Państwowa Komisja Wyborcza
http://pkw.gov.pl/
Institutul Național De Statistică
www.recensamantromania.ro/
Biroul Electoral Central
www.beclocale2012.ro/index.html
www.bec2014.ro/
www.beclocale2008.ro/rezultate.html
www.becparlamentare2012.ro/
Република Србија Републички завод за статистику
http://webrzs.stat.gov.rs/WebSite/
Република Србија Аутономна Покрајина Војводина Покрајинска изборна комисија
www.pik.skupstinavojvodine.gov.rs/

Štatistický úrad Slovenskej republiky
   www.statistics.sk/
   http://sodb.infostat.sk/scitanie/
   http://volby.statistics.sk/
Derzhavnyi Komitet Statystyky Ukrainy
   http://2001.ukrcensus.gov.ua/

Part III
Challenges in sustainable development

13
The regional dimension of migration and labour markets in Central and Eastern Europe

Jan Sucháček and Mariola Pytliková

Introduction

It is not necessary to underline that human beings create the inner basis of all relevant socioeconomic processes. From this perspective, society, economy, culture and other spheres of life represent certain external manifestations and structures of internal characteristics of human beings. Characteristics, activities and migratory movements of the population act as one of the major embodiments of regional differentiation. The highly debated migration of individuals and populations with all their social, economic, political, cultural, ethnic and other dimensions certainly affects sending and receiving societies in complex ways and in many domains.

In this chapter, we focus in particular on migration trends in the Central and Eastern European (CEE) region after 1989. We analyse and discuss the evolution and characteristics of emigration waves following the collapse of communist regimes, and in connection with the EU enlargements towards the East. We review thoroughly the existing empirical evidence to highlight the major migration determinants, socioeconomic characteristics of CEE migrants and impacts of migration. Furthermore, we discuss the changing character of the region from the typically net emigration towards the net immigration region, and we give an overview of major immigration trends to the CEE countries. In the chapter, we also draw attention to the contemporary refugee crisis and the position of the macro-region within that crisis. Later on, we discuss some major developments and characteristics of the CEE labour markets. Spatial patterns of labour markets in individual countries will be depicted in a synthetic way.

Emigration from Central and Eastern Europe after 1989

During state socialism, migration from the centrally planned economies was restricted, and those who emigrated did so primarily as political refugees. After the fall of the Iron Curtain in 1989, travelling freely became one of the newly acquired freedoms, and it was relatively easy for Central and Eastern Europeans, in particular those well educated, to search for a job abroad. Many indeed have chosen to experience that freedom of movement in order to improve their economic conditions or simply to experience living and working in another country without a fear of not being able to come back and see their relatives in home countries. Thus the CEE region became *a new source of emigration* after the collapse of communistic regimes. The emigration waves from CEE became substantial in connection with the EU enlargements towards the East, and with gaining employment rights under the free movement of workers regime.

The opening of labour markets towards workers from the new EU entrants in the 'old' EU15 countries, Iceland, Norway and Switzerland took place gradually: some countries lifted restrictions immediately after the EU enlargement rounds, and some kept the so-called 'transitional restrictions'. The overview of policy changes with respect to lifting restrictions on the access to labour market for workers from the new EU8 (the 2004 CEE entrants) and EU2 countries (the 2007 EU entrants, Bulgaria and Romania) is given in Table 13.1. We can observe that Ireland, Sweden and the

Table 13.1 Overview of dates of lifting restrictions on the free movement of labour for workers from CEE countries that entered the EU in 2004 (EU8) and in 2006 (EU2)

	2004 EU enlargement	2007 EU enlargement
	Lifting restrictions for workers from EU8	Lifting restrictions for workers from EU2
EU countries		
Austria	May 2011	January 2014
Belgium	May 2009	January 2014
Denmark	May 2009	May 2009
Finland	May 2006	January 2007
France	July 2008	January 2014
Germany	May 2011	January 2014
Greece	May 2006	January 2009
Iceland	May 2006	January 2012
Ireland	May 2004	January 2014
Italy	July 2006*	January 2012
Luxembourg	November 2007	January 2014
Netherlands	May 2007	January 2014
Norway	May 2009	January 2014
Portugal	May 2006	January 2009
Spain	May 2006	January 2009 (restrictions for Romania August 2011)
Sweden	May 2004	January 2007
Switzerland	May 2011	January 2014
UK	May 2004	January 2014
'New' EU		
Czechia	May 2004	January 2007
Estonia	May 2004	January 2007
Hungary	May 2004	January 2009
Latvia	May 2004	January 2007
Lithuania	May 2004	January 2007
Poland	May 2004	January 2007
Slovakia	May 2004	January 2007
Slovenia	May 2004	January 2007
Cyprus	May 2004	January 2007
Malta	May 2004	January 2014

Source: Authors' construction.

Note: * Italy has been open to the EU8 since 27 July 2006.

Migration and labour markets

UK were the first countries to open their labour markets to the EU8 newcomers in 2004, followed by Finland, Greece, Iceland, Portugal and Spain.

Given the historical circumstances, *the number of Central and Eastern Europeans has increased in almost all developed countries after the collapse of communism*. However, the EU enlargements and the EU transitional arrangements applied differently across the EU countries towards citizens of the new EU8 and EU2 member states, together with other factors such as linguistic and cultural proximity or economic conditions, led to *a significant variation in terms of the intensity of migration flows towards different destination countries*. This has resulted in differences in stocks of foreign population stemming from the CEE countries. According to Figure 13.1, the highest percentage of immigrants originating from the new EU8 countries relative to the host country population is in Iceland, reaching 3.9 per cent of the total Icelandic population, followed by Ireland and Austria with 3.4 per cent and 2.2 per cent, respectively. However, Iceland and Ireland experienced the largest increase in migration stocks from those new EU8 countries, from almost nothing in 1995 (0.3 per cent and 0.01 per cent, respectively) to 3.9 per cent and 3.4 per cent respectively in 2010, whereas Austria has long been the traditional destination for people from the CEE countries, to a great extent rooted in historical reasons. For instance, the largest number of CEE foreigners is of Czech origin with 54,000 Czechs living in Austria, and the majority of them (90 per cent) came during the period shortly after the Second World War (Lebhart, 2003). Norway, Great Britain and other Scandinavian countries have experienced quite substantial increases in percentages of the population coming from the new EU member states. Sweden, which has traditionally been a popular destination country for political refugees from former state socialist countries, experienced a significant outflow of citizens to their home countries after the year 1989. This return migration, together with a growing overall Swedish population during the 1990s and relatively moderate migration inflows, contributed to a relatively small increase

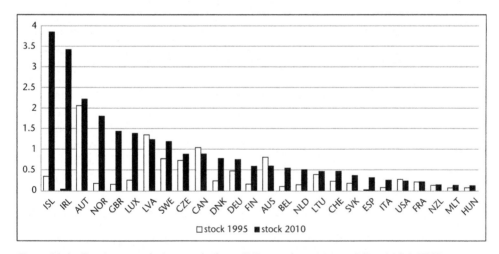

Figure 13.1 Foreign population stocks from EU8 member states residing in 26 OECD countries as a percentage of host country population, 1995 and 2010

Source: Adserà and Pytliková (2015); Authors' calculations.

Notes: Due to data availability the table shows information on: 2000 instead of 1995 for Austria; 1998 instead of 1995 for Belgium and Great Britain; 2009 instead of 2010 for Belgium and Spain; 1996 instead of 1995 for Czechia and Canada;1997 instead of 1995 for Iceland, Italy and Portugal; 2000 instead of 1995 for Luxembourg and USA; 1999 instead of 1995 for France; 2005 instead of 2010 for Greece; 2006 instead of 2010 for Canada and New Zealand; 2008 instead of 2010 for France and Malta; 2009 instead of 2010 for Australia and USA.

213

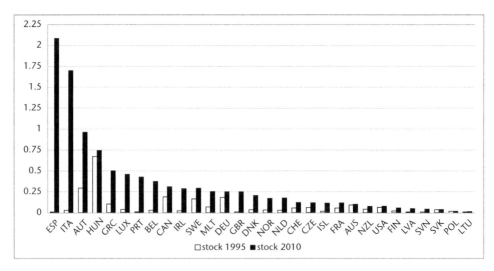

Figure 13.2 Foreign population stocks from Bulgaria and Romania residing in 26 OECD countries as a percentage of host countries population, 1995 and 2010.

Source: Dataset by Adserà and Pytlikova (2015); Authors' calculations.

Notes: Due to data availability the table shows information on: 1991 instead of 1995 for Austria; 2009 instead of 2010 for Belgium and Spain; 1996 instead of 1995 for Czechia 2000 instead of 1995 for Luxembourg; 1999 instead of 1995 for France; 2005 instead of 2010 for Greece.

in proportion of the immigrants from the new EU members from 1995 to 2010 (Figure 13.1). We can observe a similar pattern in the case of Switzerland and France.

Figure 13.2 focuses on the development of foreign population stocks from the EU2 countries, Bulgaria and Romania. Not surprisingly, due to linguistic and geographical proximity, the largest migration flow of Romanians and Bulgarians headed towards Southern European countries. In particular, Italy and Spain dominated as the most attractive destinations for the EU2 migrants, while at the other end of the range were mainly the EU8 destination countries. Specifically, the share of Romanians and Bulgarians in Spain increased from almost nothing (0.003 per cent) in 1995 to 2.08 per cent of the Spanish population in 2010. Given the massive increase in particular of Romanians in Spain at the time of harsh economic crisis and tough unemployment conditions, Spain reintroduced restrictions on the free movement of workers from Romania as of August 2011.

The migration from the new EU countries became a highly debated issue in a number of the 'old' EU15 countries. The most visible 'identity crisis' in connection with the massive post-EU enlargement inflow of workers from CEE countries could be observed in the UK in connection with the 'Brexit' referendum in June 2016. The fears from immigration, in particular the growing numbers of CEE immigrants in the UK following the 2004 EU enlargement have been a major reason for the pro-Brexit vote. The anti-migration rhetoric towards migrants from the new EU countries has been similar in a number of other EU countries.

Major migration determinants, socioeconomic characteristics of CEE migrants and impacts of CEE migration

Given the large migration wave from CEE countries, it became an important task to understand the economic consequences of CEE migration on both destinations and origins. Numerous empirical studies have documented the *mostly positive effects of CEE immigration in terms of overall fiscal effects* (see Dustmann and Frattini, 2014, for evidence from the UK, and Wadensjö, 2007, for Sweden), GDP per capita (Kahanec et al., 2013; Kahanec and Pytlikova, forthcoming) and moderation of inflationary pressures (Blanchflower and Shadforth, 2009; Kahanec et al., 2013). In line with the traditional economic debate on the effects of immigration on wages and employment among natives, empirical evidence on the effects of migration from CEE tend to find non-negative effects on wages, and show that although local adjustment may occur, *immigration does not increase the total unemployment rate* (Kahanec and Zimmermann, 2010; Elsner and Zimmermann, 2016; del Boca and Venturini, 2016; Rodríguez-Planas and Farré, 2016). Existing studies have also found no support for the welfare magnet hypothesis (Giulietti et al., 2013; Giulietti, 2014).

Regarding the impact of emigration from CEE macro-region on sending economies, the empirical evidence is rather scarce, mostly due to lack of data. Nonetheless, the existing studies tend to find increased wages and lower unemployment in connection with CEE emigration. For instance, Elsner (2013a, 2013b) documents that emigration from Lithuania, one of the countries with the highest emigration rates, has resulted in higher wages in that country. Zaiceva (2014) finds that emigration reduced excess supply of labour, lowered unemployment and increased wages in the new EU member states and led to additional positive effects through remittances and possibly brain gain. Dustmann et al. (2015) show that emigration from Poland led to a small increase in wages for high- and medium-skilled workers, which are the two groups with the largest relative outmigration rates. Polish workers at the low end of the skill distribution might have experienced wage decreases.

Empirical data show that immigration from CEE since the fall of communist regimes has been *mostly economic, i.e. labour-driven migration* (Palmer and Pytlikova, 2015), and that emigrants from CEE tend to be drawn from both ends of educational and skill distribution (Kahanec et al., 2016). For instance, Kahanec et al. (2016) compare the EU8 and EU2 migrants in the 'old' EU15 countries with the native population in terms of education and skill levels, as well as unemployment and participation rates. It seems that *Denmark* in particular attracts the most highly educated EU8 and EU2 immigrants compared to the educational composition of natives, followed by France and the Netherlands; whereas Italy, Greece and Portugal receive mostly lower-educated EU12 migrants. Compared to their natives, Cyprus, Spain and the United Kingdom also attract relatively lower-educated EU8 and EU2 migrants. Moreover, the data show that EU8 and EU2 migrants tend to work in less skilled occupations, which indicates that there is a prevalence of down-skilling among CEE migrants working in the EU15 countries, or in other words, accepting jobs in the destination that are below their qualifications (Kahanec et al., 2016).

Central and Eastern Europe – from net emigration towards net immigration

The macro-region of Central and Eastern Europe after the fall of the Iron Curtain in 1989 has experienced a changing migration character from being a net emigration region towards being a net immigration region. In particular, the relatively strong economies of Czechia, Slovenia, Poland and Hungary have experienced relatively significant immigration inflows in recent years (Figure 13.3). Czechia is receiving the most immigrants per destination population and the migration peak around the year 2007 was driven by immigration from Ukraine. On the other hand, the Baltic countries and Slovakia are receiving a relatively low number of immigrants.

We can expect that with the improving economic conditions and low fertility rates in the CEE macro-region and faster growing populations in major sending countries, together with continuing globalisation, improving communication technologies, widely available Internet connections and falling travel costs, the immigration pressure to CEE countries will be on the

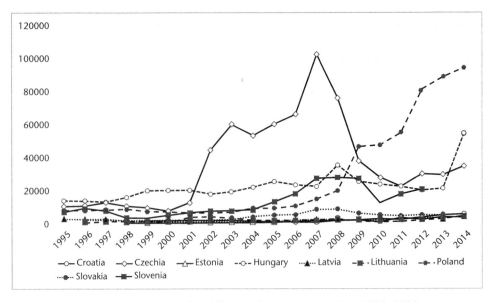

Figure 13.3 Immigration to Central and Eastern European countries, 1995–2014

Source: Data collected from national statistical offices of particular countries, and from Eurostat.

Note: Due to data availability, some data points for Croatia, Baltic countries, Slovenia and Poland have been extrapolated.

rise. In 2014, the annual migration to the CEE macro-region was around 340,000 foreigners (flow), and the number of foreigners living in CEE countries reached 2.6 million, i.e. on the average around 5 per cent of total CEE population were foreign-born. The percentage varies across the different CEE countries, between below 1 per cent foreign-born population in Bulgaria and Croatia to 11.5 per cent, 14.6 per cent and 14.8 per cent in Slovenia, Estonia and Latvia, respectively.[1] Regarding Estonia and Latvia, the majority of foreigners come from the countries of the former Soviet Union, in particular Russia, whereas for Slovenia, most foreigners originate in the former Yugoslavia, in particular Bosnia and Hercegovina and Croatia. Thus, those *relatively large percentages of foreign-born populations in some of the CEE countries are a result of historical developments in the region and country break-ups rather than immigration from other parts of the world per se.* All in all, we might say that populations in the CEE macro-region are rather homogeneous.

Regarding the regional dimension of immigration into CEE, we can observe a similar pattern of immigrant settlement as in the rest of developed world, with the majority of immigration flows heading towards the largest agglomerations.

The European refugee crisis and the position of Central and Eastern Europe

Hundreds of thousands of refugees pouring into Europe during the years 2014–2016 created an unforgettable picture of how humanitarian crises (in this case associated with the civil war in Syria) can shape an international migration. The contemporary migration crisis, which concerns Europe as a whole, represents one of the most important challenges for its societies. Uneven distribution of wealth as well as quality of life in combination with armed conflicts can be considered as the main underlying mechanisms of that crisis. In the future, demographic trends, such as an ageing population in Europe and a rapidly growing population in neighbouring Africa and the Middle East, can play an increasingly important role in driving further migration to Europe. Specifically, the regions of North Africa and the Middle East are characterised by a persistent high fertility with a large and growing population of young people looking for work and troubled by low wages and lack of stable employment prospects. Many countries in this region are suffering profound political and economic crisis, or even military conflicts. In this situation, *we can expect a continuing and growing migration pressure towards Europe, as to the neighbouring continent.* In addition, the existing communities such as Algerians in France, Moroccans in Spain, Turks in Germany and Scandinavia, and migrants from sub-Saharan Africa in Italy surely will offer a support to those newly arriving, which may further fuel so-called 'chain migration' from these countries.

So far, it is apparent that *CEE countries do not belong among typical final destinations of immigrants from Africa and the Middle East*. It concerns mainly Germany, France, Great Britain and Scandinavian countries, where both living standards and the presence of communities from Africa and the Middle East create an attractive environment for refugees. CEE countries mostly serve as specific 'changing stations' or transitionary territories for immigrants trying to get to Western Europe. However, we might expect that a part of the future migration from the less developed regions of Africa and the Middle East will sooner or later come to the CEE macro-region. The magnitude will depend on a variety of factors.

Regarding the recent refugee crisis, it is worth noticing that CEE countries adopted a different attitude than their counterparts from the Western part of Europe. Both politicians and the wider public of CEE countries have shown quite a reluctant or even hostile attitude to newly coming refugees. This might be a reflection of different economic conditions between Western and Central and Eastern Europe on the one hand, as well as relative lack of contacts and experience with immigrants in CEE on the other. Prior to the refugee crises, immigration had not been a concern of politicians, media and the public in the majority of the CEE countries. This has changed dramatically since 2014 and migration and threats of terrorism have become some of the major issues discussed, similar to Western EU countries. However, the leaders of CEE countries have started to diverge significantly from their Western counterparts regarding the willingness to adopt mandatory refugee quotas as a possible solution to the refugee crisis and refugee placement across Europe. The latest developments show that leaders of the European Union agree on solutions involving the closure of the outer borders of the European Union, together with support to countries close by the military conflicts in accommodating the refugees.

Nevertheless, the crisis has uncovered the West–East divide in attitudes. Western politicians and media have criticised the CEE countries and their leaders for a lack of solidarity and a failure to show a more humanistic face to the refugees. Particularly when, not that long ago, the Central and Eastern Europeans themselves had fled their home countries to seek asylum as political refugees, and continued to emigrate in large numbers as economic migrants to the 'old' EU15 countries. Apart from the critique of having a short memory in terms of the large outflows of Central and Eastern Europeans and the refusal of CEE leaders to cooperate on refugee placement across Europe using the refugee quotas, the Western media criticise the CEE leaders for spreading a negative picture of immigrants leading to an increase in anti-immigrant sentiments and xenophobia, instead of calling for more tolerance and acceptance, which are much needed for a smooth integration of immigrants into societies and labour markets.

Are Central and Eastern Europeans indeed more xenophobic than their Western counterparts? The recent rounds of the World Value Survey (WVS) show that the picture is more complex. The question in WVS that is usually considered to measure the degree of racial tolerance asks respondents which sorts of people they wouldn't want to have as neighbours, and offers 'people of another race' as an option. By looking at the outcomes from WVS in years 2005–2009, intolerance appears more widespread in Hungary, Czechia, Serbia and the Baltic States, but levels of intolerance are much higher in France than in those countries. Interestingly the tolerance levels seem to be about the same in Poland, Slovakia and Croatia, Bosnia and Hercegovina and Montenegro as in Germany, Italy, Finland, Greece or the UK.[2] The most recent WVS survey for the years 2010–2014 does not include the majority of European countries including France and Hungary, yet the percentage of German respondents saying they wouldn't want foreign workers or immigrants for neighbours went up from 13 per cent to 21 per cent – higher than the percentage of respondents saying the same in the supposedly xenophobic CEE countries of Poland (7 per cent) and Ukraine (19 per cent), and around the same percentage as in Romania.[3] Thus, although there is surely a space for an improvement for CEE representation in terms of informing the public on the benefits of immigration and increasing tolerance towards ethnic minorities and ethnic diversity, the task is there for their Western counterparts as well. A greater tolerance is important not only to reduce tensions in the societies, but also to ensure the smoother integration of foreigners into labour markets and societies. Furthermore, recent research suggests that natives' hostility acts as a strong deterrent of migration particularly for highly educated economic migrants, migrants from developed countries and linguistically close countries (Gorinas and Pytlikova, 2015). Growing hostility towards migrants may impose a major challenge for policymakers who would like to alleviate the fiscal burdens of ageing societies by attracting highly educated migrants.

General features of labour markets in Central and Eastern Europe

After the fall of the Iron Curtain, the move from totalitarian to democratic political systems and transition from central planning to market mechanisms was enriched by more general modernisation tendencies. Labour markets as an integral part of the economy and its characteristics in individual countries underwent profound changes as well.

Human resources are indispensable in this respect, as their quality affects the character of the labour as well as the directions and modes of developmental paths. One cannot omit that the labour market represents one of the most complex forms of market existing in the economy. It also forms an arena where individual living conditions are determined (Vogel, 2003). Labour markets simultaneously reflect and co-shape regional differentiation within CEE regions.

Due to their complexity, labour and labour markets are often researched by economists, geographers, demographers, sociologists, psychologists as well as other specialists. It is generally accepted that perfect competition in labour markets is just a theoretical construct. In reality, there exist numerous substantial barriers; those that apply to post-transition countries we focus on with even higher intensity.

As indicated by the World Economic Forum (2013), *labour market efficiency is distinctively lower in the Central and Eastern part of Europe*. Unfavourable institutional milieus, high transaction costs, networking and connections are still concomitant for labour markets within CEE. These phenomena can be treated as one of the manifestations of the deformed institutional fabric from the communist past (see also Sucháček, 2008).

The next attribute of CEE space is a relatively high affinity to home region or locality. This is palpable mainly when compared with the USA, for instance. In spite of enormous technological advances and the development of transport, the geographical segmentation of labour markets and the lower mobility of the workforce are still present. Moreover, these factors were 'enriched' by emigration from researched countries after 1989, which is discussed elsewhere in this chapter.

One of the rather more common features of CEE countries lies in the relatively *persistent gender differentials on labour markets* that have paradoxically not decreased in the last decades. What is more, the gender differentials on labour markets, e.g. in employment and wages, have even increased in some countries compared to the situation prior to the collapse of communist regimes (Filipová and Pytliková, 2016). Thus, *while the labour market position of women in Western Europe has improved over the last generation, that of women in CEE has declined.* Some of the reasons behind the trend are surely institutional, such as setting up lengthy maternity and parental leave schemes hand in hand with the destruction of the formerly well-functioning system of childcare facilities for children younger than 3 years. This has consequently forced women in their prime age to take long breaks from their careers to take care of their children. A part of the explanation for the change was to differ from communist practices, which were considered detrimental for society in terms of the destruction of the 'traditional family' features through maternal employment (Filipová and Machová, 2011). We can also observe that trend in changing gender identity attitudes, which in some of the post-communist countries became more conservative during economic transition. For instance, Mullerova (2016) shows using the European Values Study that a preference for long parental leaves in Czechia stems mostly from a purely intra-household value change in favour of higher task specialisation between men and women. The author observes a significant turn towards specialised couple preferences – among both women and men, both parents and non-parents, and both the higher and lower educated (Mullerova, 2016).

Finally, we might expect that *due to its ageing populations and in view of structural changes and shifts, the CEE macro-region is likely to need extra workers to fill job vacancies and newly emerging jobs.* The structure of present (Balcar et al., 2014) and mainly future labour demand, similarly to other developed countries, is expected to be varied and to include high- as well as low-skilled jobs. As for the possible sources available, we may mention immigration, increase in female labour supply and extending the working age. The developments should surely fill a greater space in the agendas of policymakers and researchers.

The spatial dimension of labour markets within Central and Eastern European countries

Generally speaking, we have to be pretty careful when deriving and making conclusions about the spatial pattern of labour markets at the regional level within Central and Eastern Europe. The key problem consists in the fact that the position of individual regions is determined by numerous factors showing strong spatial differentiation: sectoral economic structure, industrial and entrepreneurial traditions, geographic position, educational structure, level of technical infrastructure, state of the environment, relations between self-government and state government, institutional thickness or position in mental maps and many others (see also Lux, 2008). At the same time, one cannot ignore the ranking-territorial scale the analysis is performed at. Decreasing territorial scale leads to generally higher spatial differentiation.

Nonetheless, one can contemplate a certain polarisation between regions with the presence of large cities and agglomerations and more rural regions. The next attribute of our researched space is a certain East–West gradient with growing prosperity to the West. However, this applies to different countries with different intensity. Special attention should be devoted to the capital cities as they can be generally treated as transformational winners. Gorzelak and Smetkowski (2010) speak about a dichotomy between the capital city regions and remaining regions of a given country in Central and Eastern Europe (see also Chapters 2, 4, 8 and 15).

There are several indicators describing the character of the labour markets, which include: employment rate, unemployment rate, labour productivity and several others. These indicators present the mixture of intended regional policy and other policies and macro-structural mechanisms on the one hand and the spontaneous work of market forces on the other hand. Representativeness and the availability of different indicators naturally vary enormously.

However, *unemployment can be perceived as a relatively synthetic indicator of complex territorial changes during the transformation period* (see for example Hampl, 2001). Thus, it creates a certain point of departure for the assessment of the regional dimension of the labour markets within Central and Eastern Europe. Apart from the profound geographical differentiation of unemployment (as well as other relevant labour market indicators), long-term unemployment and the unemployment of young people can be treated as issues CEE has in common. It should be noted that the researched countries differ not only in terms of settlement hierarchies but also as to macro-economic conditions and spatially differentiated institutional conditions. We are even entitled to talk about varieties of capitalism at different territorial scales, discussed in this volume in the Introduction and Chapter 19.

CEE countries show different labour market characteristics on the regional level, influenced by their own spatial structures. Due to its size, Poland can be characterised by a genuinely polycentric settlement system. The spatial pattern of its labour market is determined by a network of relatively strong regional metropolises, and settlement hierarchy plays quite a dominant role here. At the beginning of transformation, polarisation between urban and rural Poland as well as East–West polarisation were the most pronounced features of this country from the perspective of labour market. The capital city turned out to be rather resilient in fact during the whole period after 1989. This became particularly salient after the crisis in 2008. Surprisingly, in recent years, certain backward regions of rural character have also enjoyed quite good unemployment values. According to Gorzelak (2010), this can be accounted for by the poor connection of these localities and regions with the external world.

The other three Visegrad countries differ from Poland in terms of country size as well as their settlement hierarchy. Generally speaking, they are much more monocentric, which also manifests in the spatial profiles of their respective labour markets.

In contrast to Poland, Czechia is much more monocentric. The labour market of the capital city bears certain qualitative differences in contrast with the rest of the country. The level of unemployment in Prague did not exceed the non-accelerating inflation rate of unemployment (NAIRU) level for the whole time. The position of regional metropolises is largely sound, too. Problematic territories are represented mainly by Old Industrial Regions in the northern part of the country that were compelled to undergo rather painful restructuring, as well as by some agricultural areas (Sucháček et al., 2012).

In terms of the spatial profile of the labour market, neighbouring Slovakia does not differ so much from Czechia. Nonetheless, East–West polarisation plays a more important role in Slovakia, and differences between urban and rural structures are more pronounced. The dominant position of the capital city of Bratislava is further strengthened by its proximity to Vienna. Eastern regions (with the partial exception of Košice) suffer from numerous shortcomings in the labour market. There are also many problematic micro-regions. It should be also mentioned that the Czech labour market is relatively attractive for Slovaks.

As for Hungary, the country's character is influenced by the size and weight of Budapest, which naturally induces somewhat intense centripetal forces. The north-western part of Hungary attracted fairly large Foreign Direct Investments, resulting in a dual economy (Fazekas and Osvald, 2010; Chapter 3). East–West polarisation also represents a *sui generis* inbuilt mechanism in Hungary, and affects labour markets accordingly. Again, the majority of regional metropolises are performing well, and perceivable problems are constituted by lagging agricultural regions and some problematic micro-regions. Notably, functional losses (hollowing-out) affect several small and medium-sized towns, which have lost their ability to integrate the surrounding labour markets (Chapters 2, 5 and 8).

The large area of Romania supports a more polycentric settlement system, albeit the capital city's position seems unshakeable. The country represents a peculiar mixture of regionalism and centralism. Bucharest, which has the most service-oriented economy in the country, is followed by the Western region bordering with Hungary and partly also the Central region. Constanța with maritime transportation routes is also prospering relatively well. On the contrary, the north-eastern part of the country occupies the very bottom in terms of the labour market (see also Ianos, 2010). Thus, the country's structure is despite certain East–West polarisation, of much more mosaic-like or even patchy character.

Neighbouring Bulgaria also has a rather irregular spatial profile. Nonetheless, regions within the sphere of influence of large towns enjoy the highest levels of economic development. Sofia is a clear leader, followed by Varna (see also Kirilova, 2014). The Black Sea coast strives for the development of tourism, which is tangible also in Bourgas. Rural territories represent the largest weakness of the Bulgarian economy, which also affect the country's labour market.

Last but not least, as mentioned, decreasing territorial scale leads to generally higher geographical differentiation. Thus, there are numerous micro-regions in CEE countries which perform much better or worse than the respective NUTS-2 or NUTS-3 regions where they belong. These successful or problematic micro-regions can be thus statistically 'hidden' within larger territorial units. In the future, micro-regions should definitely attract increased attention from both theoreticians and practitioners, and not only in the sphere of the labour market.

Summary

In this chapter, we discussed development in migration in Central and Eastern Europe, both emigration and immigration. It is clear that after the fall of communist regimes, the CEE macro-region became one of the major sources of emigration. The CEE emigrants headed mostly towards the EU, and the emigration wave was particularly pronounced in connection with the EU enlargements towards the East. As we discuss in this chapter, the CEE emigration clearly reacted to the opening of the labour markets in the 'old' EU15 countries to workers from CEECs, i.e. to gaining employment rights under the free movement of workers regime. Consequently, the number of Central and Eastern Europeans has increased substantially in EU countries over the last decades. However, there was a significant variation in the stocks of CEE migrants across the EU15 destinations resulting from differences in the EU transitional arrangements applied towards citizens of the new EU member states together with other factors such as linguistic and cultural proximity or economic conditions in both destinations and origins. The available empirical evidence is rather conclusive that migration from the CEE macro-region has been economic, i.e. labour-driven migration, and that it affected the EU destinations' economies positively. We might note however that the economic situation in the CEE region is improving – the region is steadily growing at a much higher pace than the rest of the 'old' EU countries and it is characterised by relatively low levels of unemployment. Therefore, we might expect CEE emigration towards the West to slow down, although it will not reduce completely due to pull factors of the existing CEE diaspora in the EU and the existing (although diminishing) economic disparities between the countries. Furthermore, any unexpected instability in the CEE region might reverse the trends.

In spite of the emigration trend, the majority of CEE countries have witnessed tremendous immigration inflows and changed from being typical net emigration countries towards net immigration countries. In particular, the relatively strong economies of Czechia, Slovenia, Poland and Hungary have experienced relatively significant immigration inflows in recent years. The largest inflows to CEE countries come from neighboring or culturally close origins; nevertheless the diversity of origins of CEE immigrants is growing over time. Due to demographic trends such as an ageing population in Europe and a rapidly growing population in Africa and the Middle East, and due to persistent economic and political crises in the African region that are often accompanied by conflicts, we may expect a continuing and growing migration pressure towards Europe. A part of the future migration from the less developed regions of Africa and the Middle East will eventually come to the CEE countries as well.

With respect to the recent refugee crisis in 2014–2016, the Central and Eastern European countries were not among typical target destinations of refugees; they mostly served as 'transit' countries towards the West. Yet, the issue of refugee migration became a very sensitive topic accompanied by increased hostility and xenophobia in a number of countries. As we show, however, the picture is more complex, and a number of CEE countries such as Poland, Slovakia and former Yugoslav countries are not more intolerant than their Western counterparts. Yet, the East–West divide has been visible in the reaction of leaders to the refugee crisis. Overall, it is important for all EU countries to manage successfully the immigration pressures and at the same time to work towards a greater tolerance and a reduction of the existing anti-immigrant sentiments. This is important in order to ease the growing social tensions as well as to ensure a smooth integration of foreigners into labour markets and societies in order to turn ethnic diversity into competitive advantage.

Labour market efficiency is still lower in CEE countries. This can be accounted for by institutional peculiarities in this part of Europe. Gender differentials are another striking feature of CEE countries. As to the spatial patterns of labour markets at the regional level in Central and Eastern Europe, they are often of a mosaic or even patchy character. However, East–West gradient, urban–rural polarisation or exclusive position of capital cities are concomitant to these countries and the situation will likely not change in the foreseeable future.

Acknowledgements

The work on this chapter was funded in part by the Operational Programme Education for Competitiveness (No. CZ.1.07/2.3.00/20.0296), SGS Research Grants No. SP2016/138 and No. SP2016/56 and by a Czech Science Foundation grant (No. GA15-24642S).

Notes

1 The numbers are based on data collected from national statistical offices and Eurostat for year 2014.
2 World Values Survey, waves 2005–2009, www.worldvaluessurvey.org/WVSOnline.jsp
3 World Values Survey, waves 2010–2014, www.worldvaluessurvey.org/WVSOnline.jsp

References

Adserà, A. and Pytliková, M., 2015. The Role of Language in Shaping International Migration. *Economic Journal*, 125(586), pp. F49–F81.

Balcar, J., Janíčková, L. and Filipová, L., 2014. What General Competencies Are Required from the Czech Labour Force? *Prague Economic Papers*, 23(2), pp. 250–265.

Blanchflower, D.G. and Shadforth, C., 2009. Fear, Unemployment and Migration. *Economic Journal*, 119(535), pp. F136–F182.

Del Boca, D. and Venturini, A., 2016. Migration in Italy Is Backing the Old Age Welfare. In: M. Kahanec and K.F. Zimmermann, eds. *Labor Migration, EU Enlargement, and the Great Recession*. Berlin: Springer, pp. 59–83.

Dustmann, Ch. and Frattini, T., 2014. The Fiscal Effects of Immigration to the UK, *Economic Journal*, 124(580), pp. F593–F643.

Dustmann, Ch., Frattini, T. and Rosso, A., 2015. The Effect of Emigration from Poland on Polish Wages. *Scandinavian Journal of Economics*, 117(2), pp. 522–564.
Elsner, B., 2013a. Does Emigration Benefit the Stayers? Evidence from EU Enlargement. *Journal of Population Economics*, 26(2), pp. 531–553.
Elsner, B., 2013b. Emigration and Wages: The EU Enlargement Experiment. *Journal of International Economics*, 91(1), pp. 154–163.
Elsner, B. and Zimmermann, K.F., 2016. Migration 10 Years After: EU Enlargement, Closed Borders, and Migration to Germany. In: M. Kahanec and K.F. Zimmermann, eds. *Labor Migration, EU Enlargement, and the Great Recession*. Berlin: Springer, pp. 85–101.
Fazekas, K. and Osvald, E., 2010. The Geography of the 2008–2009 Crisis – The Case of Hungary. In: G. Gorzelak and Ch. Goh, eds. *Financial Crisis in Central and Eastern Europe: From Similarity to Diversity*. Warsaw: Scholar, pp. 91–105.
Filipová, L. and Machová, Z., 2011. Wage Determination with Special Reference to Role in a Family. *Journal of Economic Perspectives*, 11(1), pp. 37–56.
Filipová, L. and Pytliková, M., 2016. Gender Differentials on the Czech Labour Market – Development over 25 Years since the Fall of Iron Curtain. In: G. Razzau eds. *Gender Inequality in the Eastern European Labour Market*. London: Routledge (2017).
Giulietti, C., 2014. The Welfare Magnet Hypothesis and the Welfare Take-up of Migrants. *IZA World of Labor*. http://wol.iza.org/articles/welfare-magnet-hypothesis-and-welfare-take-up-of-migrants/long
Giulietti, C., Guzi, M., Kahanec, M. and Zimmermann, K.F., 2013. Unemployment Benefits and Immigration: Evidence from the EU. *International Journal of Manpower*, 34(1), pp. 24–38.
Gorinas, C. and Pytliková, M., 2015. The Influence of Attitudes toward Immigrants on International Migration. *International Migration Review*. doi:10.1111/imre.12232
Gorzelak, G., 2010. The (Non-existing?) Polish Crisis. In: G. Gorzelak and Ch. Goh, eds. *Financial Crisis in Central and Eastern Europe: From Similarity to Diversity*. Warsaw: Scholar, pp. 139–149.
Gorzelak, G. and Smetkowski, M., 2010. Regional Development Dynamics in Central and Eastern European Countries. In: G. Gorzelak, J. Bachtler and M. Smetkowski, eds. *Regional Development in Central and Eastern Europe*. London: Routledge, pp. 34–58.
Hampl, M. ed., 2001. *Regionální vývoj: specifika české transformace, evropská integrace a obecná teorie*. Prague: Charles University.
Ianos, I., 2010. Spatial Pattern of Romania's Uneven Territorial Development. *Revista Romana de Geografie Politica*, 12(1), pp. 5–17.
Kahanec, M. and Zimmermann, K.F. eds., 2010. *EU Labor Markets after Post-enlargement Migration*. Berlin: Springer.
Kahanec, M. and Pytliková, M. (forthcoming). The Economic Impact of East–West Migration on the European Union. *Empirica*.
Kahanec, M., Zimmermann, K.F., Kurekova, L. and Biavaschi, C., 2013. Labour Migration from EaP Countries to the EU – Assessment of Costs and Benefits and Proposals for Better Labour Market Matching. *IZA Research Report*, 56.
Kahanec, M., Pytliková, M. and Zimmermann, K.F., 2016. The Free Movement of Workers in an Enlarged European Union: Institutional Underpinnings of Economic Adjustment. In: M. Kahanec and K.F. Zimmermann, eds. *Labor Migration, EU Enlargement, and the Great Recession*. Berlin: Springer, pp. 1–34.
Kirilova, Y., 2014. *Regional Disparities in Bulgaria – Evaluations and Policy Implications*. Sofia: Institute of Economic Studies, Bulgarian Academy of Sciences.
Lebhart, G., 2003. Volkszählung 2001: Geburtsland und Staatsangehörigkeit. *Statistische Nachrichten*, 4, pp. 258–265.
Lux, G., 2008. *Industrial Development, Public Policy and Spatial Differentiation in Central Europe: Continuities and Change*. Pécs: Centre for Regional Studies. (Discussion Papers, 62).
Mullerova, A., 2016. Czech Welfare and Gender Role Preferences in Transition. PhD thesis. University of Paris West Nanterre la Defense.
Palmer, J. and Pytliková, M., 2015. Labor Market Laws and Intra-European Migration: The Role of the State in Shaping Destination Choices. *European Journal of Population*, 31(2), pp. 127–153.
Rodríguez-Planas, N. and Farré, L., 2016. Migration, Crisis and Adjustment in an Enlarged EU: The Spanish Perspective. In: M. Kahanec and K.F. Zimmermann, eds. *Labor Migration, EU Enlargement, and the Great Recession*. Berlin: Springer, pp. 163–188.
Sucháček, J., 2008. *Territorial Development Reconsidered*. Ostrava: VŠB-Technical University of Ostrava.
Sucháček, J., Wink, R. and Drobniak, A., 2012. *New Processes in Old Industrial Regions. The Case of Leipzig-Halle Agglomeration, Upper Silesian Agglomeration and Ostrava Agglomeration*. Saarbrücken: LAP Publishing.

Vogel, J., 2003. The Labour Market. *Social Indicators Research*, 64(3), pp. 349–372.

Wadensjö, E., 2007. Migration to Sweden from the New EU Member States. *IZA Discussion Paper*, No. 3190.

World Economic Forum, 2013. The Global Competitiveness Report 2012–2013. www3.weforum.org/docs/WEF_GlobalCompetitivenessReport_2012-13.pdf [Accessed 5 August 2016].

Zaiceva, A., 2014. Post-enlargement Emigration and New EU Members' Labor Markets. *IZA World of Labour*, 40, pp. 1–10.

14
The changing role of universities and the innovation performance of regions in Central and Eastern Europe

Zoltán Gál and Balázs Páger

Introduction

Knowledge, learning, innovation activities and their relations became increasingly important in regional development during the last third of the twentieth century. There is an extensive literature, which contains several models on how innovations are created, how new products affect development and which factors have an influence on these processes. The national and regional issues of research and development (R&D) as well as of innovation have become one of the most important topics in the last decades. As Gál (2005) wrote, until the early 1990s, innovation policy was primarily oriented towards national growth targets, and its spatial implications were mostly neglected. Since the emergence of the knowledge-based economy, there have been innovations relating not only to new technologies and products, but also to systemic and network approaches which emphasise the importance of spatial proximity and regionally organised production (Koschatzky, 2003). Innovation and research had already played an important role in the Central and East European (CEE) countries in the era of planned economy. During that period research and technological development (RTD) were given high political priority, particularly in some specific industrial sectors. R&D activities were carried out mainly in public industrial research centres. Although the activities of these research centres were dedicated to supporting development in specific industrial branches, they most resembled the 'Fordist' innovation systems in that they had little interaction with industry as a whole. R&D activities declined significantly during transition on account of the drastic reduction of both public and private R&D funding. After halving the number of R&D units, the number of people employed in this sector decreased radically (Gál, 2005).

In this chapter, we focus on the innovation performance of Central and East European regions. Our research relates to the concentration of high-tech firms and employees in the CEE regions characterised by high innovation performance. The chapter is structured as follows. First, we give a short summary of the innovation processes, and the theoretical background of innovation systems. Then we shall deal with higher education, as universities may play an important role in knowledge-based development and innovation. The innovation performance of CEE regions will be evaluated by using the Regional Innovation Scoreboard. This scoreboard, its methodological background and the investigation results will be summarised in the third section. In the fourth section we identify the regions where high-tech firms and employees are concentrated, and compare their location to our results about regional innovation performance. The chapter ends with a summary and conclusions.

Restructuring the R&D funding system in the transition period

Expenditures on R&D as a percentage of GDP (GERD) in the communist (in CEE terminology 'socialist') regimes were not much below similar expenditures in the Western countries (Lepori et al., 2009). For example, in Russia, in 1990, such expenditures reached 2.98 per cent which was the original target of the Lisbon strategy from 2000. A large share of research and development in the communist economies was a type of 'reinventing the wheel', that is, it related to excessive import substitution (Yegorov, 2009; Radosevic and Lepori, 2009). On the other hand, there were only a few R&D units (particularly in military and space research, but also in heavy engineering, chemistry and pharmaceutical research) in the socialist countries that were able to keep pace with their counterparts in the Western countries. Some industry branches and their research institutes (like the garment or the food chemistry research institutes) were fully comparable with the world leading research base. At the same time, there were neglected R&D branches (like IT, biotechnology, etc.). Nevertheless, despite the cutting-edge military and space research, the number of spin-off consumer goods remained very limited.

Research funding in CEE has undergone significant changes in terms of funding sources, performers and instruments. The overall transformation was shaped by economic restructuring and by two key systemic changes: the reopening of previously closed research systems and the gradual introduction of the quality principle as a funding criterion (Radosevic and Lepori, 2009). In the first phase of transition, the decline in economic development came largely as a result of institutional uncertainty, not only in terms of the institutional system, but also of disrupted production, technology and trade linkages (Havrylyshyn, 2006; Mickiewicz, 2005). Applied research institutes of large companies were shuttered by foreign investors with the support of IMF and World Bank programmes. Domestic demand for modern technology applicable in production also declined substantially because of the general economic deterioration. This mismanaged economic transformation made the Central and East European countries more dependent on external technologies than ever before (Gowan, 1995; Gál, 2005).

The general trend was a very sharp decline in the funding of R&D as a percentage of GDP, as the average gross expenditure on R&D (GERD/GDP ratio) in the CEE-10 fell from well above 1 per cent to 0.8 per cent in 1994, declining further to 0.52 per cent by 2004. The GERD to GDP ratio grew to 0.92 per cent by 2013. In the case of Hungary and Czechia, the minimum value (0.7 per cent of GDP) was reached in 1996–1999 (Gál, 2005). The economic growth in all CEE-10 countries from 2000 led to a stabilisation of relative R&D expenditures. Continuing growth and, even more importantly, accession to the EU resulted in an increase in GERD/GDP ratios, which could be attributed mainly to EU funding. The average of the CEE-10 GERD/GDP increased from 0.75 per cent in 2000 to 0.90 per cent by 2007, although there were great differences among the countries. GERD/GDP exceeded 1.5 per cent in Czechia and Estonia, thus approaching the level of some of the West European countries, while it was still below 0.5 per cent in Slovakia and Bulgaria (Gál and Ptáček, 2011).

In the CEE-10, the main source of R&D funding is the government (49 per cent in 2007), closely followed by industry (40 per cent). This is typical of countries at this level of development; in the two most advanced countries (Slovenia and Czechia) the share of industry is above 50 per cent (Radosevic and Lepori, 2009).

Changes in the funding bodies and the sources since the early 2000s have reflected the Europeanisation of the R&D systems in the CEE-10 (Radosevic and Lepori, 2009). Many R&D institutes adopted various adjustment strategies and/or opted for moving upstream towards basic research, as they knew that the state would preserve its funding responsibility in this field. In some countries the R&D systems have lost their function as a funding agency, like in Poland, where the institutes are entirely funded by the research ministry (Jabłecka and Lepori, 2009). In other cases their size has been reduced, but they have kept their organisations and functions (as in Czechia; Lepori et al., 2009) or have even remained dominant in the public research sector as in Bulgaria (Simeonova, 2006) and Hungary (Havas, 2009; Mosoni-Fried, 2004). Higher education institutions (HEIs) have emerged as major actors in public research in all countries, but there are significant differences among them in this respect.

The slowly changing role of higher education: universities' innovation performance

This section gives an overview of both the regional distribution of HEIs and their share in the number of students, as well as of the regional breakdown of R&D indicators from the perspective of HEIs' performance. It also examines to what extent regional, mid-range universities may enhance economic development in a lagging area, and to what extent the European models of the universities' third role may be relevant in these regions. The concept of mid-range universities in Central and East Europe is used here as developed by Gál and Ptáček (2011).

Higher education is an important factor which attracts capital not only by creating competitive advantages in the local labour market, but also by absorbing innovative capacities. Determined by the R&D units of big firms, advanced technology systems have been concentrated in metropolitan agglomerations all over Europe (see Chapter 2). At the same time, technology innovations of the SMEs and the organisation of local and regional technology clusters have been initiated by institutes of higher education in the majority of cases. Regional higher education has been a driving force in the development of West European core regions (Bennett and Krebs, 1991).

In the CEE countries, the role of universities in the R&D sector is different and relatively weaker than in the Western countries. This is primarily due to the different organisational system of basic and applied research in these countries before the transformation in 1989, and to a large extent also two decades later. Both basic and applied research was mostly concentrated in the academies of sciences or in industrial research institutes and not in universities. In CEE, universities traditionally did not play a central role in basic and applied research, because they focused on teaching. This situation was but little changed during the transition in the 1990s, as most governments in the region at the time stressed the universities' educational role. During that decade, the finance system did not encourage them to search for new contacts and collaboration with industry, and it was much easier to survive by having a rising number of students. Basic research was carried out at the institutes of the Academy of Sciences, and it was not their primary role to focus on collaborating with industry.

The spatial concentration of universities in capitals or large cities, such as Budapest or Prague, did not help knowledge spillover from here to the mid-range universities located mostly outside these agglomerations. Therefore, it can be argued that the highly concentrated networks of the institutes of the Academy of Sciences have not played a significant role in regional innovation (Gál and Ptáček, 2011).

Traditionally, university education was elite education and universities primarily focused on teaching, with research being but secondary. In the last 50 years there has been a gradual expansion of higher education in many European countries. The appearance of mass education and lifelong learning supports the traditional education-oriented mission of the universities. Their second mission is to generate research-oriented knowledge (the generative role of universities). Universities have been forced to perform (basic as well as applied) research activities on a profit-oriented basis (Clark, 1998; Chatterton and Goddard, 2000). Later, in addition to teaching and research, universities started to adopt a third mission, their developmental role, which can be described as 'regional engagement' in Europe (Holland, 2001), or 'regional innovation organisation' or 'academic entrepreneurialism' (OECD, 1999).

Beside the shift of higher education from elite to mass education and the prevalence of lifelong learning, universities are also required to educate graduates in compliance with the needs of the regional labour market. By this, universities play the role of an interface between graduates and the regional labour market.

With the development of universities in Central and East Europe in the past 10 years and with the development of the infrastructure of provincial university centres created by integrating universities and colleges, the number of students, the training infrastructure and the potential of the universities have all increased exceptionally. However, research funding at provincial university centres had not increased significantly until the turn of the millennium.

The expansion of universities was most noticeable through the increasing number of students and the growth of university centres. By 2009, the number of full-time students increased to nearly 4.5 million in the region, which is almost twice as high as it was in 1998. However, their number has been declining since 2009. Between 2009 and 2014 the rate of decline was more than 20 per cent (from 4.8 to 3.8 million), that is, a decrease of 1 million. In 2009 Poland alone had more than 2.1 million students (Table 14.1). The highest concentration of universities in terms of the number of students can be found in the capital cities and in significant urban centres.

Table 14.1 Number of students and HE institutes in CEE countries and capital city regions

	Number of students			Number of HE institutes (2013)
	2009	2014	2009 = 100	
Bulgaria	274,247	283,294	103.3	45
BG41	123,176	135,827	110.3	24
Czechia	416,847	418,509	100.4	64
CZ01	154,319	149,539	96.9	32
Croatia	139,069	166,061	119.4	24
HR04	98,025	114,213	116.5	19
Hungary	397,679	329,455	82.8	70
HU10	179,915	172,832	96.1	38
Poland	2,149,998	1,762,666	82.0	431
PL12	441,115	379,007	85.9	104
Romania	1,098,188	578,705	52.7	147
RO32	470,100	181,344	38.6	49
Slovakia	234,997	197,854	84.2	34
SK01	77,355	74,681	96.5	13
Slovenia	114,391	90,622	79.2	29
SI02	81,228	61,039	75.1	16
CEE Total	4,825,416	3,827,166	79.3	844

Source: Authors' calculations and construction based on Eurostat data.

Concentration in Bucharest, Warsaw and some secondary city regions in Poland (Cracow, Poznań and Wrocław) is higher than, for instance, in Budapest or Prague. Despite the general expansion of regional HEI centres during the 1990s, these centres started to decline rapidly in many countries from the end of the first decade of the 2000s. Due to ageing and outmigration, the Bucharest region lost more than half of its students during the crisis years. In other countries, like Hungary, only the central capital city region managed to stabilise its position to the detriment of the less developed regions where the number of students rapidly declined. The fastest contraction took place in Romania and Hungary where government funding for higher education decreased the most rapidly and HEI policies failed to mitigate these negative trends. Horváth (2003) argues that the size and weight of provincial university centres in Hungarian higher education are lagging behind the European standards. The four provincial university centres together have only one-fourth of the total number of students, while this figure is one-third or one-half in other European countries with similar population size (Horváth, 2003). The Czech higher education is the positive extreme in CEE. All regions with significant HEIs in Czechia increased their higher educational capacities (by 3–30 per cent). This is a clear consequence of successful decentralisation and of the major governmental and EU funds granted to the higher education sector. Similar expansion can be found in Croatia and in a few regions in Bulgaria (Sofia, Pleven), where the HEIs' expansion started later.

The gradual 'marketisation' of the higher education sector in the CEE transition countries started after 2000 as a result of several factors. In general, it was the recognition of knowledge as a source of economic growth. In the marketisation process, universities started to use standard tools borrowed from West Europe, but the result could not be the same, because both the history and the position of universities in the regional or national innovation systems are different.

Many empirical studies on the universities' role in regional and economic development derive their findings from investigations at large, world-class research universities located in highly developed economic environments. These findings are not necessarily relevant to all universities, especially not to mid-range universities. The term 'mid-range' universities was first described in a study by Wright et al. (2008). It was defined as all universities except the top and the new (post-1992) ones. However, in the UK and other European countries, there are many first-ranked universities located in non-metropolitan regions. Gál and Ptáček (2011) argue that due to the high concentration of top universities in CEE almost exclusively in metropolitan areas, mid-range universities in this region are most often located in non-metropolitan regions.

Mid-range, regional universities are characterised by being located in secondary cities where regional demand for innovation is moderate, the R&D potential as well as the 'density of contacts' are much lower and possible spillover effects emerge more rarely. They seldom possess a world-class research base; their academics work in smaller, local scientific communities where they might interact with industry, but the creation of spin-off companies here is quite infrequent (Wright et al., 2008).

EU accession and the possibility of using EU development funds (such as cohesion funds) for building knowledge infrastructure induced the universities to adopt an active approach. Supporting innovation infrastructure (like science and technology parks) continued to be established and developed at the universities thanks to the role of intermediaries (mostly technology transfer offices or R&D services). These were engaged in building ties with industry on the one hand and in gaining EU funds for infrastructure building on the other. At that time the requirements of incoming foreign direct investments shifted from low-paid routine labour towards skilled and/or university-educated labour. In this sense, multinational companies had a pioneering role in promoting knowledge spillover from universities to industry (Ptáček, 2009). At the same time, mid-range universities located mostly outside the metropolitan areas had to face problems and disadvantages similar to those of their Western counterparts, namely, less intensive university–industry linkages, weak local R&D networks, etc. (Gál and Ptáček, 2011). The regional impacts of these processes have led to increasing disparities in the R&D potential between metropolitan and non-metropolitan areas; that is, R&D resources and research capacities have become more and more unequally distributed among the regions (Ptáček, 2009; Gál, 2005).

The differences between the advanced core regions of metropolitan agglomerations and the most backward regions are reflected in the relationship between universities and their region (Ács et al., 2000). Varga (2000) argues that a 'critical mass' of agglomeration must be present in a metropolitan area to expect that academic research can have substantial local economic impacts. Only metropolitan areas with about 3 million population, 40,000 employees in high-tech industry and with a professional research staff of at least 3,000 have the 'critical mass' of local economic activities that can absorb university-generated spillovers in the most efficient manner.

There are huge differences among the regions as regards R&D expenditure per capita. Statistics indicate the inferior role of HEIs in peripheral CEE regions. The capital city regions are the clear winners: Prague, Ljubljana and Bratislava have the largest per capita euro spending on research. (The second largest Czech region where Brno is located is ahead of many capital cities.) In terms of higher education expenditure, once again the order is Prague, Brno and Bratislava. On the other hand, Bulgarian, Romanian and recently Hungarian regions are underperformers even compared to the CEE average. This demonstrates what happens when regional decline is accompanied by decreasing funding for higher education R&D. The average proportion of HERD (Higher Education Expenditure on R&D) was 32 per cent between 2011 and 2013 in CEE. The largest share (50–75 per cent) can be found in the less developed regions, indicating the lack of strong SMEs and the very low level of business expenditure on R&D. The HERD to GDP ratio grew moderately from 0.13 to 0.24 per cent between 2004 and 2013 in Central and East Europe, while business expenditure doubled, reaching 60.0 per cent (Table 14.2). The low level of business and higher education R&D expenditures (BERD, HERD), as well as the larger share of universities as a performing sector from HERD means less opportunity to catch up with the most advanced regions.

It is often argued that universities are able to generate economic impacts based on knowledge spillovers and innovation transfers to businesses (Etzkowitz et al., 2000). Chatterton and Goddard (2000) as well as Arbo and Benneworth (2007) contend that universities are engaged in their regions not only in the fields of education and research but also in regional institutions and governance systems. Benneworth and Hospers (2007) review the literature describing the ways universities can play an integrative role in the regional innovation systems of less advantageous regions. For example, universities can help build large-scale excellence in research attracting new external partners; be an additional body/institution in governance networks, thereby increasing network connections; and train educated and informed citizens for public institutions. Furthermore, universities can promote the inflow of new ideas to old industries; act as globally focused actors raising demands for new kinds of planning arrangements; and actively shape through their consultancy the development of programmes that address and represent the cornerstones of regional innovation systems. Consequently, universities' regional engagement is a key factor in the innovation-based economic development of the peripheral regions (Gál and Zsibók, 2012).

The differences between the advanced regions of metropolitan agglomerations and the most backward regions are reflected in the relationship between universities and their regions (Ács, et al., 2000). This means that in most of the non-metropolitan Central and East European regions, where both the regional innovation systems and the university–industry linkages are still weak, the role of universities in local development has to be revised and, consequently, the

Table 14.2 The HE and business sector R&D expenditures as a percentage of GDP*

	HE R&D expenditure in per cent of GDP			Business R&D expenditure in per cent of GDP		
	2004–2007	*2008–2010*	*2011–2013*	*2004–2007*	*2008–2010*	*2011–2013*
Bulgaria	0.04	0.06	0.06	0.12	0.20	0.36
BG41	0.10	0.09	0.08	0.20	0.31	0.64
BG31	0.00	0.01	n.a.	0.04	0.04	n.a.
Czechia	0.21	0.25	0.46	0.73	0.74	0.95
CZ01	0.41	0.44	0.55	0.92	0.82	0.93
CZ04	0.02	0.04	0.05	0.20	0.20	0.28
Croatia	0.31	0.25	0.20	0.34	0.35	0.36
HR04	n.a.	0.31	0.24	n.a.	0.48	0.45
HR03	n.a.	0.12	0.13	n.a.	0.08	0.17
Hungary	0.23	0.23	0.22	0.43	0.62	0.86
HU10	0.23	0.23	0.20	0.68	0.86	1.16
HU31	0.18	0.17	0.16	0.15	0.37	0.47
Poland	0.18	0.24	0.27	0.17	0.19	0.31
PL12	0.19	0.24	0.27	0.33	0.34	0.50
PL31	0.17	0.25	0.42	0.11	0.14	0.17
Romania	0.08	0.13	0.10	0.21	0.18	0.16
RO32	0.14	0.23	0.11	0.46	0.34	0.28
RO21	0.09	0.17	0.12	0.09	0.06	0.07
Slovakia	0.11	0.13	0.26	0.22	0.22	0.32
SK01	0.24	0.26	0.41	0.22	0.25	0.63
SK04	0.10	0.23	0.27	0.12	0.13	0.18
Slovenia	0.22	0.26	0.28	0.88	1.21	1.91
SI02	0.33	0.38	0.41	1.05	1.33	2.04
SI01	0.08	0.10	0.11	0.67	1.05	1.74
CEE average	0.17	0.19	0.23	0.39	0.46	0.65

Source: Authors' calculations and construction based on Eurostat data.

Note: * The values of the country, the capital city region and the most underdeveloped region (according to GDP per capita) are involved.

expected economic impact of universities cannot be unambiguously extended to the transition economies. Therefore, universities' collaboration with industry faces both organisational and administrative constraints. Namely, in the post-socialist countries, applied research was concentrated in companies and branch R&D institutes which had limited connections with (mostly technical) universities.

For example, a Hungarian study concluded that the knowledge-producing ability of the academic sector did not increase the knowledge-exploitation ability of the local business sector, and both the universities and the less developed local economy might have been responsible for the several hindering factors in the intraregional knowledge transfer between the universities and industries (Gál and Csonka, 2007). Similarly, Bajmócy and Lukovics (2009) showed that university research might be an outstanding instrument for local economic development in the case of advanced regions, but not necessarily in the less developed regions where the lack of an appropriate industrial base was one of the main constraints. They measured the contribution of Hungarian universities to regional economic and innovation performance between 1998 and 2004. According to their results, the presence of universities did not affect the growth rate of the per capita gross value added, neither the gross tax base per tax payer. Therefore, the general economic impacts of the universities and the related R&D investments are hardly visible in transition economies; that is, in many Central and East European regions.

Radosevic (2011) investigated science–industry links in CEE countries and concluded that the innovation policies focused much more on connecting organisations than on supporting the relationship between science and industry. Kravtsova and Radosevic (2012) studied the efficiency of innovation systems in the East European countries. According to their results, East European countries lost some advantages concerning the size of their R&D. The analysis showed that economic growth in these countries was driven by production and not by innovation. They drew attention to three points related to innovation policy in the East European countries: (1) the importance of production capability, (2) the capabilities of firms in both production and technology, and (3) the change of focus in the East European R&D systems from exclusive knowledge generation to knowledge diffusion and absorption (Kravtsova and Radosevic, 2012).

An EU-wide empirical survey was conducted on the innovation performance of 20 selected EU regions, examining 92 time-series indicators in these 20 regions (Gál and Csonka, 2007). This research analysed the state of the regional techno-economic systems and the related research and technological development investments and policies, with special regard to university-based knowledge transfers. The main objective of this project was to find the key factors conducive to increasing R&D investment, and to identify the main barriers to knowledge transfer from academia to economy. The research focused on the constraints of knowledge transfer in the less developed transition regions having traditional, non-research mid-range universities. The survey managed to identify the main causes of the poorer performance in RTD transfers in these regions. A mismatch was found between the economic and research specialisations, combined with a low share of the business sector in RTD investment, a high proportion of the traditional lower-tech sectors, the small size of local SMEs, and the consequent shortage of resources to be invested in RTD and to absorb its results. The survey also identified the lack of demand for research results on the part of larger (mainly foreign-owned) companies[1] and, to a lesser extent, the lack of necessary knowledge supply in the region for some sectors and in particular disciplines (Gál and Csonka 2007; Dőry, 2008).

Investments in RTD activities have had the greatest impact in the developed regions having high absorption capacities and an economic structure where innovative actors are able to exploit transferred research results. In light of the examined cases, even increased RTD investments in selected areas might have only limited impact on economic performance in a region, at least in the short to medium term. This is particularly true for the less developed transition regions with traditional, less knowledge-intensive sectors (e.g. agriculture, food processing or tourism). It should be accepted that these regions are specialised in activities that are not highly research intensive, therefore, increased R&D expenditures cannot be easily exploited by the local businesses. In such situations, setting up a new research base not linked to the needs of the regional economy is like building 'cathedrals in the desert', as such research units are unlikely to be able to effect spillovers or knowledge transfer to the local economic actors (Dőry, 2008; Gál, 2010).

Measuring innovation performance

Regional innovation performance is a complex phenomenon that can be tracked via several soft and hard indicators. These measure various factors with different roles depending on their position in the innovation process (like the input or output of innovation). Since regional innovation performance involves many different indicators, its measurement requires a complex approach. To evaluate the innovation performance of CEE regions in a European context, we have used the assessment of the Regional Innovation Scoreboard (RIS). The different indicator groups and dimensions of the RIS will be introduced shortly. Then we summarise the methods and the results of a cluster analysis built on the RIS indicators.

The theoretical framework of the RIS is based on the concept of the Innovation Union Scoreboard (IUS). This latter was first published in 2001 as the European Innovation Scoreboard, its methodology and structure being more or less the same since its first publication. The IUS indicators are clustered into three groups: enablers, firm activities and outputs. The first group contains the input factors ('main drivers') of innovation. It has three dimensions: 'human resources', 'open, excellent research systems' and 'finance and support'. The second group of indicators also has three dimensions: 'firm investments', 'linkages and entrepreneurship' and 'intellectual assets'. They cover the firm-related indicators that describe the financial background (like investments), the collaborations and the outputs (like patents or trademarks) of the business innovation processes. The third group includes two dimensions relating to the outputs of innovation. They measure the number of innovative SMEs on the one hand, and the economic impacts (like employment, export of products and sales) on the other. There are altogether 25 IUS indicators. The difference between IUS and RIS lies in the number of applied indicators. Since some of the IUS indicators are not available at regional level, RIS contains only those which are measured at subnational level, too.

The RIS reports include 11 indicators. Human resources are measured by the proportion of people with higher education compared to the 25–64 year olds, 'finance and support' is assessed by the public sector R&D expenditure. These two indicators belong to the 'enablers' factors. Firm investments are measured by the R&D and non-R&D innovation expenditures of companies. Linkages are estimated by the proportions of in-house innovating SMEs and those working in collaboration. The indicator for patent applications measures the intellectual assets. These five indicators show the 'firm activities' factors. The 'output' attributes are captured by four indicators. Within 'outputs', 'innovators' are indicated by the proportion of innovative SMEs, while economic impacts are measured by the share of employment in knowledge-intensive activities and the share of sales of innovations in the turnover. The RIS was published in 2006 for the first time (with data from 2004), the latest data were collected from 2013 and published in 2014. Our data were collected from three RIS reports published in 2009, 2012 and 2014. These data refer to 2007, 2009, 2011 and 2013. Since the different scores were rescaled to a 0–1 scale for each year, we do not know the real values of the pillars behind the normalised score. Hence, we can only compare the positions of the regions, without seeing the development of a given pillar. Therefore, we shall investigate the changes in the rank of the CEE regions during the mentioned years.

First of all, we attempted to determine the level of regional innovation performance in CEE comparing it to all of the European regions. In order to make this comparison, we collected the RIS data on 168 European regions from 21 countries. Some countries are missing from the analysis (like the three Baltic states), but most of the EU members are included. Some of the countries were measured at NUTS-1 level (Austria, Belgium, Bulgaria, France, Germany, Greece and the UK), the others at NUTS-2 level. Although we are investigating the CEE regions, we compare them to all of the European regions. For this analysis, we applied data from 2007 and 2013 only. The score for regional innovation performance was computed as the average value of pillar scores. According to their regional innovation performance, we divided the regions into six groups by using the 17th, 33rd, 50th, 67th and 83rd percentiles (Figure 14.1). Since the normalised scores may have had different real values, the mentioned percentiles are also different in the case of 2007 and 2013.

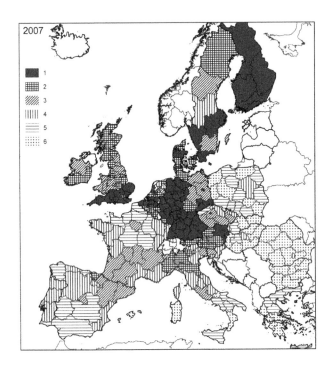
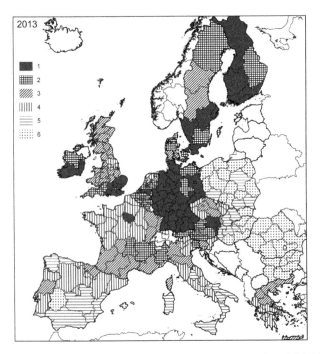

Figure 14.1 Regional innovation performance in Europe in (a) 2007 and (b) 2013

Source: Authors' calculations and construction.

Note: The darker a region, the higher its innovation performance.

The core regions in innovation performance are in Germany (only some East German regions have a little lower performance), in southern Britain, and in some of the Scandinavian and Western European capital regions (like Vienna or Paris). The CEE countries have significantly lower performance. Even the most innovative territories in the CEE countries have moderate performance in EU comparison, and several CEE regions belong to the under-average or the lowest performing ones. Compared to 2007, the positions of the CEE regions deteriorated by 2013. It may be due to the recovery of innovation performance in the CEE regions having been slower than in the West European countries after the economic crisis.

The innovation profile of the CEE regions

Our cluster analysis was performed in two steps. First, a hierarchical cluster analysis was conducted on different years and groups of variables. This helped us to determine the number of groups. We applied Ward's method for creating groups in the hierarchical cluster analysis. Its results suggested dividing the sample into 6 groups. On performing the k-means cluster analysis, we checked the F values in the case of 5 or 7 groups. Finally, we decided to use 6 groups in the k-means cluster analysis, as this was the most appropriate solution and it could be applied for all of the three dimensions. After accomplishing the cluster analysis, the created groups were identified according to their performance level in each dimension. In order to determine the level of performance of a cluster, we took into account the average values of the different variables within a dimension and also the differences among them. These levels of performance relate to the results of all included regions. Then the CEE regions were classified into these categories according to their clusters. Since our data were normalised, we conducted the analysis for both 2007 and 2013 (Tables 14.3 and 14.4).

'Enablers' can be identified as the input factors of innovation. This dimension contains the public R&D expenditures and the proportion of people having higher education. According to the scores from 2007, two groups can be clearly identified. The first includes those CEE regions which include the capital city (or are the capital city regions themselves). They perform much better than the other regions, and their performance is above average. Most of the CEE territories belong to the second group, having low-level performance in an all-European comparison. In the case of Bulgaria, Romania, Slovakia and Slovenia, all of their regions belong to this group, except the region including the capital city. There are few CEE regions belonging to the so-called 'under average' and 'moderate low' group. Most of these have a traditional university city and this may explain their relatively better performance among the non-capital CEE regions.

The category 'firm activities' contains the indicators that measure the innovation potential of firms, like patents, business R&D expenditures or cooperation in innovation. The CEE regions do not belong to the high or above average performing territories. The best performances in this respect can be found in Czech and Slovenian regions as well as in the Hungarian capital city region. All the other CEE regions have low performance.

The 'output' dimension includes innovations, the sale of new products and high-tech employment. Here the Czech regions are the best. Four of them, the agglomeration of Prague and the higher education centre Brno among them, reach the high performance level. Other four Czech regions also perform above the average. Some CEE regions have moderate (around average) performance. Here belong the Romanian, Hungarian and Slovak capital city regions as well as some Polish regions (primarily the ones having relatively strong industries, like Silesia or Pomerania). However, most of the CEE regions have the lowest performance among the European regions in this dimension, too.

Table 14.3 Innovation performance of the CEE regions, 2007

Performance level	Enablers	Firm activities	Outputs
High	CZ: 01		CZ: 01, 02, 03, 06
Above average			CZ: 07, 08
Moderate high	BG: 4	CZ: 01, 02, 03, 04, 05, 06, 07, 08	CZ: 04, 05
	HU: 10		PL: 22, 51, 52, 63
	PL: 12		RO: 32
	RO: 32		SI: 01, 02
	SI: 02		SK: 01
	SK: 01		
Moderate low	PL: 31, 34, 51, 63	HU: 10	HU: 10, 21
		SI: 01, 02	SK: 02
Below average	CZ06		PL: 21, 31, 32, 33, 34, 61
	HU: 23, 32, 33		RO: 21, 22
	PL: 11, 21, 41		
Low	BG: 3	BG: 3, 4	BG: 3, 4
	CZ: 02, 03, 04, 05, 07, 08	HU: 21, 22, 23, 31, 32, 33	HU: 22, 23, 31, 32, 33
	HU: 21, 22, 31	PL: 11, 12, 21, 22, 31, 32, 33, 34, 41, 42, 43, 51, 52, 61, 62, 63	PL: 11, 12, 41, 42, 43, 62
	PL: 22, 32, 33, 42, 43, 52, 61, 62		RO: 11, 12, 31, 41, 42
	RO: 11, 12, 21, 22, 31, 41, 42	RO: 11, 12, 21, 22, 31, 32, 41, 42	SK: 03, 04
	SI: 01	SK: 01, 02, 03, 04	
	SK: 02, 03, 04		

Source: Authors' calculations and construction based on data from the Regional Innovation Scoreboard.

Note: The regions are given by their NUTS Codes in Tables 14.1 and 14.2.

Table 14.4 Innovation performance of the CEE regions, 2013

Performance level	Enablers	Firm activities	Outputs
High	CZ: 01		
Above average	PL: 12		
	SK: 01		
Moderate high	CZ: 06	CZ: 01, 02, 06	CZ: 01, 02, 03, 04, 05, 06, 07, 08
	PL: 21	SI: 01, 02	SI: 01, 02
	SI: 02		SK: 01
Moderate low	BG: 4	CZ: 04, 05, 07	
	HU: 10	SK: 03	
	PL: 31		
	RO: 32		
Below average	BG:3	CZ: 03, 08	HU: 10, 21, 22
	PL: 11, 22, 32, 33, 34, 41, 42, 43, 51, 52, 61, 63	SK: 01	PL: 51
			RO: 31, 32
	SI: 01		SK: 02, 03

Performance level	Enablers	Firm activities	Outputs
Low	CZ: 02, 03, 04, 05, 07, 08 HU: 21, 22, 23, 31, 32, 33 PL: 62 RO: 11, 12, 21, 22, 31, 41, 42 SK: 02, 03, 04	BG: 3, 4 HU: 10, 21, 22, 23, 31, 32, 33 PL: 11, 12, 21, 22, 31, 32, 33, 34, 41, 42, 43, 51, 52, 61, 62, 63 RO: 11, 12, 21, 22, 31, 32, 41, 42 SK: 02, 04	BG: 3, 4 HU: 23, 31, 32, 33 PL: 11, 12, 21, 22, 31, 32, 33, 34, 41, 42, 43, 52, 61, 62, 63 RO: 11, 12, 21, 22, 41, 42 SK: 04

Source: Authors' calculations and construction based on data from the Regional Innovation Scoreboard.

Note: The regions are given by their NUTS Codes in Tables 14.1 and 14.2.

The innovation performances of the CEE regions did not change much: in 2013 they were similar to those in 2007. However, there were some important differences. In the case of 'enablers', the Czech, Polish, Slovak and Slovenian regions having research and higher education centres (e.g. Prague, Brno, Warsaw, Cracow, Bratislava and Ljubljana) were ranked to a higher performance level in 2013 than they were in 2007. At the same time, the Hungarian non-capital regions with traditional university centres (like HU23, HU32 and HU33) shifted lower during the same period. The dimension of 'firm activities' shows the most similar values in 2013 and 2007. Higher ranking could be observed in the case of the Czech regions and some Slovak ones, while the region involving the Hungarian capital city ranked lower in 2013 than in 2007. There is a clear divide in the 'output' dimension between the Czech and Slovenian and the other CEE regions. The first subgroup performed at the moderate high level, whereas the latter territories had below average and low performance. The Bulgarian regions, almost all of the Polish ones, and several Hungarian and Romanian regions belong to the low performance group. It may be explained by the low performance of firm activities, but this should be verified, which would require deeper investigation.

Conclusion

Our study investigated the relationship between innovation performance and higher education in the CEE regions. The transition period changed the economic environment, which caused serious problems in both the economic structure and the social dimension of these regions. However, knowledge and innovation might play a crucial role in their socio-economic development. We studied the innovation performance and its background in two ways: we reviewed the restructuring and evolution process of R&D institutes from transition up to now; and we examined higher education and its development during the last 25 years.

There has been an expansion in higher education since the transition. Several new institutes have been founded and the number of students has increased as well. Most of the regions have attributed an important role to higher education institutes in their economic strategies, since universities may attract people, investments and new firms. However, the shift towards the marketisation and entrepreneurial turn in universities started much later in Central and East Europe than in West Europe and, furthermore, the result cannot be the same anyway, because of the different legacy, the fragile local economies and the weaker position of universities in the regional or national innovation systems. The regional impact of these processes has led to the still ongoing polarisation of the R&D potential between the metropolitan and the non-metropolitan areas (Gál 2005; Ptáček, 2009). The best universities and research institutes are concentrated in the capital city regions, and they attract many qualified students or graduates from other (especially peripheral) regions, which do not have any universities, or if they do, their economy is relatively weak, so they cannot keep the qualified graduates and employees.

Having reviewed the innovation profile of the CEE regions we can state that the so-called input factors of innovation ('enablers') can be found in most of them. However, the innovative activities of firms do not reflect this. This dimension was relatively weak in several regions and only some capital city regions as well as Czech and Slovenian non-capital regions showed a relatively higher performance in all-European comparison. This may be explained by the migration of qualified graduates and by the lack of innovative small and medium-sized enterprises in these regions. The 'output' dimension shows similar results. Therefore, it would be important to support emerging innovative SMEs in the regions which have at least moderate positions regarding the input factors of innovation. Their presence may also help in keeping qualified employees in these semi-peripheral regions and strengthening regional economic performance in the longer term.

Acknowledgement

This research has been supported by project #104985 of the Hungarian National Research, Development and Innovation Office.

Notes

1 There have been a few large enterprises in high-tech electronics engaged in high-tech activities, but their influence on the local RTD sector is marginal as they usually rely on the in-house RTD activities of their parent companies and import the technology from outside the region.

References

Ács, Z., Anselin, L. and Varga, A., 2000. Geographic and Sectoral Characteristics of Academic Knowledge Externalities. *Papers in Regional Science*, 79(4), pp. 435–445.

Arbo, P. and Benneworth, P., 2007. *Understanding the Regional Contribution of Higher Education Institutions: A Literature Review*. OECD Education Working Papers, No. 9, OECD Publishing.

Bajmócy, Z. and Lukovics, M., 2009. Subregional Economic and Innovation Contribution of Hungarian Universities. In: Z. Bajmócy and I. Lengyel, eds. *Regional Competitiveness, Innovation and Environment*. Szeged: JATE Press, pp. 142–161.

Bennett, R.J. and Krebs, G., 1991. *Local Economic Development. Public–Private Partnership Initiation in Britain and Germany*. London: Belhaven Press.

Benneworth, P. and Hospers, G.-J., 2007. Urban Competitiveness in the Knowledge Economy: Universities as New Planning Animateurs. *Progress in Planning*, 67(2), pp. 105–197.

Chatterton, P. and Goddard, J., 2000. The Response of Higher Education Institutions to Regional Needs. *European Journal of Education*, 35(4), pp. 475–496.

Clark, B.R., 1998. *Creating Entrepreneurial Universities: Organizational Pathways of Transformation*. Issues in Higher Education. Oxford: IAU Press/Pergamon.

Dőry, T., 2008. RTD Policy Approaches in Different Types of European Regions. JRC and ERAWATCH Report. Available at http://cordis.europa.eu/erawatch/index.cfm?fuseaction=intService.display&topicID=588&countryCode=AT [Accessed 7 October 2010].

Etzkowitz, H., Webster, A., Gebhardt, C. and Terra, B., 2000. The Future of the University, the University of the Future: Evolution of Ivory Tower to Entrepreneurial Paradigm. *Research Policy*, 29(2), pp. 313–330.

Gál, Z., 2005. A New Tool for Economic Growth: The Role of Innovation in the Transformation and Regional Development of Hungary. *Geographia Polonica*, 78(2), pp. 31–52.

Gál, Z. 2010. The Role of Research Universities in Regional Innovation: The Case of Southern Transdanubia, Hungary. In: N. Longworth and M. Osborne, eds. *Perspectives on Learning Cities and Regions: Policy Practice and Participation*. Leicester: National Institute of Continuing Adult Education, pp. 84–106.

Gál, Z. and Csonka, L., 2007. Specific Analysis on the Regional Dimension of Investment in Research – Case Study Report and Database on the South Transdanubian Region (Hungary). Brussels: ERAWATCH.

Gál, Z. and Ptáček, P., 2011. The Role of Mid-range Universities in Knowledge Transfer in Non-metropolitan Regions in Central Eastern Europe. *European Planning Studies*, 19(9), pp. 1669–1690.

Gál, Z. and Zsibók, Zs., 2012. Regional Engagement of Mid-range Universities: Adopting European Models and Best Practices in Hungary. *AUDEM: International Journal of Higher Education and Democracy*, 2(1), pp. 94–120.

Gowan, P., 1995. Neo-liberal Theory and Practice for Eastern Europe. *New Left Review*, 213, pp. 3–60.

Havas, A., 2009. *Erawatch Country Reports 2009: Hungary. Analysis of Policy Mixes to Foster R&D Investment and to Contribute to the ERA*. ERAWATCH Network.

Havrylyshyn, O., 2006. *Divergent Paths in Post-Communist Transformation, Capitalism for All or Capitalism for the Few?* Basingstoke: Palgrave Macmillan.

Holland, B.A., 2001. Toward a Definition and Characterization of the Engaged University. *Metropolitan Universities*, 2(3), pp. 20–29.

Horváth, Gy., 2003. Egyetem és regionális átalakulás [University and regional restructuring]. In: F.-né Nagy and J. Vonyó, eds. *Város és egyeteme*. Pécs: Pécsi Tudományegyetem.

Jabłecka, J. and Lepori, B., 2009. Between Historical Heritage and Policy Learning. The Reform of Public Research Funding Systems in Poland 1989–2007. *Science and Public Policy*, 36(9), pp. 697–708.

Koschatzky, K., 2003. The Regionalization of Innovation Policy: New Options for Regional Change? In: G. Fuchs and P. Shapira, eds. *Rethinking Regional Innovation: Path Dependency or Regional Breakthrough?* London: Kluwer, pp. 291–312.

Kravtsova, V. and Radosevic, S., 2012. Are Systems of Innovation in Eastern Europe Efficient? *Economic Systems*, 36(1), pp. 109–126.

Lepori, B., Masso, J., Jablecka, J., Sima, K. and Ukrainski, K., 2009. Comparing the Organization of Public Research Funding in Central and Eastern European Countries. *Science and Public Policy*, 36(9), pp. 667–681.

Mickiewicz, T., 2005. *Economic Transition in Central Europe and the Commonwealth of Independent States*. Basingstoke: Palgrave Macmillan.

Mosoni-Fried, J., 2004. Hungary: From Transformation to European Integration. In: W. Meske, ed. *From System Transformation to European Integration. S&T in CEE at the Beginning of the 21st Century*. Munster: Lit Verlag, pp. 235–258.

OECD, 1999. *The Response of Higher Education Institutions to Regional Needs*. Centre for Educational Research and Innovation (CERI/IMHE/DG(96)10/REVI). Paris: OECD.

Ptáček, P., 2009. The Role of Foreign Direct Investment (FDI) in Establishing of Knowledge Economy in the Czech Republic: The Case of Knowledge Intensive Business Services. In: Z. Ziolo and T. Rachwal, eds. *Problems in the Formation of Industrial Spatial Structures and Their Surrounding*. Vol. 14, pp. 22–30. Warszawa-Krakow: Prace Komisja Geografii Przemysłu Polskiego Towarzystwa Geograficznego.

Radosevic, S., 2011. Science–Industry Links in Central and Eastern Europe and the Commonwealth of Independent States: Conventional Policy Wisdom Facing Reality. *Science and Public Policy*, 38(5), pp. 365–378.

Radosevic, S. and Lepori, B., 2009. Public Research Funding Systems in Central and Eastern Europe: Between Excellence and Relevance: Introduction to Special Section. *Science and Public Policy*, 36(9), pp. 659–666.

Simeonova, K., 2006. Research and Innovation in Bulgaria. *Science and Public Policy*, 33(5), pp. 351–363.

Varga, A., 2000. Local Academic Knowledge Spillovers and the Concentration of Economic Activity. *Journal of Regional Science*, 40(2), pp. 289–309.

Wright, M., Clarysse, B., Lockett, A. and Knockaert, M. 2008. Mid-range Universities' Linkages with Industry: Knowledge Types and the Role of Intermediaries. *Research Policy*, 37(8), pp. 1205–1223.

Yegorov, I. 2009. Post-Soviet Science: Difficulties in the Transformation of the R&D Systems in Russia and Ukraine. *Research Policy*, 38(4), pp. 600–609.

15
Transition and resilience in Central and Eastern European regions

Adam Drobniak, Adam Polko and Jan Sucháček

Theoretical background

The concept of resilience was introduced to regional sciences by the debate about sustainable development linking the concept with climate change and exogenous shocks (Simme and Martin, 2009). Resilience has quite a long list of definitions in regional studies, which all emphasise different aspects of this phenomenon. For example, one stresses that resilience is a reaction to specific extraordinary events (Simme and Martin, 2009), the next explains resilience as the stability of a system against interference (Welter-Enderlin, 2006; Lang, 2011) and another underlines a system's capacity to avoid or manage natural and human-induced hazards (Bosher and Coaffee, 2008).

Reviewing resilience surveys conducted in the first decade of the twenty-first century that were applied to regions and cities, we find few that are devoted directly to the different socio-economic factors strengthening resilience. Gleaser and Saiz (2004) argue that a region's human capital is the major driving force of resilience. Briuglio et al. (2006) note that concentration of a nation's export into a few industries inhibits regional resilience. Feyrer et al. (2007) show that the population and level of employment in regions dominated by traditional industries have failed to grow even after approximately two decades following certain exogenous disturbances, such as an economic crisis. However, metropolitan areas located near regions of traditional industries have a positive impact on the level of resilience. Nunn (2009) claims that institutions, behavioural norms, knowledge and technology specific to the region have a long-lasting effect on regional resilience. According to Desmet and Rossi-Hansberg (2009), regional economies may be restored after experiencing exogenous disruptions if their firms can introduce new goods for export or use new technologies to produce such goods in a relatively quick way. Gerst et al. (2009) show that after experiencing exogenous disruptions, both the impact of the decline and the path to recovery vary considerably among regions. Service-based economies perform better than those dealing with manufacturing. Direct reference to regional resilience can also be found in Hill's research on resilience in metropolitan areas of the USA (Hill et al., 2010). According to him, a metropolitan area's industrial structure affects the likelihood of the region experiencing a downturn, assuming that an older economic structure means a higher probability of recession.

Both earlier and current studies of the concept of resilience allow us to draw the conclusion that a resilient system – a region or a city – tolerates exogenous disruptions through some general attributes that restrict their impact. According to Godschalk (2003), Walker and Salt (2006) and Taşan-Kok et al. (2013), attributes defining a region's or a city's resilience include: *redundancy*, *diversity*, *efficiency*, *autonomy*, *adaptability*, *collaboration* and *interdependence*.

These attributes can be arranged further for the purpose of regional resilience diagnosis according to basic regional structures, such as: economic–technological, socio-cultural, environmental–spatial and institutional–political aspects (Drobniak, 2014). The economic–technological dimension corresponds with a region's structure for creativity and manufacturing goods and services for internal and external markets. The socio-cultural aspect is linked with a region's demographic and community behaviour patterns, beliefs and attitudes. The environmental–spatial dimension refers to a region's method of land use for various functions, including the impact of these decisions. The institutional–political framework describes a region's institutional and system planning capacity, political values and power structure.

Taking into account the scope of the chapter, the analysis of the CEE regions' transition and resilience was restricted to the economic–technological, environmental–spatial and institutional–political dimensions. According to Eraydin and Taşan-Kok (2013), economic resilience is about coping with slow and (or) radical change in a city's or a region's economy that results from the interactions of endogenous and exogenous conditions.

Referring to the attributes mentioned above regarding a region's resilience, one may find a set of specific factors determining regional economic resilience. Among them, we will find: the level of economic activity and entrepreneurial skills, networking patterns among small businesses and large companies, the level of diversity in a region's economic base (mix of specialisation), business support infrastructure and economic development patterns (Lansford et al., 2010). Important aspects of enhancing regional economic resilience are connected to innovations and new technologies (Hess, 2013). Cooke (2008) highlighted that a good adaptation strategy is to develop a system of innovation in order to create new technologies and products that can quickly find demand in external markets. This means that a region is resilient if it is able to develop new value chains after the disruptions that correspond with the external global market.

Including technological aspects in a region's economic resilience opens up a new perspective in research on factors determining the resilience level – especially those related to knowledge, clusters and networks. According to Wolfe, 'During the early phase of technology development many different cities and regions have the potential to emerge as a location where a technology and its corresponding industry take root and develop' (2013). But the potential to attract new technologies is the consequence of many factors determining the advantages of a particular location (such as: labour force capacities, particular place image, tax incentives, capacity of local firms to cooperate).

To conclude, a region's resilience and transition capacity in the economic–technological dimension may be defined as successful foresight of technology trends along with efficient and effective capitalisation of these changes according to the regional economic base (Drobniak, 2014).

Factors determining a region's resilience in environmental and spatial structures can be analysed in a variety of ways. According to Cruz (Cruz et al., 2013), the following general aspects influence the dynamics of the environmental–spatial dimension of a city or a region: urban sprawl, polycentric development, shrinkage and compactness. All of them can impact resilience in different ways and can influence more than the environmental and spatial structures of a particular space.

For example, in the case of urban sprawl, the negative impact on environmental structures is linked with air pollution and the inefficiency of public infrastructure and social services delivery

(Cruz et al., 2013). Another effect of urban sprawl is frequently connected with the lack of urban space continuity, which leads to the fragmentation of space as well as the fragmentation of socio-economic activity. The polycentric development approach offers another interesting perspective for strengthening a city's or a region's resilience. This concept holds that an urban or a regional space has multiple centres, and polycentric development surveys are usually focused on the functions performed by these centres (Capello, 2000). Thus, polycentric development in regional space includes a few centres such as regional cities, edge cities, slow cities and network cities. According to others, polycentric development is connected with diversity, and results from different dynamics in demography, a region's population attitudes and lifestyles (Champion, 2001).

The popular concept of shrinkage, related to the loss of a city's or a region's population, also affects the dimension of environmental and space development. According to Cruz, shrinkage effects manifest themselves in the slow growth of a few districts, and stagnation or decline in the majority of others (Cruz et al., 2013). However, shrinkage effects may also increase the pressure for growth, for rebuilding structures or for releasing areas in order to increase green spaces.

Another concept of environmental and space development, in the context of resilience, is connected with compactness. Neuman (2005) argues that compactness stands in opposition to urban sprawl, because it assumes linkages among agglomerations of urban and regional functions, activities and inhabitants, as well as physical proximity and continuity. Compactness results in less pollution because of the proximity of the places of residence, work and services.

In the last decade, important determinants of regional resilience and transition have also been perceived in the integrated development approach, connected with a green economy. This development concept stems from the complex relationship among the elements of an ecosystem. Improving links between the ecosystems and the economic–technological structures of a region requires not only reduction of carbon dioxide (CO_2), but also the increase of energy efficiency in transportation and housing, attractive landscape development, appropriate management of water supply and waste management (Ayres and van der Lugt, 2011).

Last but not least, the institutional and political dimension of transition and resilience is of a versatile nature, and involves various categories, such as: laws, rules, organisations operating at different territorial levels, civic institutions, norms of behaviour, conventions, self-imposed codes of conduct, culture of society, institutional density and many other factors (Drobniak, 2014). Thus, the selection of categories that are truly relevant seems necessary.

The notion of institution is often connected with an establishment of the 'rules of the game' in society. North (1990) perceives institutions as the formal and informal constraints on political, economic and social interactions. According to him, the major role of institutions consists of reducing uncertainty by establishing a stable structure to frame human interactions.

It is useful to conceptualise the whole notion by distinguishing formal and informal institutions. Formal institutions are represented by artificial constraints invented by humankind: they comprise mainly existing legislation and organisations. They concern economic, political, administrative and many other realms of life.

Territorial administration is of utmost importance in this context, as the structure and the performance of public administration define the possibilities and constraints of other local and regional processes to a large extent. Competences and financial resources should be balanced between both of the fundamental components of public administration, i.e. state administration and self-administration.

Informal institutions form conventions and codes of human behaviour, and present a spectrum of constraints creating 'the culture of society'. These institutions come from socially transmitted information and are part of the culture. Boyd and Richerson (1985) perceive culture as the transmission of knowledge, values and other factors that determine behaviour across generations.

Informal institutions are submitted to evolutionary processes, and they have been developed over the course of many centuries. Although formal institutions may be changed overnight as the outcome of political or judicial decisions, informal institutions embodied in customs, traditions and codes of conduct are much more persistent and can hardly be influenced by deliberate policies in the short term. These cultural constraints connect the past with the present and the future.

Methods and data used for the quantitative and qualitative analyses of resilience in CEE regions

Conducting research on the subject of resilience and transition in the CEE regions – in the diagnostic dimension – faces several limitations. The first relates to the diversity of these regions, which is connected to their basic development attributes, such as: demographic, social, economic, technological, special, environmental, political and institutional dimensions. For example, the demographic dimension of each CEE region is characterised by a different scale of inhabitants. In the social dimension they are distinguished by varying educational structures, cultural patterns, creativity, etc.

The second limitation, essential for quantitative research, refers to the accessibility of statistical data, along with their continuity. Both the availability of completed data series, as well as their continuity (regarding each of the CEE regions) are the basic prerequisite for conducting even a simple examination of the changes that have occurred in these regions over the last decade. The diversity of the statistical systems of individual CEE countries affects possibilities for any quantitative analysis, in principle, with the exception of data from Eurostat.

Finally, taking into account these limitations, our quantitative analysis of resilience and transition of the CEE regions was conducted with the following assumptions:

- the calculation was based on Eurostat data for eight CEE countries (i.e. Bulgaria, Czechia, Croatia, Hungary, Poland, Romania, Slovenia, Slovakia). The selection of the CEE countries and regions was determined by their EU membership (data accessibility). Resilience indicator calculations were prepared for all 53 NUTS-2 regions of the CEE area;
- the calculation of resilience indicators was based on Hill's research methodology (Hill et al., 2010), i.e. dynamic indexes were calculated with a fixed baseline value;
- economic resilience of the CEE countries and regions was measured within the *efficiency* attribute by the index: dynamics of GDP at current market prices in million euro (baseline year: 2000 = 100) for the period 2000–2011. To capture the scale of changes, the economic portfolio of the CEE countries and regions was used. It is based on the following indexes: GDP per inhabitant (2011), dynamics of GDP (2011, 2000 = 100), gross value of GDP (2011). In the regional analysis of economic resilience, reference is made to the three best and three worst regions of CEE (the best and the worst case method). For these extreme cases the factors of resilience or vulnerability were identified;
- technological resilience of the CEE countries and regions was measured within the *adaptability* attribute of the index: dynamics of patent applications to the EPO[1] calculated for the period of 2000–2012 in the case of countries (2000 = 100), and for 2001–2011 in the case of the CEE regions (2001 = 100). Due to the problems associated with the baseline value (i.e. often very low or even zero), the average value of patent applications to the EPO

between 2001 and 2011 was assumed as the baseline value for the dynamics index. In addition, due to the low baseline number of patents in 2001, the index of patents was based on the ratio between the patents' growth rate in 2011 (2001 = 100) and the number of patents in 2011. To capture the scale of changes, the technological portfolio of the CEE countries and regions was used. This was based on the following indexes: patent applications to the EPO per million inhabitants (2012), dynamics of patent applications to the EPO (2012, 2000 = 100), number of patent applications to the EPO (2012). In the regional analysis of technological resilience, reference is made to the three best and three worst regions of CEE (the best and the worst case method);

- resilience of the CEE countries and their regions in the environmental and spatial dimension has been evaluated using a set of efficiency and diversity attributes. The research focused on issues related to the efficiency of energy systems and their influence on the quality of the natural environment. The first indicator shows the dynamics of energy intensity of the national economies. It is measured by the gross inland consumption of energy divided by the GDP, between 2002 and 2013 (2002 = 100). The second indicator presents the dynamics of the share of electricity generated from renewable sources, for the period of 2004–2013 (2004 = 100). The third indicator shows the dynamics of the increase of solar collector surfaces per 1,000 square metres, between 2004 and 2013 (2004 = 100). All these indicators show changes both in efficiency and diversity regarding the energy systems in Central and Eastern European countries during the last decade.

The next section of the chapter delves into the processes of territorial transition and resilience examined from institutional and political perspectives. The institutional and political fabrics of individual countries create a wider framework for single processes at the regional and/or local levels. It should be noted that the institutional–political dimension of vulnerability and resilience is somehow of a different nature. In the case of evaluating territorial resilience and vulnerability within institutional–political structures, the quantitative approach is severely constrained. These hindrances relate mainly to the relatively small range of available statistical data and the short time span they cover. The objective lack of data forces us to adopt a different – strongly qualitative – approach to examine territorial resilience and vulnerability. There are no doubts that institutional and political structures and activities represent intensely interlinked domains. In order to conceptualise this part of our research, it is useful to divide institutional–political structures and activities into previously depicted formal and informal classes.

Findings – the economic dimension of resilience and transition in CEE countries

CEE countries show similar trends in the dynamics of GDP between 2001 and 2011. Instead, differences can be found in the scale of these dynamics. It is clear that there is a group of 'rapidly emerging' countries, such as Romania, Slovakia, Bulgaria and Czechia, where the GDP growth rate is significantly higher than the average GDP growth rate measured in the EU28 countries (the EU28 dynamics is 138) – see Figure 15.1a.

Other CEE countries, such as Croatia, Hungary, Poland and Slovenia are also characterised by higher GDP growth than the EU28 average, but this rate is lower than the growth recorded in countries labelled as 'rapidly emerging'. The GDP growth dynamics of the CEE countries were disrupted in 2009 by the effects of the global financial crisis. A large decline in the GDP's dynamics was recorded mainly in Romania, Czechia, Hungary, Poland and Croatia. The economies of these countries, in terms of resilience, can therefore be considered as more sensitive

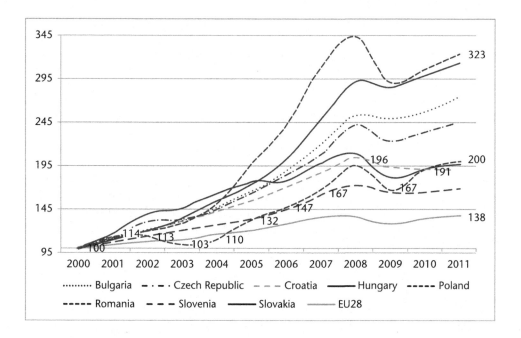

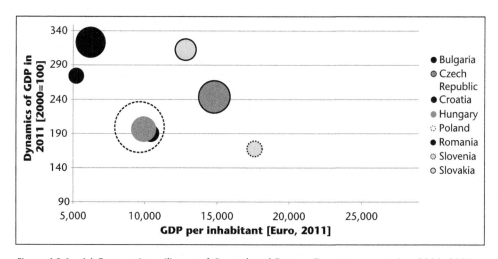

Figure 15.1 (a) Economic resilience of Central and Eastern European countries, 2001–2011 – dynamics of the GDP at current market prices in million euro (2000 = 100); and (b) Central and Eastern European countries' economic portfolio – GDP per inhabitant (2011), dynamics of GDP (2011, 2000 = 100), value of GDP (2011)

Source: Author's calculations based on Eurostat data.

to changes in the world economic situation, also as compared to the average GDP dynamics of the EU28. However, with the exception of Croatia, GDP dynamics in these countries again experienced a positive turn after 2009.

Despite the CEE countries' significantly higher GDP growth in relation to the EU28, these areas – in the years between 2001 and 2011 – were characterised by continued low value of GDP per inhabitant. The share of their joint GDP in the total GDP of the EU28 accounted for a mere 7.4 per cent in 2011. This is confirmed by the analysis of the economic portfolio of individual CEE countries (see Figure 15.1b).

The analysis of the 53 regions of the CEE countries (NUTS-2 level) points to significant differences in their growth rates measured by the dynamics of GDP in the years 2000–2011 (see Figure 15.2a). Despite the existing differences in the level of development of these regions, it should be stressed that none of them developed slower than the average GDP dynamics measured in the EU28 during the period considered. This means that even in the case of the three CEE regions that showed the lowest growth dynamics, i.e. HU23: Dél-Dunántúl (167), BG31: Severozapaden (164) and SI01: Vzhodna Slovenija (161), we find that none of them experienced a lower GDP dynamic than the average measured in the EU28 (138). However, in the case of these three regions, their growth rate slowed down and moved into a horizontal trend after 2008 (impact of the global financial crisis).

In the case of *HU23: Dél-Dunántúl* region, the lower GDP dynamics may be a consequence of the following factors analysed in the context of resilience (South, 2007):

- *regional economic development*: the large number of micro-villages; micro-regions falling behind; internal and external peripheries; functional weakness of the urban network; weak regional economy;
- *R&D, innovation*: low R&D expenditures; little FDI inflow; low employment rate; underutilised industrial parks; weak market orientation of education and training.

Similarly, the low growth dynamics of *BG31: Severozapaden* region (in comparison to other CEE regions), result from factors such as:[2]

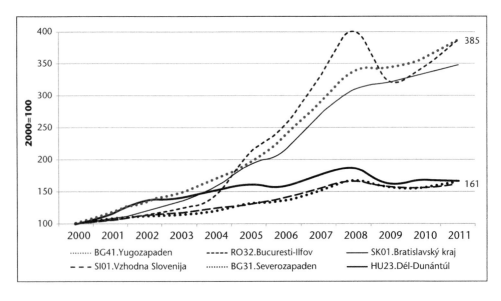

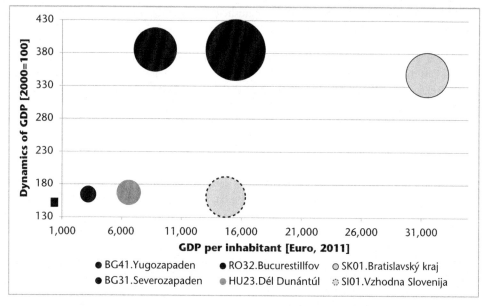

Figure 15.2 (a) Economic resilience of the best and worst performing regions of Central and Eastern Europe, 2000–2011 – dynamics of GDP at current market prices in million euro (2000 = 100); and (b) Economic portfolio of the best and the worst performing regions of Central and Eastern Europe – GDP per inhabitant (2011), dynamics of GDP (2011, 2000 = 100), value of the GDP (2011)

Source: Author's calculation based on Eurostat data.

- *regional economic development*: the lowest GDP per capita in Bulgaria, industrial sectors are connected with old branches (textile and clothing, food and beverage production, metallurgy, metal processing, machine building);
- *R&D, innovation*: low employment in high-tech and knowledge-intensive activities (less than 2 per cent); lack of critical mass regarding national universities and R&D; weak links between science and industry.

Almost the same factors apply to *SI01: Vzhodna Slovenija*, limiting the resilience of the region and enhancing its sensitivity to exogenous disruptions. These include (Operational, 2007):

- *regional economic development*: the poorest region in Slovenia (but the GDP per capita is 72.5 per cent of the EU28 average); a low level of employment in services; high employment in the secondary sector and in the agriculture sector;
- *R&D, innovation*: a weak position of innovation firms; low level of employment in knowledge-intensive activities; lack of plots equipped for investments; lack of cooperation patterns among education, research institutions and business.

Examining the three CEE regions with very high dynamics of GDP between 2000 and 2011, i.e. BG41: Yugozapaden (385), RO32: București-Ilfov (385) and SK01: Bratislavský kraj (348), we find that their resilience and successful transition factors are undoubtedly linked with the advantages of a socio-economic and institutional–political concentration of activities generated by a central city that is simultaneously the capital (i.e. Sofia, Bucharest, Bratislava). This is expressed in the absolute volume of GDP generated by these regions (see Figure 15.2b).

The *BG41: Yugozapaden* region is distinguished by the following factors reinforcing its resilience:[3]

- *regional economic development*: the most developed region with the capital city; the largest share of the population; the highest employment rate; the region contributes 48.6 per cent of Bulgaria's GDP; its former industrial sector was replaced in importance by the services sector;
- *R&D, innovation*: the region with the highest concentration of universities and R&D infrastructure; the highest employment in high-tech industries and knowledge-intensive activities.

When analysing the details of the GDP dynamics trends in Bulgaria's Yugozapaden's region it should also be noted that the path of its development was not weakened after the 2008 global financial crisis.

Regarding the *RO32: București-Ilfov* region, we have to note that its GDP dynamics index was the highest among all 53 surveyed regions of the CEE in 2005–2008. There was a sharp decline of the region's GDP dynamics in 2009, but it increased again in 2010–2011. The economic resilience of the București-Ilfov region results from the following facts (Regional, 2007):

- *regional economic development*: a high concentration of population within Bucharest – the biggest urban agglomeration in Romania; a high employment rate; a high level of educational attainment; very low unemployment rate; the share of the service sector workplaces grew from 53.1 per cent in 1995 to 75.4 per cent in 2005; the highest density of SMEs in the country;
- *R&D, innovation*: highly attractive techno-economic environment due to the existing institutional structure; specialised human capital; the highest FDI value (60.6 per cent of total value in the country in 2005).

The third CEE region with very high GDP dynamics in the years 2000–2011 is *SK01: Bratislavský kraj* in Slovakia. Its path of growing GDP is similar to BG41: Yugozapaden, which means that despite the global financial crisis in 2008, GDP dynamics were maintained. The factors that determine the high economic resilience of the region include (Program hospodárskeho, 2006):

- *regional economic development*: a diverse economic base including chemical and automotive industries as well as agri-food production (gardening, fruits, wine); the highest economic values in relation to other regions (level of employment and GDP); a significant concentration of FDI; a high share of services;
- *R&D, innovation*: advanced ICT in public and private sectors; a well-developed network of institutions supporting the economy; a high proportion of population with higher education; significant potential in universities and R&D institutions.

Undoubtedly, the region also benefits from the proximity and the trans-border location of the Vienna and Budapest agglomerations.

Findings – the technological dimension of resilience and transition in CEE countries

The analysis of technological resilience and transition within the CEE countries was carried out using the dynamics index of patent applications submitted to the EPO (see Figure 15.3a).

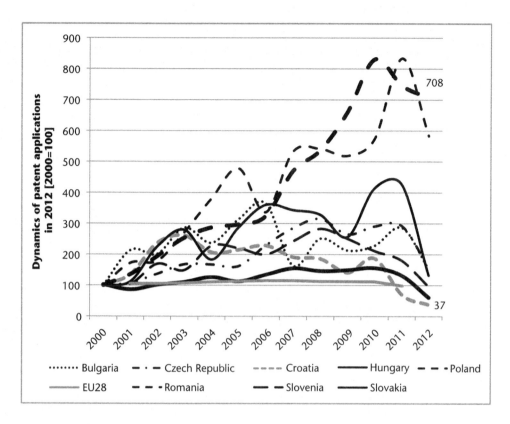
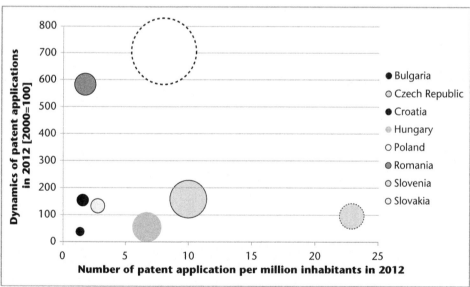

Figure 15.3 (a) Technological resilience of Central and Eastern European countries, 2000–2012 – dynamics of patent applications to the EPO (2000 = 100) and (b) Central and Eastern European countries' technological portfolio – patent applications to the EPO per million inhabitants (2012), dynamics of patent application to the EPO (2012, 2000 = 100), number of patent applications to the EPO (2012)

Source: Author's calculations based on Eurostat data.

According to its findings, countries can be divided into two groups. The first group, including Poland and Romania, is characterised by very high dynamics regarding the number of patent applications to the EPO in the period 2000–2012. Their dynamics figures stayed high even through the collapse in 2010–2012.

The second group of countries, i.e. Bulgaria, Czechia, Croatia, Hungary, Slovenia and Slovakia, is characterised by significant volatility regarding the dynamics of the patents submitted. In Czechia, Slovakia and Bulgaria the number of patent applications to the EPO fell sharply in 2010–2012. In 2012, the growth rate of the number of such applications submitted by Croatia, Hungary and Slovenia was lower than their 2000 rate. Despite the above, these countries recorded high (Croatia, Hungary, Slovenia) or very high (Slovakia, Bulgaria and Czechia) dynamics of GDP in relation to the average dynamics of its value in the EU28 in the years 2000–2011 (see Figure 15.1a).

Analysing the technological portfolio of CEE countries demonstrates not only significant differences in the technological resilience of the CEE countries, but also a large gap between these countries and the EU28 in terms of the absolute number of patents.

The analysis of the regional dimension of technological resilience in the CEE region (see Figure 15.4) clearly shows the high level of diversification regarding the dynamics of patents submitted to the EPO in the years between 2001 and 2011. Analysis of the three CEE regions with the highest dynamics and the highest number of patents submitted shows that a leader in this area is the PL21: Małopolskie region (the index of patents[4]: 16,237), followed by the PL12: Mazowieckie region (the index of patents: 11,203) and HU10: Közép-Magyarország (the index of patents: 10,138).

In the case of the Polish *PL21: Małopolskie*, the trend of the strengthening technological potential results from the following factors (Małopolski, 2007):

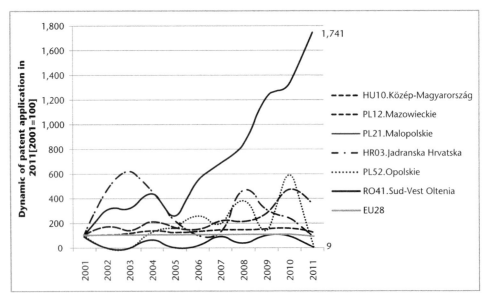

Figure 15.4 Technological resilience of the Central and Eastern European regions with the best and worst performers, 2001–2011 – dynamics of patent applications to the EPO (2001 = 100)

Source: Author's calculations based on Eurostat data.

- *regional economic development*: high investment attractiveness, good availability of qualified staff; active business environment institutions; a high level and a wide range of cultural events;
- *R&D, innovation*: high position of Krakow academic environment; good conditions for development of online services; strong and competitive enterprises within the ICT industry; highest investment and employment growth in R&D in the country.

The high position of the second region, *PL12: Mazowieckie* is linked to the following attributes (Regionalny, 2011):

- *regional economic development*: high (and growing) level of economic development; Warsaw's outstanding economic potential; large concentration of FDI in the Warsaw agglomeration; strong businesses of national and international importance; a well-developed market services sector; the highest level of exports and imports in the country;
- *R&D, innovation*: the highest national share of expenditure on R&D; high-value research equipment; the highest share of people employed in R&D in the country.

The third fastest growing region in CEE in the technological dimension is *HU10: Közép-Magyarország*. Its high technological position is due to the occurrence of the following factors (Közép-magyarországi, 2006; Viturka et al., 2009):

- *regional economic development*: the strong economic position of the capital city (Budapest); national and international centre of finance, trade and services, and government institutions; a high proportion of residents with higher education; cultural diversity;
- *R&D, innovation*: a high concentration of knowledge-base in the region; higher education, R&D and innovation capacity; a significant share of creative industries.

An extremely different (i.e. the most unfavourable) situation, in terms of technological resilience and transition capacity, can be experienced in the following regions of CEE: HR03: Jadranska Hrvatska (the index of patents: 35), PL52: Opolskie (the index of patents: 3) and RO41: Sud-Vest Oltenia (the index of patents: 4).

The first of them, *HR03: Jadranska Hrvatska* has a weak resilience position in the technological dimension mainly due to the following factors (Regional Competitiveness, 2007):

- *regional economic development*: lack of high-value added production of SMEs; lack of cooperation and networking among firms; insufficient and inadequate soft services for SMEs; an absence of vertical and horizontal institutional coordination of SMEs and regional development policies;
- *R&D, innovation*: SMEs: low investment in R&D; low share of labour and capital in sectors which drive competitive economies (like biotech, ICT); undeveloped seed and venture capital market.

Among the surveyed regions of the CEE, *PL52 Opolskie* has shown the lowest dynamics of patent applications to the EPO (4 in 2011, 2001 = 100) and the lowest value of the index of patents (10). The poor technological resilience of this region may result from the following factors (Regional Operational, 2007):

- *regional economic development*: low rate of economic growth; underdeveloped business support institutions; low level of entrepreneurship; underdeveloped service sector;

- *R&D, innovation*: very low level of expenditures on R&D; low potential of the high-tech sector; low competitiveness of scientific and research offers.

RO41: Sud-Vest Oltenia is the last of the CEE regions with the worst indicators of technological resilience. The region showed significant fluctuations in the dynamics of the number of patents between 2001 and 2011. The factors responsible for the low resilience of this region are as follows (Regional, 2007):

- *regional economic development*: a relatively low share of population in urban areas; low employment rate; a high share of employment in agriculture and forestry; migration from urban to rural areas and abroad; the old energy and mining sectors; mono-industrial centres (Balş, Tg. Cărbuneşti, Rovinari Motru);
- *R&D, innovation*: agriculture, as the predominant sector of the region's economy, is based on less advanced technology; a weak business support infrastructure; a low potential of R&D infrastructure.

Findings – the environmental–spatial dimension of resilience and transition in CEE countries

Environmental and spatial factors are important components of regional resilience in CEE countries and their regions. In the case of the environmental dimension, the great challenge is to improve energy efficiency and to improve the quality of the natural environment; while in the spatial dimension the biggest claim is to boost the efficiency of land use and ameliorate the quality of the built environment. These challenges mainly result from global competition and environmental policies carried out at the EU level. In this section of this chapter we focus on issues related to the energy system, and how these compound the efficiency of economies. There is a gap between CEE and Western European countries in terms of the efficiency of energy systems. During the last few years the CEE countries have tried to reduce the gap in this field.

The indicator measuring the efficiency of an energy system in the most synthetic way is the ratio between the gross inland consumption of energy and the gross domestic product (GDP) for a given calendar year. The gross inland consumption of energy is calculated as the sum of the gross inland consumption of five energy types: coal, electricity, oil, natural gas and renewable energy sources. Since gross inland consumption is measured in kgoe (kilogram of oil equivalent) and GDP in 1,000 euro, this ratio is measured in kgoe per 1,000 euro. In 2013 the average value of the ratio for the EU28 countries was 141.6 kg of oil equivalent per 1,000 euro. All eight CEE countries had higher rates than the EU28 average. The highest value of indicators throughout the period of 2002–2013 was measured in Bulgaria, and it was significantly higher than in other CEE countries (610.6). Czechia (353.8), Slovakia (337.2) and Romania (334.7) were the countries with a relatively high index level, while the lowest indexes were measured in Croatia (219.5) and Slovenia (225.8). The gap between the CEE and Western European countries is significant. The lowest energy intensity of the economy can be observed in the UK, Ireland and the Scandinavian countries (for example in 2013: UK – 102.7; Ireland – 82.4; Denmark – 86.6). The energy efficiency of economies improved in all eight CEE countries during the period 2002–2013. A change of course can be observed between 2008 and 2010, when energy intensity increased in all of the CEE countries. The degree of change was generally the same in all countries, and it was mainly caused by the lower GDP. After 2010 we can observe a decrease of the energy intensity ratio. During 2002–2013 the greatest progress in reducing energy intensity was observed in Slovenia and Romania.

The second interesting indicator measuring adaptability in the field of environmental resilience is the changes in energy sources, which tend to be more and more environmentally friendly. In the period 2004–2013 all the CEE countries increased their share of electricity generated from renewable sources. Croatia, Romania and Slovenia are countries where more than 30 per cent of electricity is produced from renewable sources. In these countries the share of renewable resources in the energy system is higher than the average for the EU28 countries. In Hungary, Poland and Czechia the majority of electricity is generated by non-renewable sources. However, these countries have made the biggest progress to shift from fossil fuels to renewable energy. For instance, Poland, in comparison to 2004 data, increased its share of electricity generated from renewable sources by 509 per cent.

The third indicator measures the diversity of energy sources. The CEE countries invest more and more in solar collectors as one of the renewable sources. According to incomplete data, one can observe the highest rate of growth in Croatia (increase from 11,000 sq.m in 2007 to 143,000 sq.m in 2013), Bulgaria (increase from 28,000 sq.m in 2007 to 309,000 sq.m in 2013). The rest of the countries (except Slovenia) also present stable growth of solar collector surface. This indicator, together with the general ratio of electricity generated from renewable sources, shows that CEE countries are gradually following the path of adaptability of energy systems, focusing their efforts on higher efficiency using more diverse renewable sources.

Findings – the institutional–political dimension of resilience and transition in CEE countries

In order to prevent problems concerning the performance of administrative functions, a multi-tier structure of public administration appears to be a necessity. Taylor (1993) states that governments as bureaucratic organisations have to decentralise some of their functions along the geographical scale in order to attain higher efficiency.

Central governments decentralise some of their functions to subnational governments, i.e. to self-governing structures. Decentralisation means the devolution of state functions to autonomous regional governments which can act on their own behalf. Local or regional autonomy is introduced, referring sometimes to the existence and traditions of local or regional societies or communities, which are then institutionalised by what some have called a 'local or regional state' and are represented by a local or regional government (Cockburn, 1977). In this case, we are dealing with regional self-administration.

For state administration, a hierarchical 'top-down' system is typical. State policies and preferences are accomplished via public administration only. In contrast to state administration, self-administration applies a 'bottom-up' approach, based on the needs of territories and entities present in the area in question.

The power of self-administration essentially expresses the rate of democracy in the given country. We are dealing with the reflection of the distribution of political power alongside the hierarchical axis state–region–locality. In the case that self-governing structures are not strong enough to act as partners or to balance the power of state administration, the threat of the emergence of power centralisation becomes rather pressing. In this context, the principle of subsidiarity should be called to attention.

There is no doubt that political and institutional factors influenced the development of the examined countries enormously. What Poland, Hungary, Czechia, Slovakia, Romania and Bulgaria have in common is their shared past. After the Second World War they all followed the path of a totalitarian political system and a controlled economy based on central planning. Democratic political systems and market economies did not develop in the countries in focus, or, more precisely, the Visegrad 4 together with Romania and Bulgaria avoided that economic–political mainstream. Effects of these developments are palpable even today. On the other hand, these countries are relatively heterogeneous in terms of municipal systems as well as population and area size. The same applies to the economies of these countries (Lux, 2008).

Naturally, these parameters also influence the administrative–political structure of the individual countries, as well as the performance of regional administration. Typically, administrative divisions within the countries only loosely comply with natural regions. Moreover, quick changes and lack of coordination are often experienced symptoms of administrative–political developments in virtually all of these countries.

Vivid debates related to the political–administrative division of the country were concomitant in Poland in the first decade after the regime change. Questions related mainly to decentralisation mechanisms and to the number and organisation of administrative units. In 1998, it was decided that the country would be divided into communities (*gminy*), districts (*powiaty*) and voivodeships (*województwa*). However, the amount of 49 voivodeships enfeebled self-governing power, which culminated in political disputes. Thus, the number of voivodeships was reduced to 16, which are in compliance with historical regions within the country (Dolata, 1998). In 2000, Poland managed to harmonise the division of NUTS and LAU territorial units in accordance with EU needs. With the exception of NUTS-3, all of the European regional administrative levels are in consonance with the current administrative division of the country.

In Hungary, the traditional self-governing administrative regions are called *megyék*. However, their stability was not solid during certain periods. In 1990, altogether 19 *megyék* were renewed and Budapest received a special administrative status. Still, the manoeuvring space for regional self-governments was very constrained at the time: the self-government emphasis was put on the level of municipalities. In 1998, seven statistical and planning regions (*tervezési-statisztikai régiók*) composed of *megyék* came into existence. These regions act as the NUTS-2 level of administration; however, these *megyék*-level regions have virtually no space for influencing their development. From 1 January 2013 a new administrative division of the country was introduced. New districts (*járások*) were added within the *megyék*.

As to Czechia, self-government for a long time existed merely at the level of communities. District offices (*okresy*) that covered state administration in a way represented 'little state kingdoms'. After long debates and several postponements, the reform of public administration took place in 2000. From 2001 on the structure of the administrative division of the country has been transformed to communities (*obce*), districts (*okresy*) and regions (*kraje*). In 2003 district offices ceased to exist, and that is why *okresy* serve mainly as statistical units, although some of their functions remained untouched. The number of regions (*kraje*) is 14. As for the NUTS-2 level, there are altogether eight units composed of 14 regions.

Slovakia has traditionally been the part of other, larger states, which naturally affected the Slovak political–administrative regional division. The peculiar feature of Slovakia was that the administrative centres managing Slovak territory were often located beyond the current Slovakian border. Decentralisation and a new administrative arrangement of the country took place in 1996 when eight new regions (*kraje*) and 79 districts (*okresy*) were set up. NUTS classification was applied in 1998. NUTS-2 regions were formed under the name *oblasti*, which are composed of self-governing regions. Similar to Hungary and Czechia, there are distinct centripetal tendencies in the relation between the capital city and other parts of the country.

Romania faces numerous obstacles stemming from weak administrative–political decentralisation. The local administrative level can be considered very integrated (centralised) thanks to the systemic integration process taking place in the second half of the twentieth century (Illés, 2002). The country is divided into 42 counties and the self-governing status of the whole system is rather disputable (Horváth, 2003). The eight development regions were established in 1995 with the help of the PHARE programme. The entire conception was consulted on with foreign professionals. Since functions performed by the regions turned out to be formal rather than real, embeddedness as well as networking represent non-negligible moments of the country's regional development (Pálné Kovács, 2007).

Last but not least, Bulgaria has some specific features. After 1990, nine administrative regions were created. These were replaced in 1998 in the framework of an administrative–political reform by 28 exclusively state-managed public administrative units led by governors. The role of regional self-government is performed at the level of the strongly integrated 262 local governmental units (Drumeva, 2001). The establishment of six NUTS-2 regions was accomplished with resources from the European Union. Their role, however, remains rather formal; they exist only for the sake of built-in, centralised distribution of public funds (Alexandrova, 2005).

The affinity between informal institutions and local and regional development is seldom discussed; nonetheless it cannot be ignored, as 'internal' characteristics of the population form the basis of 'external' processes that can be directly contemplated. That is why values, modes of behaviour, routines and habits should be studied carefully.

The immense spatial differentiation of informal institutions has been confirmed, for instance, by the notorious Inglehart values scheme, which provides us with the basic overview of values in different cultures. It is worth noting that although the countries of Central and Eastern Europe are usually perceived as one homogeneous set, in reality their values show quite a large heterogeneity. This can be treated as a point of departure for further assessments.

It is natural that the long-lasting socialist system also affected, or more precisely deformed, the informal institutions of the populations of the countries in question. Obviously, territoriality represents one of the most relevant components of human lives. Consequently, it is hardly surprising that the consciousness of local or regional development as an integral part of public affairs in developed countries was high compared to countries that were much more subdued. This dismal situation was further aggravated by ubiquitous central planning, which virtually eliminated the notion of regional development. In this socialist milieu, isolated from world developmental trends, people were barely able to realise the importance of regional development for the improvement of their living standards.

Regional identity is of utmost importance in this context (Sucháček, 2005). It can be generally explained as the positive relationship of the individual to the area in which he or she lives. Essentially, regional identity also reflects the readiness of the population to act on behalf of the given region. It acts as an immaterial regional clue, *sui generis*.

One cannot omit that people that belong to a particular region are bound to it by interests, emotions and feelings. In that case we can delimit territory according to a seemingly hidden layer of (common) interests and emotional ties. This concept does not reveal its finality but, compared to cultural identity, it is much more concrete, as a particular area has finite borders that meet the borders of another territory. Contrary to the traditional approaches in local and regional studies, we concentrate on inherent, inborn characteristics. Regional identity can be measured at different territorial ranks. It is important to focus namely upon local and national levels. We found a strongly differentiated picture of local and national identities in Central and Eastern European countries (see also Plecitá, 2004).

The largest gap between mean local and national identities can be observed in Czechia and Hungary. In the case of other countries, the difference is not so palpable, but national identities are generally stronger than local ones. As already indicated, during the socialist era, regional or local problems were silently swept under the rug. After 1989, debates concerning local or regional development were generally avoided. National identity consequently served as a certain substitute for suppressed local and regional identities. The only country that had larger mean local identity than that at the national level was Bulgaria.

However, regional identities tell us nothing about national pride. In every nation, national pride is related to national achievements and failures. Should people bear genuine emotional ties to their country, they are proud of their nationality. The lowest national pride among the countries in question was found in Bulgaria.

As Plecitá (2004) underlines, national pride is low in countries experiencing discontinuity of states, regimes and economic systems. Historical experience might be the source of the low national pride of the Czech population. Moreover, correlation analysis showed that national pride significantly correlates with satisfaction with the development of democracy, satisfaction with the development of the economic situation of the country and, last but not least, the evaluation of justice in the societies of the Central Eastern European countries.

Final remarks and conclusions on CEE regions' resilience and transition

In the years between 2000 and 2011, CEE countries were characterised by meaningful differences in terms of GDP growth, but for the entire group the dynamics of GDP were significantly higher than the EU28 average. All of the CEE countries have shown a high sensitivity to the impact of the global financial crisis, while after 2009 there has been a renewed increase in their GDP growth. A greater level of sensitivity characterises the economy of Romania, Czechia, Slovakia, Hungary and Slovenia.

Despite their significant growth dynamics, transition in the CEE countries – in relation to the size of the GDP and GDP per capita calculated for the EU28 – is still very low. This indicates the existence of a serious gap in the socio-economic development of the CEE countries.

CEE countries have different dynamics in technological development. Poland and Romania developed rapidly in this field over the period 2000–2012. In other CEE countries (Bulgaria, Czechia, Croatia, Hungary, Slovenia and Slovakia) dynamics of patent applications to the EPO significantly collapsed after 2010.

In regional terms, in the years 2000–2011, the weakest economic regions of the CEE were growing much faster than the average growth of the GDP in the EU28. However, this group of regions proved to be sensitive to negative changes caused by the global financial crisis, since after 2008 their growth slowed down. This points to the weakness of their economic bases, which do not have sufficient capacities to adapt to the changing external economic environment quickly enough.

Detailed analysis shows that the growth paths of CEE regions varied considerably after exogenous changes (such as: EU accession, decay of the old industries, global financial crisis). In some cases external disruptions made a radical positive impact (BG41: Yugozapaden, RO32 Bucureşti-Ilfov, SK01: Bratislavký, PL21: Małopolskie, PL12: Mazowieckie, HU10: Közép-Magyarország), while in others they led to negative outcomes (HU23: Dél-Dunántúl, BG41: Severozapaden, SI01: Vzhodna Slovenija, HR03: Jadranska Hrvatska, PL52: Opolskie, RO41: Sud-Vest Oltenia).

The CEE regions whose economies are based on services usually perform better than those mainly dealing with industry or agricultural activities. In a few CEE regions, especially in those with a central city which is often the capital of the state (Yugozapaden – Sofia, Bucureşti-Ilfov – Bucharest, Bratislavký – Bratislava, Mazowiecki – Warsaw, Közép-Magyarország – Budapest), a desirable combination of resilience attributes has begun to take shape (*diversity, efficiency, adaptation*). In general words, metropolitan areas – because of their higher economic–technological and social capacities – usually cope with external disruptions more successfully, and this also applies to the CEE regions. However, in some cases when economic–technological and social capacities are not rooted in a particular region, a metropolitan area may be more exposed to exogenous disruptions (like Bucharest in the București-Ilfov region, which suffered a severe drop in growth dynamics after 2008).

There are a few other factors that have also been strengthening the CEE regions' economic–technological resilience in the first decade of the twenty-first century. These include: quality of human capital and its high concentration, entrepreneurship attitudes, networking patterns, diversity of economic base, business support infrastructure and R&D infrastructure, along with linkages between R&D and the business sector. They facilitate a capacity for transition, and enhance the development of new regional value chains adjusted to external market changes. However, the capacity to create or attract new technologies along with the regional value chains is impacted by many other factors, such as: advantages of location, local knowledge assets, image of a particular place, cultural offerings and openness to cooperation.

By analogy, the regional economic–technological resilience of the CEE regions is weakening mainly due to factors such as: a weak urban network, an underdeveloped R&D infrastructure, low level of FDI, low employment rate, high share of employment in agriculture, low level of employment in R&D and knowledge-intensive sectors, lack of a critical mass of universities and R&D institutions, weak links between science and businesses, and lack of awareness of external economic trends (which are not considered in regional development strategies). The list of the factors weakening the resilience of the CEE regions should be supplemented by at least one other: the socialist 'heritage' of old industrial districts (RO41: Sud-Vest Oltenia). The concentration of un-restructured old industries, because of their low efficiency and low level of knowledge-intensive activities, exposes CEE regions to greater vulnerability in the context of global market turmoil.

Environmental and spatial factors also play a crucial role in the resilience and transition of the CEE countries and their regions. Both efficiency and diversity of energy sources proved to increase the level of resilience in the face of environmental pressures. Selected data collected for eight CEE countries show high dynamics of changes that focus on the shift from traditional to more renewable sources of energy. Nevertheless, we still find countries like Poland, Czechia or Hungary where energy systems are mainly based on fossil fuels. This causes negative effects on the natural environment and the quality of life, which manifests in increased emission of greenhouse gases and a high level of particulate matter (PM10). The efforts to transform the energy system result in lower and lower energy intensity of economies. The pace of transformation has been constant; minor fluctuations were experienced only during the economic crisis in 2008–2010.

As the chapter discusses, the countries we analysed generally suffer from ill-fitting administrative and political structures, which considerably increases transaction costs. Totalitarian regimes centralised these important administrative and political structures, which also affected the social fabric shaping the development of the Central and Eastern European economies. During the transformation shock, both physical and intangible structures reproduced themselves, and the centripetal spatial pattern of the transition/post-transition countries remained basically unchanged. However, some of the larger countries also show some signs of benign decentralisation, which applies for instance to Poland.

The concept of resilience was born in the context of matured, Western institutions, and should be understood as a natural continuation of neo-endogenous conceptions of regional development, which apply largely to the economically advanced world. Thus, this conception has only limited validity for transition/post-transition economies, which institutionally and evolutionarily constitute a completely different story (Sucháček, 2008, 2014).

The evolutionary conceptualisation of post-socialist transformation proposed by Grabher and Stark (1997) draws on the above-mentioned arguments. Pendall et al. (2007) also depict slow burns or slow-moving crises. This also holds true for previous systems in transition/post-transition countries embodying a peculiar blend of shocks and slow burns. Transition/post-transition economies were exposed to innumerable shocks of various kinds combined with slow burns during both the socialist era and the last 20 years of transformation. These shocks markedly differed from the Western world. Moreover, they were aggravated by the totalitarian political system and the absence of a market during socialism, and by the painful return to a natural developmental trajectory and general external pressure for modernisation in the last 20 years. We are thus entitled to talk about a peculiar series of resilience or responses to almost permanent, and at the same time specific, shocks in Central and Eastern Europe. Paradoxically, although deformed historical trajectories are closely interconnected with these peculiar 'series of resilience', contemporary projections of these previous occurrences are seldom considered.

Thus, resilience and adaptability applying to more affluent Western counterparts can barely provide us with satisfactory explanations of contemporary processes in Central and Eastern Europe. Transition/post-transition economies continue to face the great challenge concerning the creation of original, non-copied and tailored approaches to regional development in Central and Eastern Europe.

Acknowledgements

The publication of this research has been supported by Project #104985 of the Hungarian National Research, Development and Innovation Office; and has been prepared according to the methodology of the research project titled *Urban Resilience Concept and Post-industrial Cities in Europe* (2011–2014), financed by the Polish National Science Centre, No. DEC-2011/01/B/HS5/03257.

Notes

1. European Patent Office.
2. https://ec.europa.eu/growth/tools-databases/regional-innovation-monitor/base-profile/severozapaden
3. https://ec.europa.eu/growth/tools-databases/regional-innovation-monitor/base-profile/yugozapaden
4. The index of patents = dynamics of patent application to the EPO in 2011 (2001 = 100) multiplied by the number of patents per 1 million inhabitants in 2011.

References

Alexandrova, S., 2005. Grant Transfers and Financial Supervision in Bulgaria: Principles and Practice. In: Šević, Ž., ed. *Fiscal Decentralisation and Grant Transfers in Transition Countries*. Bratislava: NISPAcee, pp. 212–232.
Ayres, R. and van der Lugt, C., 2011. Manufacturing. Investing in Energy and Resource Efficiency. In: Green Economy Report, Towards a Green Economy: Pathways to Sustainable Development and Poverty Eradication, UNEP.
Bosher, L. and Coaffee, N., 2008. Editorial: International Perspective on Urban Resilience. *Urban Design and Planning*, 61(DP4), pp. 145–146.
Boyd, R. and Richerson, P.J., 1985. *Culture and the Evolutionary Process*. Chicago, IL: University of Chicago Press.
Briguglio, L., Cordina, G., Bugeja, S. et al., 2006. *Conceptualizing and Measuring Economic Resilience*. Department of Economics, University of Malta. Mimeo.
Capello, R., 2000. The City Network Paradigm: Measuring Urban Network Externalities. *Urban Studies*, 37(11), pp. 1925–1945.

Champion, A.G., 2001. A Changing Demographic Regime and Evolving Polycentric Urban Regions: Consequences for the Size, Composition and Distribution of City Population. *Urban Studies*, 38(4), pp. 657–677.
Cockburn, C., 1977. *The Local State*. London: Pluto Press.
Cooke, Ph., 2008. Regional Innovation Systems, Clean Technology and Jacobian Cluster-platform Policies. *Regional Sciences, Policy and Practice*, 1(1), pp. 23–45.
Cruz, S.S., Costa, J.P., de Sousa, S.A. and Pinho, P., 2013. Urban Resilience and Spatial Dynamics. In: A. Eraydin and T. Taşan-Kok, eds. *Resilience Thinking in Urban Planning*. Dordrecht: Springer, pp. 53–69.
Desmet, K. and Rossi-Hansberg, E., 2009. Spatial Growth and Industry Age. *Journal of Economics Theory*, 144, pp. 2477–2502.
Dolata, S., 1998. Problémy teritoriálního členění státu a společenského uspořádání jako determinanty regionální politiky na příkladu Polska', *Regio 98. Sborník referátů z mezinárodní vědecké konference v mariánských Lázních a v Erfurtu*. Erfurt: Universität Erfurt, pp. 17–20.
Drobniak, A., 2014. Factors Determining City's Development Dynamics. In: A. Drobniak, ed. *Urban Resilience Concept and Post-industrial Cities in Europe*. Katowice-Gliwice: University of Economics in Katowice, Helion.
Drumeva, E., 2001. Local Government in Bulgaria. In: E. Kandeva, ed. *Stabilization of Local Governments*. Budapest: OSI/LG, pp. 141–179.
Eraydin, A. and Taşan-Kok, T., 2013. Introduction: Resilience Thinking in Urban Planning. In: A. Eraydin and T. Taşan-Kok, eds. *Resilience Thinking in Urban Planning*. Dordrecht: Springer, pp. 1–16.
Feyrer, J., Sacerdote, B. and Stern, A.D., 2007. Did the Rustbelt Become Shiny? A Study of Cities and Counties that Lost Steel and Auto Jobs in the 1980s. *Brookings Wharton Paper on Urban Affairs*, pp. 41–102.
Gerst, J., Doms, M. and Daly, M.C., 2009. Regional Growth and Resilience: Evidence from Urban IT Centers. *FRBSF Economic Review*, pp. 1–11.
Gleaser, E.L. and Saiz, A., 2004. The Rise of the Skilled City. *Brookings Wharton Paper on Urban Affairs*, pp. 47–94.
Godschalk, D.R., 2003. Urban Hazard Mitigation: Creating Resilient Cities. *Natural Hazards Review*, 4(3), pp. 136–143.
Grabher, G. and Stark, D., 1997. Organizing Diversity: Evolutionary Theory, Network Analysis and Postsocialism. *Regional Studies*, 31(5), pp. 533–544.
Hess, D., 2013. Sustainable Consumption, Energy and Failed Transitions: The Problem of Adaptation. In: M.J. Cohen, H. Szejnwald Brown and Ph.J. Vergragt, eds. *Innovations in Sustainable Consumption. New Economies, Socio-technical Transitions and Social Practices*. Cheltenham: Edward Elgar, pp. 159–178.
Hill, E., Clair, T., Wial, H. et al., 2010. *Economic Shocks and Regional Economic Resilience*. George Washington, Urban Institute. Building Resilience Region Project. Conference on Urban and Regional Policy and Its Effects: Building Resilience Regions, Washington, DC, May.
Horváth, Gy., ed., 2003. *Székelyföld* [Székely Land]. Budapest–Pécs: Dialóg Campus.
Illés, I., 2002. *Közép- és Délkelet-Európa az ezredfordulón. Átalakulás, integráció, régiók* [South and Southeast Europe at the turn of the millennium. Transformation, integration, regions]. Budapest–Pécs: Dialóg Campus Kiadó.
Közép-magyarországi 2006. operatív program 2007–2013 [Operative programme 2007–2013]. Budapest: A Magyar Köztársaság Kormánya, 6 December.
Lang, T., 2011. Urban Resilience and New Institutional Theory – A Happy Couple for Urban and Regional Studies? In: B. Müller, ed. *German Annual of Spatial Research and Policy 2010, Urban Regional Resilience: How Do Cities and Regions Deal with Change?* Berlin: Springer-Verlag, pp. 15–24.
Lansford, T., Covarrubias, J., Carriere, B. and Miller, J., 2010. *Fostering Community Resilience. Homeland Security and Hurricane Katrina*. Farnham: Ashgate.
Lux, G., 2008. *Industrial Development, Public Policy and Spatial Differentiation in Central Europe: Continuities and Change*. Pécs: Centre for Regional Studies. (Discussion Papers, 62).
Małopolski 2007. Regionalny Program Operacyjny na lata 2007–2013. Zarząd Wojewodztwa Małopolskiego, Kraków, październik.
Neuman, M., 2005. Compact City Fallacy. *Journal of Planning Education and Research*, 25(1), pp. 11–26.
North, D., 1990. *Institutions, Institutional Change and Economic Performance*. Cambridge: Cambridge University Press.
Nunn, N., 2009. *The Importance of History for Economic Development*. NBER Working Paper 14899. Cambridge, MA: National Bureau of Economic Research.

Operational, 2007. *Programme for Strengthening Regional Development Potentials for Period 2007–2013*. Ljubljana: Republic of Slovenia, Government Office for Local Self-Government and Regional Policy, 6 March.

Pálné Kovács, I., 2007. Fragile Structures of Public Administration in Southeast-Europe. In: Z. Hajdú, I. Illés and Z. Raffay, eds. *Southeast-Europe: State Borders, Cross-Border Relations, Spatial Structures*. Pécs: Centre for Regional Studies, pp. 99–118.

Pendall, R., Foster, K.A. and Cowell, M., 2007. *Resilience and Regions: Building Understanding of the Metaphor*. Working Paper 2007-12. Berkeley, CA: MacArthur Foundation Research Network on Building Resilient Regions, Institute for Urban and Regional Development, University of California.

Plecitá, K., 2004. *National Identity in the Eleven CEE Nations: Where Do We Belong?* Institute of Sociology, Academy of Sciences of the Czech Republic. www.soc.cas.cz/projekty/ewelcome/pdf/PP_nat_ident_WA_02.pdf [Accessed December 2015].

Program hospodárskeho 2006. *a sociálneho rozvoja Bratislavského samosprávneho kraja na roky 2007–2013*. Cast 'Analýza ekonomických a sociálnych východísk'. Bratislavský samosprávny kraj, Bratislava, 2006. www.region-bsk.sk/clanok/program-hospodarskeho-a-socialneho-rozvoja-na-roky-2007-2013-892123.aspx

Regional 2007. *Operational Programme 2007–2013*. Government of Romania, Ministry of Development, Public Works and Housing, Bucharest.

Regional Competitiveness 2007. *Operational Programme 2007–2009*, Republic of Croatia, September. www.regionalna-konkurentnost.hr/default.aspx?id=3583

Regional Operational 2007. *Programme of the Opolskie Voivodeship 2007–2013*, Executive Board of the Opolskie Voivodeship, Opole, October.

Regionalny 2011. *Program Operacyjny Województwa Mazowieckiego 2007–2013*. Urząd Marszałkowski Województwa Mazowieckiego, Warszawa, grudzień.

Simme, J. and Martin, R., 2009. The Economic Resilience of Regions: Towards an Evolutionary Approach. *Cambridge Journal of Regions, Economy and Society*, pp. 1–17.

South, 2007. *Transdanubia Operational Programme 2007–2013*. Budapest: Government of the Republic of Hungary.

Sucháček, J., 2005. *Restrukturalizace tradičních průmyslových regionů v tranzitivních ekonomikách*. Ostrava: VŠB-TU.

Sucháček, J., 2008. *Territorial Development Reconsidered*. Ostrava: VŠB-TU.

Sucháček, J., 2014. Regional Development in Central and Eastern Europe: A New Approach. *Actual Problems of Economics*, 154(4), pp. 106–115.

Taşan-Kok, T., Stead, D. and Lu, P., 2013. Conceptual Overview of Resilience: History and Context. In: A. Eraydin and T. Taşan-Kok, eds. *Resilience Thinking in Urban Planning*. Dordrecht: Springer, pp. 39–51.

Taylor, P.J., 1993. *Political Geography*. Harlow: Longman.

Viturka, M., Žítek, V., Klímová, V. and Tonev P., 2009. *Regional Analysis of New EU Member States in the Context of Cohesion Policy*. Národohospodářský Obzor, 9(2), pp. 71–90.

Walker, B. and Salt, D., 2006. *Resilience Thinking: Sustaining Ecosystems and People Changing World*. Washington, DC: Island Press.

Welter-Enderlin, R., 2006. *Resilienz – Gedeihen trotz widriger Umstände*. Heidelberg: Carl-Auer-Systeme.

Wolfe, D.A., 2013. Resilience and Growth in Canadian City-regions. In: Ph. Cooke, ed. *Re-framing Regional Development. Evolution, Innovation and Transition*. London: Routledge, pp. 54–72.

16
New trends in Central and Eastern European transport space

Ferenc Erdősi

Introduction

In the past quarter-century Central and Eastern Europe's transport has been affected by the deepening of globalisation, which had begun earlier (and is seen significantly in technology and in continuous changes in modal split), but even more so by new processes and phenomena that were direct and indirect results of political/economic regime changes, which in and of themselves contributed to globalisation.

The author of this study has published close to one and a half thousand pages of research on CEE transportation (on general and branch characteristics and on the two grand regions) in a half dozen books (Erdősi, 2005a, 2005b, 2007, 2008, 2009, 2010) and numerous articles. Based on these and on the most important phases of the past few years this short study will summarise the most important processes and consequences seen in the sector (and the regions thereof) in the given transport space (in Central and Eastern Europe and the states of the former Soviet Union) from the early 1990s on.

Regarding the period before the regime changes, we will refer only to those processes which began earlier, or where a lack of knowledge of antecedents would make the interpretation of new processes too difficult. Due to space limitations we will only make use of data tables and map-based illustrations where necessary.

A poor heritage and the partial nature of technology change

At the time of planned economies in *CEE all transportation branches lagged behind those of the West* in terms of both quality and effectiveness, whereas in terms of transportation technology (particularly vehicles) the gap was enormous. Differences, however, varied among groups of countries or among countries. Unlike China, South Korea, etc., the developments of the transportation network (and partly the vehicle stock) to this day are of a reactionary nature, i.e. they lag far behind growing demand, compared to the situation in Western Europe. In most branches the gap is most evident in quality.

From the 1990s to the recent past, efforts taken to close the gap on the whole have been moderate or have produced partial success in technology and infrastructure quality. The (mostly developed) transportation modes and vehicle stock success stories happened mostly in terms of engineering level, the demand for which increased the most after the regime change, where the engineering foundation was typical of the developed world.

The 'socialist camp' strove to execute land, water and air transportation with *vehicles it produced itself*. Vehicle producers were helped through licensing agreements, but the quality of automobiles produced through such contracts usually lagged behind Western European models by 15–20 years, with a generation's worth of quality difference appearing between CEE and the West. A common characteristic of vehicles in all transportation branches was their high level of fuel consumption and emissions. In some Western European countries lorries and aeroplanes arriving from CEE were made to pay environmental duties. Environmental protection elements had appeared in the state and regional public spheres of socialist countries at the time, but support for them was weak (see Chapter 17).

Since the regime changes, the most advantageous change in the vehicle stock has taken place in the once 'sovietised' stock of aeroplanes. The output of post-soviet aeroplanes production fell to a fraction of its former self after losing its markets, given that it was unable to close the technology gap. For this reason, it took only a few years for Russian- and Ukrainian-produced 'kerosene gulping' aircraft to be replaced by Boeing and Airbus models. (In the Commonwealth of Independent States (CIS) region the process of replacing aeroplanes reached a proportion of 60–80 per cent, with the largest changes taking place on international routes.)

The gradual exchange of the public road vehicle stock took place through acquisitions from developed countries and started with the elimination of the environmentally most damaging two-stroke engine vehicles. The automobile factories built by transnational corporations produce for export at a rate of 85–97 per cent; however, their developed technology and high level of organisation have had a positive effect on the labour culture and professional expertise of the workforce (Fekete, 2015). With their high technological levels these factories serve as model islands (innovation centres) in their wider environment and through their networks of suppliers have become factors in regional development (e.g. through increasing employment) that cannot be ignored, from the Volga region to the Carpathian Basin and Poland (see Chapter 3).

Compared to 1990 levels *the gap in railway technology has hardly closed at all*. While in the western half of Europe (and the Far East), high-speed (200–350 km/h) technology has created a renaissance in long-distance rail transportation, in CEE advanced-speed (140–160 km/h) technology is available and used by railway companies on a few lines only, given a lack of appropriate vehicles and tracks. *CEE also lags behind in the use of fast and economical automatic gauge change technology* on the border of the CIS region (Erdősi, 2009). However, almost everywhere the *length of the electrified railway network* has grown. Bulgaria leads in electrifying main lines, but CEE as a whole has not reached Western levels. Further, unifying power types across countries or even within countries has not yet succeeded – the Czech railway network alone uses three different types of power. The lack of inter-compatibility is one of the factors that has increased costs in rail and thus adversely affected its competitiveness (Erdősi, 2010).

Soviet recommendations put sliding shipping technology, which was able to move shipments of several thousand tons, at the forefront of inland waterway shipping from the 1970s on. Given the significant decrease in inland water traffic, the most appropriate boats for most circumstances (especially for using the Danube–Main canal system) are self-propelled ships, but the stock of such is lacking. Not only has the composition of the stock of ships been unable to meet the needs of the EU, but the depth of waterways has also been insufficient. Contradicting the tendency that forces adaptation to new-style loads in globalising shipping, CEE shipping companies significantly lag behind in the use of container and roll-on roll-off (RoRo) ships, which can seriously increase efficiency (Erdősi, 2008).

The largest *gap in transportation infrastructure*, and the one that is hardest to bear for transporters, is the *very low quality of the road network in rural areas*, along with the embryonic state of the highway system in the CIS region. Efforts to close the gap have been partially successful at best. There have been few advantageous changes in the utilisation value of low-grade roads. At the same time the western countries of CEE have undergone remarkable development of their highway systems, often adopting western technology for collecting road fees.

The direct consequence of political regime change

In the 1970s and 1980s, a number of public road border crossings were established between the socialist countries of CEE, but up until the 1990s international traffic among these countries increased, despite the long distances between border crossings. In contrast, the strict conditions for crossing borders between the Soviet Union and other socialist countries hardly changed, and only a few new border crossings were opened. Only a few railway crossings functioned across this 'second Iron Curtain'. (For example, there was only one border crossing connecting the Soviet Union and Finland – that of the Leningrad-Helsinki railway line and highway – despite a 1,500 km border.)

Within a short period of time after the regime changes economic/social relations between the western and eastern halves of Europe increased in intensity, all of which weakened within CEE itself. With the change in orientation to foreign economies came a need to establish infrastructure that could meet new needs: the largest tasks had to be dealt with by the countries on the borders of the old Iron Curtain, although the establishment of new border crossings and border area infrastructure development were supported by European communal CBC, PHARE and other funds. As such, new connection modes between Bratislava and Vienna (in the Petrzalka–Kittsee–Pandorf direction) were constructed in Slovakia to assist shuttle services, while on the Czech–German border, five one-time border crossings were reconstructed and reopened (Erdősi, 2010).

The new ease in travelling across the former Iron Curtain, followed by the full freedom of movement after European Union enlargement, with its inherent generation of extensive business travel and shopping tourism and resultant networks of interpersonal relations, and an increased demand for mobility, all contributed to the multiplication of group and individual modes of transportation.

In a political sense, mobility in the areas along the former Iron Curtain that separated Western from Central and Eastern Europe is of a different nature than the lands along the 'inner Iron Curtain' by the borders of the former Soviet Union. The most visible result of improvement in conditions for crossing borders (Figure 16.1) and the new semi-permeable condition of borders has been the 'bundling' of the population. With official trade with the post-Soviet region having fallen to one-third or one-quarter of its former level, the massively developed reloading complexes built at the transfer points where wide and narrow gauge tracks met lost much of their regional significance. These workplace–residential complexes acted as development nodes ('seeds of crystallisation') in underdeveloped regions. Their degradation not only led to an underutilisation of production and service capacities, but significant drops in population as well (Erdősi, 2009). However, in east–west goods flow, in contrast to routes going through the Carpathian Basin and the Balkans, the magistral between the CIS region and Western Europe, flowing through the Polish–German plains, gained significant advantage. This is indicated by a strong growth of goods traffic at the Małaszewicze reloading complex on the Polish–Belarusian border along the Moscow–Berlin corridor (Erdősi, 2010).

One of the necessary conditions of interregional cooperation between the countries of CEE – even that coming to life within Euroregion spatial units – is the establishment of transportation relations across state borders that have become porous. As an example, we may consider the south-eastern part of the Carpathian Basin (the areas around the cities of Szeged, Timișoara and

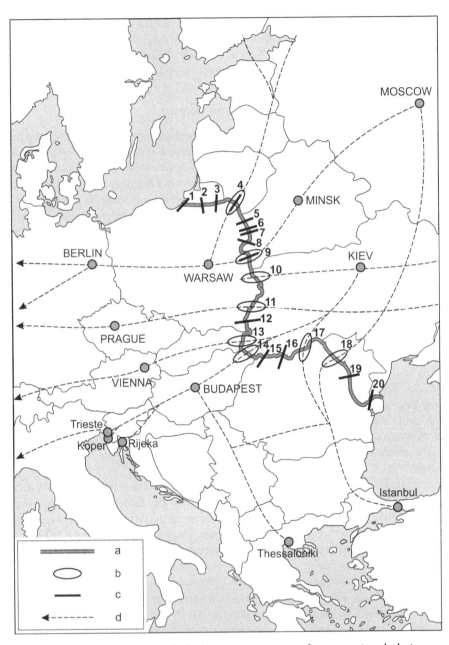

Figure 16.1 Reloading stations/belts between two types of gauge networks between East Central Europe and the CIS region, 2014

Source: Author's construction based on *Jane's World Railways* yearbook, 2014 data.

Legend: a – state border; b – exceptionally important reloading belts in trans-European corridors; c – other international border crossings; d – key directions of flow in the largest reloading belts; 1 – Mamonovo–Braniewo; 2 – Bagrationovsk–Bartoszyce; 3 – Zheleznodorozhny–Skandawa; 4 – Šeštokai–Trakiszki; 5 – Łosośnia–Kuźnica-B.; 6 – Bol'šaja Berestovica–Zubki Białostockie; 7 – Svislach–Siemianówka; 8 – Vysokaje (Visoko Litovsk); 9 – Brest–Terespol/Małaszewicze; 10 – Jagodin–Dorohusk; 11 – Mostyska–Medyka; 12 – Kirov–Krościenko L.; 13 – Chop–Čierna nad Tisou; 14 – Chop–Záhony; 15 – Korolevo–Halmeu; 16 – Yasinya–Valea Vişeului; 17 – Vadul Siret–Vicşani; 18 – Iaşi–Ungheni; 19 – Suceava–Paşcani; 20 – Buzău– Galaţi–Reni–Bendery.

Subotica), where before the First World War the close economic relations between regions were served by main railway lines between the given regional centres. One of the goals of the Danube–Kriş–Mureş–Tisza Euroregion established in 1997 is to revitalise the railway link between Szeged and Timişoara.

The 'Rail Baltica' plan aims to re-establish rail transportation between the three Baltic countries, which ceased in the period 1991–1993. The plan envisions the construction of a track that can manage advanced speed trains, which would serve transit between the Finland/St Petersburg region and Poland (Robinson and Kängsepp, 2013). The 'Via Baltica' expressway serves a similar purpose: it would contribute to cohesion among the Baltic States, as well as transit between Finland and CEE, bypassing the CIS countries.

The consequences of the disintegration of synthetic state formations on transportation

The multinational state forms that were established after the First World War (Yugoslavia, Czechoslovakia, Soviet Union) disintegrated in 1990–1991, and as such the *political–economic space segmented. The drawing of new borders fragmented transportation networks*. The multiplicity of new state borders cut across infrastructure lines, and from one day to the next the ability to cross one-time political spatial units worsened dramatically. Depending on the number of former members of earlier federations and the development of relations between successor states (and to a degree on the form of successor states), difficulties in transportation appeared. The rural populations of new border areas were placed in situations of disadvantage depending on their distance from new border crossings. In the end the inner peripheries of former federations (e.g. between Bosnia and Hercegovina and Serbia, or between Belarus and Ukraine) became outer peripheries. The bilateral spatial division of production within and among successor states, along with traffic, as factors that formerly encouraged mobility ('imperial' scale official travel, supply needs aimed at raw materials from other members' states as a basis for industrial concerns), declined significantly or ceased, or fundamentally lost their significance. The backward development of transport supply was a visible response to the strong drop in demand.

By far the most drastic consequences in terms of the degree of segmentation in all aspects arose from the disintegration of Yugoslavia, affecting not only the number of successor republics on the given territory, but transportation in the entire Balkan Peninsula and finally all of Europe – even in peacetime – thanks to the worsening ability to cross borders (Erdősi, 2001). The transportation corridor connecting Turkey to Western Europe was 'diced up' and as a result transit times increased alarmingly, three- to four-fold (Voivodic, 1997). Beyond the segmentation of political space, for years the war fully blocked SE–NW directed trans-Balkan transportation.

To relieve transportation difficulties arising from new borders cutting across lines various solutions were implemented, which depended on the scale and weight of consequences as dictated by concrete situations (e.g. geographic relations, engineering options, the nature of political relations between two countries). The simplest modus vivendi was the acquisition of transit-use rights for a fee on short sections of tracks that ran on the territory of neighbouring countries (e.g. the Bosnian stretch of the Belgrade–Bar railway). Much more expensive was the construction of new connections (e.g. the Hungarian–Slovenian railway line; Erdősi, 2005c).

The inability to traverse the post-Yugoslav territories due to war, and the high expense of using transportation on routes that avoided the region in the north (leading through Bulgaria, Romania and Hungary) forced the traders of South-Eastern Europe (largely Greece and Turkey) to move an ever larger portion of their export–import shipments onto sea routes. A small proportion of RoRo ships formed a connection with Southern Italy, while a larger proportion connected directly to the ports of the northern Adriatic. A significant portion of this sea/land combined transportation functions to this day, as it turned out to be competitive in peacetime and as a result became permanent, in contrast to traditional trans-Balkan railways and motorways that stretched across several borders.

One of the factors that makes transit across CEE – *particularly that related to Western Europe – difficult is the lack of bridges across major rivers* (especially the Danube).

Beyond the ability to traverse, the relative length of domestic shipments is also influenced by the *shape of new countries*. Unlike the most compact countries, the most disadvantageous shape is that of Croatia, which has diverse effects. The distance between South Dalmatia and East Slavonia (i.e. Dubrovnik and Osijek) is, as the crow flies, 330 km; however, using the Croatian motorway system, which leads one on a significant detour to the west, the distance is over 700 km. For this reason the average shipment distance in Croatia – which incorporates supplying the coast and islands, which are of great significance for tourism – is 20 per cent higher than it is for Hungary, despite Croatia having less territory (World Bank Transport Data & Statistics, 2013).

Beyond transit lines being broken up by state borders and time delays, the coming into existence of ethnically marked and semi-independent entities further fragmented the already poor railway network in Bosnia and Hercegovina, where (in terms of Annex 9 of the Dayton Peace Accord document of December 1995) three unviable railway companies were established in accordance with ethnicities in Bosnian and Serbian ethnic territories. This impossible situation is damaging not only for Bosnia and Hercegovina itself, but for Croatia – which is highly dependent on transit goods shipping across its neighbour – and other countries that need to make use of such transit (Komlós, 2001).

European Union transportation policy in Central and Eastern Europe and the development of the modal split

The EU expects the renewing transportation policies of former socialist countries to follow the pan-European canon. Community transportation policy demands transitions that are in line with a market economy for transportation (in ownership structure and organisational structure) in the interest of establishing regulated conditions for competition. The support of environmentally friendly sub-branches is also prioritised in EU transportation policy. Despite this, one of the worrying paradoxes of current community transportation policy is that the favoured liberalisation/deregulation process happens to favour transportation modes that are the most damaging to the environment (competitive advantages created for automobiles), while it has a braking effect on the maintenance of railway transportation, which is more advantageous in environmental terms, and has made the option of placing rail and inland waterway transportation in a better position a pipe dream. In the end all the countries of CEE, in response to EU transportation policy expectations, have seen motorways and air transportation, which are the most stressful for the environment, grow at the expense of rail and waterway transport. Realistic forecasts calculate a continuation of this trend (e.g. White Paper, 2011).

However, the validity of environmental protection aspects has been maintained in many cases where local/regional, and to a degree national, special authorities have been able to delay the construction of motorways, force the redrafting of their routes or cancel them altogether. In Hungary the route of a motorway to Vienna was moved further south from the Szigetköz area, while in Czechia environmentalists managed to reroute the Prague–Dresden motorway through an area that was less ecologically sensitive (Autobahn Dresden–Prag 2000), and in Poland the construction of regional airports in environmentally protected lands was obstructed.

One of the most difficult to execute intentions of the European Union calls for the level of growth of transportation performance (tonne/kilometre, passenger/kilometre) to not exceed GDP growth (White Paper, 2011). In this area the situation of freight transport is worst, despite logistical innovations.

In terms of both point of view and structure the transportation policy documents of CEE countries are particularly stereotypical, given that they use phraseology adapted from the EU to outline European transportation policy, whose contents and concepts are automatically projected to national levels – holding out for organisation reforms to be undertaken and the modes, directions and structures of development. However, *the unique characteristics and special features* of given countries are only *rarely mentioned in national transportation policies*. We do not discover how conditions that veer from EU stereotypes are to be modified in the interest of goals to be met, what measures are to be taken, what tools are to be utilised and how the latter will affect development budgets. The other extremity is when declared transportation policy contains detailed, registry-like developmental programmes that ignore conceptual approaches (e.g. Lithuanian plans, as interpreted by Butkiovicius and Musteikis, 2005).

Rarely do we find instances where priorities in later planned transportation development strategies grossly diverge from fundamental principles announced in national transportation policy designed in the spirit of EU policy. This took place in Croatia, where the national policy announced the prioritisation of rail, while the strategy concentrated on the construction of the motorway network as soon as possible (Croatian Transport Strategy, 2014).

The lack of harmonisation in transportation infrastructure development and the disproportions in sub-branches and networks-space are areas worth criticising when reviewing the gap between investments and their low effectiveness and competitiveness. The authors of national transportation policy are forced to manoeuvre their way between various interest groups and the forces behind them. Most ruling political regimes ignore professional documents and respond to public pressure, putting unilateral emphasis on constructing motorway networks that meet the needs of '*automorph*' societies, at the expense of an important area of spatial development, namely the road networks of rural areas.

After the political 'outset' the majority of developed countries (avoiding a situation whereby the East would destabilise the West) made available large sums of financial aid, foremost for countries of CEE in the 'waiting room' for EU accession, to contribute to their transportation development, especially regarding infrastructure that crossed borders. For this reason the proportion of certain transportation branches within all of transportation performance (modal split) was increasingly influenced by community, and partially national transportation policy, as opposed to natural geographic features and production structure.

In CEE after the regime changes most countries saw passenger traffic decrease less than freight traffic. From the demand side, social transformation and the change in the structure of the economy had a disadvantageous effect mainly for freight transport – where decrease was dramatic in some cases. Within passenger traffic, as a result of mass unemployment, mainly just 'professional traffic' (executed largely by means of public roads) was acutely affected, while locality changes connected to other activities of the population, largely conducted with automobiles, became more common.

In those countries, however, where unemployment reached a high rate but where the bulk of freight transport was conducted through increasing international transit, public road passenger traffic decrease was stronger than freight traffic decrease.

To conduct an accurate evaluation of the temporal changes in modal split by country, we must make use of a unified methodology in making calculations. In many cases, however, statistics do not allow us to do so. In Table 16.1, which among other things conveys inland water transport performance proportions for given countries and years, in the cases of Slovakia

Table 16.1 Changes in freighting modal split in the countries of CEE, 1990–2013 (based on tonne/kilometre performance)

Country grouping/country	Public road transport 1990/1992* %	2005 %	2013 %	Rail transport 1990/1992* %	2005 %	2013 %	Inland water transport 1990/1992* %	2005 %	2013 %
Visegrád countries									
Poland	32.1	69.5	73.5	67.8	30.3	26.5	0.1	0.2	0.0
Czechia	68.0	74.4	71.7	31.9	25.5	28.3	0.1	0.1	0.1
Slovakia	12.5	54.4	56.4	76.4	40.9	39.0	11.1	4.7	4.6
Hungary	13.4	66.5	63.3	46.1	28.8	30.7	40.5	4.7	6.0
Balkan countries									
Slovenia	43.2	73.7	65.3	56.8	26.3	34.7	–	–	–
Croatia	58.0	75.0	73.1	39.5	22.7	19.7	2.3	2.3	7.2
Bosnia-Hercegovina	28.0	34.7	70.6	78.0	65.3	28.4	2.0	0.0	1.0
Serbia-Montenegro**	27.5	18.8	42.4	72.2	80.8	47.3	0.3	0.4	10.3
Macedonia	68.0	75.0	87.4	31.5	24.7	12.5	0.5	0.3	1.0
Albania	47.9	72.8	83.1	52.1	17.2	16.9	–	–	–
Bulgaria	40.1	57.3	56.0	59.9	41.8	16.6	0.0	0.9	27.4
Romania	32.6	64.0	40.3	65.1	29.0	30.7	2.3	7.0	29.0
Baltic countries									
Estonia	28.1	41.9	36.3	71.8	58.0	63.7	0.1	0.1	–
Latvia	24.3	30.1	33.4	75.4	63.7	66.6	0.3	0.2	0.0
Lithuania	40.9	56.1	18.8	59.0	43.9	81.2	0.1	0.0	–
CIS countries									
Russia	2.3	1.4	78.9	90.1	94.4	19.2	7.6	4.2	1.9
Ukraine	15.5	28.0	23.9	82.7	71.0	76.1	1.8	1.0	0.6
Belarus	22.5	25.7	31.0	75.7	74.0	68.5	1.8	0.0	0.5
Moldavia	31.0	42.0	75.0	68.9	58.0	24.9	0.1	0.0	0.1

Source: Data for 1990/1992 and 2005 were assembled from national statistical yearbooks for given countries by the author; data for the year 2013 are taken from original ratios in Eurostat.

Notes: * For some countries 1990 data are unknown, hence 1992 data are used instead.
** In the interest of comparability data after Montenegro's split from Serbia, (2013) are the combination of Montenegrin and Serbian results.

and Hungary in 1990/1992 (national statistics), and further for Serbia–Montenegro, Bulgaria and Romania for 2013 (data published by Eurostat) it is impossible to compare the branch to the other two transportation modes and to the branch's results in earlier years. The cases of Lithuania and Ukraine, where the proportion of rail increased, are similarly problematic.

The small states that were established as a result of political fragmentation saw given transportation branches operated by organisational units that were small enough to threaten viability: from an efficiency standpoint this phenomenon is rife with problems. The companies that operate services in the Baltic States lag far behind European economies of scale. For this and other reasons the companies that sunk below the threshold of economic viability have been unable to maintain pre-1990 levels.

Beyond such problems with viability the drastic changes that took place in the branch structure of transportation, in terms of tendencies, can be embedded within general modernisation processes and evaluated as such. However, a general experience in CEE is that although the further development and general maintenance of the road network cannot keep step with the pace of the growing vehicle stock, traffic on public roads is inexorably increasing, while the share of rail and inland water (in light of EU expectations) has been continuously and dramatically decreasing.

The branch structure of transportation has significantly changed, at a rate faster than that in Western Europe, to favour public road and air transport, or to remove rail from its former leading position. New expectations have arisen in passenger transportation and freight shipping spurred by the liberalisation and deregulation of transportation policy and restructuring of the economy, and their new predominance has played a significant role. The modal split has transformed to the extent that within the sum freight transport performance, the *public road sector of CEE has a larger proportion than it does in the EU15, while the opposite is true of rail*.

Corridor-focused development and its consequences

In Helsinki in 1997 the 10 pan-European transportation corridors (PEN; now known as the Trans-European Network – Transport (TEN-T)) were marked with only approximate directions (by marking the biggest cities they were to reach), as the exact routes would be determined by the affected countries themselves. This liberty, however, was abused by given countries in that they strove to incorporate relations that diverged from the outlined direction into the network in the hope of attaining more EU support.

The corridors widely diverge from one another in terms of their degree of completion. The 'completion curve' generally moves from the west toward the east (Howkins, 2005). In theory no matter how equal the corridors are in community transportation policy, the divergence apparent in execution budgets and rate of completion reveal significant differences unrelated with physical conditions in their political/economic importance – for example, some corridors have a special economic interest for the economic great power, Germany.

From the beginning of the 1990s, *transportation development in CEE has to too large a degree been oriented toward corridor infrastructure* in the hope of accession and hoped-for EU support. This has been the case to the point where, until stricter provisions were introduced in the Cohesion and Structural Funds in 2008, an insignificant level of EU support in other areas led to a situation where a shockingly high proportion of sums from the state budget for transportation infrastructure development was taken up by corridors (80–85 per cent for rail, 70–75 per cent for public roads). Despite this, in many aspects the corridor system supplemented by national-level TINA corridors were unable to meet the needs of connecting CEE internationally. After the political changes in CEE the main aspect became the high-level connection of Western and Eastern European networks, and for this reason the majority of corridors were placed along W–E or NW–SE axes. The network contains few N–S elements, which of themselves do not constitute a coherent subsystem.

The beneficiaries of the corridor supplementations proposed by Molnár (2007) would be regions with similar culture and history – but which are economically underdeveloped and situated in a geographic periphery (e.g. Transylvania, Moldavia, East Poland and West Belarus, Podolia). The supplemental recommendations, for the most part, rest on geographic considerations.

Beyond being sources of national versus European (community) interest conflicts, corridors also have a strong population and economic draining effect that contributes to increases in regional differences. The cities they reach have a concentration of the best trained labour and attract the most modern and highest added value greenfield investments, industrial parks, services and research and development institutions, while in the periphery the processes of drain and deconstruction pick up momentum (Reggiani et al., 1995). With the exception of the Baltic States, the establishment of 'sea motorways' in CEE serve to relieve traffic in congested land corridors, where lines of ferries carrying freight lorries are created.

The losers of modal change – rail and inland water transport

In CEE the largest decrease in the scale and proportion of *rail and inland water transportation*, which earlier played a leading role in the generation of bulk goods transport, took place in those countries or regions which at one time were outstanding in terms of heavy industry, which strongly required transport, and where, as a result of de-industrialisation, the transport needs of other branches (commerce, services) were unable to brake the decrease. The Balkan countries counted among these, where mostly ore mining products were moved on rail (and on the Danube and Sava rivers) in bulk. In Poland and Czechoslovakia coal mining as well as iron and steel milling were large customers of transport.

The development of rail passenger transportation from the 1990s was polarised to the degree where:

- 'suburban' services aiming at relieving road congestion in capital city agglomerations (max. 80–90 km distances) were given priority, with intensive movement provided by modern wagons running on refurbished tracks;
- domestic long distance line conditions were improved through the use of comfortable Inter-City trains travelling between capital cities and large rural centres (and Poland, Romania and Ukraine, with their large areas and more decentralised structures, between given rural cities) – with few stops in between, and in most cases travelling at anachronistically low speeds;
- on some corridor rail lines Euro-City (e.g. Rail Jet) trains, which approach Western European comfort levels, are used at speeds somewhat higher than those of Inter-Cities;
- for other types of rail travel – 'medium' distances and branch lines of rural areas – the conditions for travel are worse from all points of view. In countries with motorways bus transportation is more competitive than rail services that run on parallel routes.

While European community transportation policy has been pro-rail for decades, since 1990 only symbolic railways have been constructed, while in some countries, to a much higher degree, passenger transport (and in rare cases freight transport) has been curbed. Beyond some short lines reopened close to the western borders, re-industrialisation has seen the network connection needs of some large new plants served through the construction of locally significant rail lines (Erdősi, 2010).

Back when environmental and sustainability factors and regional political aspects hardly had a role in micro- and macro-economies, in the period from the 1950s to the 1980s, the most developed countries of Western Europe terminated a number of rail branch lines and second-class lines that were deemed uncompetitive when compared to public road transportation (Erdősi, 2004). In the area of railway termination the countries of CEE related to the western modus vivendi in extreme ways. In those CEE countries with more liberal economic policies (Yugoslavia, Hungary, Poland) the western model was followed (despite a much lower rate of motorisation) and, following the western model, large-scale 'amputation' of railway networks took place from the 1960s, with reference made to a strong decrease in traffic and uneconomical operations. In contrast to this, the other socialist countries using orthodox political principles defied the decimation of their railway networks, which they deemed to be of strategic importance. This sharp contrast in viewpoint stayed in place after the regime changes, e.g. in Slovakia, Czechia and Romania (Erdősi, 2010).

In a paradoxical fashion the termination of railways affected Poland – which is comprised of three very diverse regions – in the most dramatic way, even though Poland had significantly expanded its network between the two world wars and after 1945. Between 1920 and 1980 close to 3,380 km of track was constructed. The 1990 passenger-carrying network total length of 26,228 km was reduced to 13,615 km by 2013. Of the still operating railways, half exclusively conduct freight transport (Statistical Yearbook of Poland, editions from 1990 to 2013). The government executed the amputation, unparalleled in CEE, in a country where the capacity of public road transport to make up for railway loss was inadequate (Taylor, 1998). This step taken by the PKP railway company was due to short-sightedness and resulted in a frantic search for solutions that could be delivered quickly. The railway terminations of the past decades did not focus on sparsely populated peripheries, but were executed across the territory of the entire country – for this reason they cannot be connected to the economic and demographic situations of affected regions. In areas that are distant from railways a significant portion of the poorly trained population (which is predestined to earn a minimum wage) became unemployed, as they could not afford the costs of commuting by automobile (Taylor, 2004).

The crisis of rail earlier had a singularly strong effect on passenger rail. At the same time, the low volume of goods transported on an increasing number of branch lines is reaching the point where arguments to sustain the service lose their force. For this reason we can anticipate the mid-term physical termination of 5,000–6,000 km of rail lines.

The seemingly unstoppable decrease in the market share of railway is connected to the question of 'collapse or renaissance' when arguing the alternative future views of rail. The 'automobile technocrats' feel that inevitably in the process of switching transportation technology rail will make way for modes that better respond to new needs. Given that this is not a tragedy but an unavoidable shift in periods, it would be wasteful to spend resources on slowing the process. The other (European Union) view holds that the further reduction in the role of rail must be stopped. A third view sees combined systems based on rail as the best solution, and thus the 'renaissance' of rail as inevitable. (This is the opinion of some optimistic environmentalists and biased rail professional circles.) There is some truth to each of these opinions. What is certain is that only an objective review of characteristics can serve as a foundation for mapping out the future in unique areas (countries, groups of countries).

The production structure transformation that accompanied the political/economic regime change brought with it a drop in demand for transport which affected not only rail but led to *inland water transport losing its significance and to a change in its functional structure*. The drop in this transport mode is connected not only to economic factors, but also the neglect of waterways (and locks in several places). These days freight traffic has stabilised at about 10–30 per cent of what it was half a century ago (Erdősi, 2008). Boats today carry mainly sand, pebbles and rock over short distances and cement and wheat over medium distances, with rare orders for long-distance transport of containers, iron ore and coke. For these reasons there is a strong will in most CEE countries to find new functions for freight ports working with minimal or low volume that previously dealt mostly with bulk goods. Most often we see plans for logistical centres that might not even have boat traffic, but contain large warehouses and storage facilities for the collection and distribution of goods, the completion of certain manufacturing processes, or the organisation and trans-shipment of combined rail and road long-distance transport.

For centuries, large rivers have been capable of serving as axes of regional development. Despite this, opinions on the future of inland water freight in CEE are remarkably polarised. The public roads lobby, which is completely insensitive to environmental aspects, considers inland water transport an 'economic historical' category whose time has passed. EU transportation policy, however, has a degree of budgetary optimism when it advertises the belief that inland water transport can be competitive through reconstruction/development supported by the community (Gudmundsson, 2003).

Both extreme opinions are anachronistic in their own right, but through differing perspectives. Instead of leaning on preconceptions, the processes influencing water freight (post-industrial society, climate change) need to be studied in more detail and from there a holistic approach should be used to work out a prognosis.

With the facts in mind, we can state that today (and even more so in the future) CEE inland waterways, in the best cases, can only have a *structure-stabilising role, as water transport in itself is not a true force in the development of the economy or regions*. Characteristically, the 'EU Strategy for the Danube Region' (2013) centres not on water transportation but in a wide sense on the socio-economic problems and processes in the Danube Valley.

The winners of modal change – motorway networks and air traffic

From the aspect of the speed of motorway network construction the drastic differences between the countries of the western half of CEE can hardly be traced back to only the average level of performance of national economies and foreign trade orientation, but instead to unique preferences in national transportation policy.

The density of motorways and economic performance and vehicle supply are most clearly harmonised in *Slovenia* (which is the 'leader' in all aspects).

Croatia, which joined the EU in 2014, was capable of moving to the forefront at the beginning of the 2000s without significant EU support in the interest of making the highest revenue sector of the country, namely coastal tourism, more competitive by ensuring faster transportation access. Although the Dalmatian motorway (which reached Ploče in 2015) is not part of the TEN-T network, the road was built to Split earlier than the V/b corridor to Rijeka. The Croatian motorway network increased from a length of 875 km in 2005 to 1,254 km in 2012 (Statistical Yearbook of Croatia, 2013) and in 2021 it will reach 1,935 km. Croatia, compared to its GDP and vehicle supply, seems to have used super strength and exceptional goal-orientation to achieve these investments.

The situation in *Hungary* is similar, where motorway density indicators are the highest among all Visegrád countries, despite the fact that its vehicle supply is the weakest and its GDP is only third.

Given its material conditions and stock of vehicles, until its accession to the EU *Poland* was exceptionally slow to expand its network, even though long distances and high transit had made motorways necessary even earlier. After having joined the EU (and thus gaining access to EU funds, making financing easier) construction picked up pace. Of the 1,970 km of motorway and 5,200 km of expressway anticipated in the 2004 long-term national plan, only 522 km of motorway and 1,000 km of expressway existed in 2005. But on 1 January 2015 the national motorway-expressway network had reached a length of 3,100 km, but despite this, among the Visegrád Four and Slovenia and Croatia, the supply of specific motorways is weakest in Poland.

Behind the Visegrad Four countries in development are the *Baltic States*, where despite strong public road transit, the short motorways from before 1990 have not been or have hardly been lengthened.

The successor states of *Yugoslavia* (Bosnia and Hercegovina, Serbia, Macedonia) have seen their economies retract over the last few years and were thus able to expand the motorway networks they inherited to a minute degree.

Compared to the country's size, distances and even income situation and vehicle supply, *Romania*'s deficit in network construction is quite strong – taking into account the fact that in 2014 three-quarters of the motorway stock had been built in the previous 8–10 years.

Today's CEE motorway network is connected to the European network at several points, and is indeed an integral part of the latter. As such, moving through the chain of motorways in the corridor starting from Western Europe and moving through the Carpathian Basin and the Morava–Vardar Valley, one can reach the Greek motorway network. This NW–SE oriented magistral is connected to a W–E path from Austria and through Slovenia and Croatia in the Belgrade area. At the same time, deficiencies include the lack of a connection between Polish ports and the North Adriatic through Slovakia, Western Hungary and Slovenia, i.e. a N–S-oriented transcontinental motorway (of which only a few elements have been constructed, despite being present in the TEM project in the 1970s).

In CEE the construction rate of the motorway network is presently lower than it was for railway construction in the past. There are more theoretical assumptions about the regional economic development benefits of motorways than there are verified experiences. It is clear today that a motorway is not a panacea, but a necessary yet insufficient condition for spatial and municipal development. It is most effective where it carries significant foreign (transit) traffic.

Perhaps the biggest winner of the regime change, the other victor in the last quarter-century among transportation branches, is *international air travel*. In contrast to domestic flights, *international air traffic has increased to a high degree* (several-fold in most cases) in every CEE country, especially to points in Western Europe and well-known tourist holiday destinations on other continents. This process has, through travel becoming more liberal and business relations gaining depth, at once helped increase the demand for travel to distant points and create a rich, highly diversified supply of new airlines. The appearance of low cost carriers in CEE (especially in the Visegrad Four) and the Baltic States has been a particularly important moment in the generation of international air traffic (Groß and Schröder, 2007). Through their flights to Western European metropolises the supply-heavy labour market of CEE has been moved closer to the demand-heavy labour market of Western Europe, and such flights have physically begun to unify the European market through a slow equalisation process.

Some 60–70 per cent of the millions of labourers from Poland, the Baltic States and to a lesser degree Slovakia and Hungary working in the British Isles, the Benelux countries and Scandinavia use air flights to travel to Western Europe and (on occasion) home. An even larger proportion of guest workers from Romania and Bulgaria choose air flights to travel to the Iberian Peninsula and Italy.

The most controversial process to have played out in the CEE air transportation sector is seen in the differentiation in function and traffic intensity of regional airports. Public traffic airports in CEE function at much greater distances from one another (sometimes several-fold) compared to Western Europe – thanks in part to smaller specific air transit demand, and partly to the underdevelopment (poor supply) of air service. The traffic and transportation significance of regional airports in the past 25 years has been differentiated according to the size of countries and the intensity of their international relations.

- In countries with small territory there are either no regularly scheduled domestic air flights or merely a handful between the capital city and one or two significant urban centres in the countryside. Regional airports in such countries serve not only domestic destinations but in some cases professional traffic to international destinations (e.g. Debrecen, Košice, Moravska Ostrava, Niš) or (through charter flights) offer international tourism services (e.g. Sármellék, Sliač, Karlovy Vary, Ohrid). Among small countries *Croatia*'s situation is exceptional, as in the 1970s/1980s eight regional airports operated.
- In larger countries with areas of several hundred thousand sq.km and a more equalised distribution of urban centres, where the capital city's 'over-significance' is lower and significant regional centres exist in the countryside, regional air traffic has developed in a varied way since 1990:
- In *Romania* the network of airports numbers about half a dozen and serves mainly domestic flights between the capital city and the most important urban centres in the countryside. Of these, Timișoara stands out (given frequent travel by engineering experts working abroad and partly by businesspersons, university lecturers and scientific researchers), as does Cluj-Napoca (where the above-mentioned groups are supplemented by foreign tourists). Iași and Oradea conduct some international traffic. However, 80 per cent of international traffic is centred in the capital city of Bucharest.
- An unparalleled increase in air traffic occurred in Poland's stock of regional airports. Since entering the EU the outflow of labour to Atlantic Europe would be too slow by land transportation, with the British Isles being a destination of importance. The majority of Polish guest workers are from the country's southern, most densely populated and industrialised area, which has led to strong territorial differentiation in the increase of airport traffic.

To a lesser degree Czechia and Hungary, which have much smaller territories, have seen air traffic concentration in their capitals moderated by the launch of flights from Brno and Moravska Ostrava and Debrecen and Sármellék to points in Western Europe and the Mediterranean.

The specific airport passenger traffic of given countries (Figure 16.2) is only partly dependent on economic development. In many cases the inadequacy of land long-distance international transportation connections and isolation (an inevitably compelling factor) have made air travel not only a supplemental, but a substitute mode. (For example, in small Montenegro, with international airports in the capital and in Tivat, or Bosnia and Hercegovina to conduct the movement of approximately 1,000,000 guest workers going abroad and visiting home.)

In contrast to functionally successful airports with growing traffic, the fate of (regional) airports with unused capacity and which from an environmental point of view are inadequately allocated, is a source of contention.

While in Western Europe it was mainly demand for tourism and business air travel that spurred the gradual construction of a dense network of regional airports in countryside areas, in CEE the regions see regional airports as one of the most effective tools for spatial development (referring to more equalised spatial development and dissatisfaction with overland distance transportation). The will to close the gap does not always go hand in hand with a guarantee of making full use of capacity and with a detailed consideration of allocations. For this reason today the European Commission will only contribute to the financing of regional airports when all

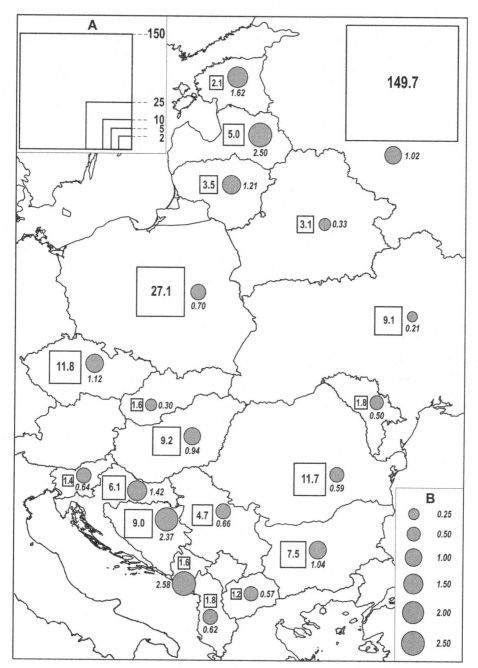

Figure 16.2 The sum traffic of airports (squares, millions of passengers) and specific passenger traffic (circles, passengers/residents) in the countries of Central and Eastern Europe, 2014

Source: National statistical yearbooks and airport statistical data as edited by the author.

aspects are substantiated. EU experts were soberingly factual when they warned Poland about creating unnecessary capacity for its smaller airports. EU, national and local opinions (that is to say interests directed from various levels) clashed over several airports – including among others the cases of Kielce, Opole and Zielona Góra (EU's Pie in the Sky, 2012). Among the airport establishment recommendations, several were criticised on environmental/natural-protection grounds, especially when these noisy institutions were planned to be located near protected wildlife areas which house large bird populations. Such aspects came to the fore regarding the Tykocin and the Warsaw Modlin airport investments, which were located near the Natura 2000 certified Bielrza National Park and Narew National Park (Warsaw Modlin Airport … n.d.)

The macro-region dislocation of sea port traffic

Partly due to political/security reasons, and partly to economic considerations ('sparing hard currency') the socialist countries strove to conduct as low a proportion of their international trade as possible through the ports of capitalist countries. For this reason the export of bulk goods (and in some cases imports) took place through further developed domestic and nearby Council for Mutual Economic Assistance (COMECON) ports. From the 1980s and 1990s, however, an increasing proportion of CEE sea-based foreign trade made use of the much more advanced services in Western European mega-ports. For CEE (sea) trade up to the 1990s the world economy mega-regions of Atlantic Western Europe and North America were of special significance. From this point of view the best situation was that of Poland and (partially) the Baltic States, as well as the ports of the Adriatic, where the worst-off ports were those of the Black Sea and the Azov Sea.

Ever since the world economy became tri-polar the Far East global power centre emerged (and in terms of volume moved into the lead). The geographic position of certain port regions changed in terms of distance, but in reality the new situation has not yet brought fundamental change to the amount of port traffic and their orientation (Figure 16.3).

Water depth at the entrance to the Baltic Sea is such that ships of a maximum of 10,000–15,000 dwt can cross. For this reason the Baltic Sea, in terms of opportunities for intercontinental shipping, is a 'victim' of economies of scale, given that it is unreachable with direct routes using ocean-going ships. By necessity shipping occurs with the mediation of Western European hubs, even for Far Eastern trade.

Although the Black Sea and the Adriatic Sea can be reached by large ships through their oceanic channels, up until recently, from the aspect of the execution of logistics of shipping efficiency, it was common practice to make use of smaller feeder ships to link to ships travelling between the hub ports of the Far East and Western Europe (Piraeus, Gioia Tauro, etc.). Recently, however, the volume of oceanic traffic between the Far East (mainly China) and Europe has reached a scale even in the inner seas of CEE's groupings of ports which in numerous cases makes direct routes profitable without making use of Mediterranean and North Sea hubs.

When considering the role of transportation in regional processes, one of the strongest indicators is the change in performance of sea ports (or the economic development power of such) over space and time. As installations ports are too expensive for merely economic considerations to change their significance (to one another and relatively) in the domestic weighting and valuation in given countries. Geographic changes in emphasis among ports in CEE have arisen mainly due to political events, especially border changes. Some examples of such changes in inland seas in CEE are as follows.

Before and after the regime change (with the exception of Russia) the raw material (mineral) export of CEE countries dropped dramatically owing to the change in the structure of the

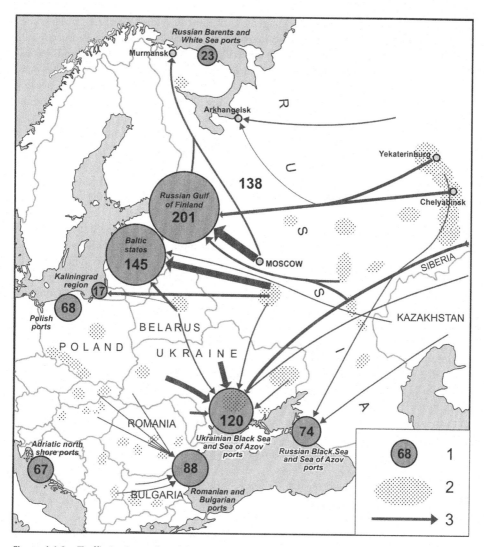

Figure 16.3 Traffic in Central and Eastern Europe's port regions, 2013, with the key directions of flow

Source: Data from web pages of given ports, national statistical yearbooks and other sources as compiled and edited by the author.

Legend: 1 – loading traffic in million tonnes; 2 – largest industrial regions; 3 – key directions of flow.

economy and to uncompetitiveness in the world economy. There was less need for domestic ports for the export and import of valuable materials, while the modern, well-equipped capital-rich mega-ports on the Atlantic used varying methods (with varying degrees of success) to increase their attractiveness to the ports of socialist and then post-socialist countries. The syphoning away of traffic led to a state where Poland's total port traffic dropped by 40 per cent in the period 1972–1989 – but with significant spatial differentiation. Given the syphoning away of traffic was strongest in the ports of the Vistula Delta, the relative weight of ports in the Oder Delta temporarily grew. The political changes created a new opportunity for the ports of the Oder

Delta, but the processes were contradictory, given that in the development of relations between German and Polish ports, international treaties have no role and as such the relationships are regulated by contracts between the ports and especially by competition. Compared to Rostock, Szczecin for a few years was capable of conquering a part of Northern Germany's market share (Radloff, 2007). However, the conditions for competition between the ports became disadvantageous for the ports of the Oder Delta given the development of German ports and the transformation of the allocation of transit traffic. The sum Polish port traffic level shifted toward the ports of the Vistula Delta. (In 2013 two-thirds of transit was concentrated in Gdańsk – Statistical Yearbook of Poland, 2013.)

The *Black Sea* holds enormous potential for the countries of its coasts, given that they can count on not only their own foreign trade, but in many cases the transit of landlocked countries behind them. Half the traffic of Ukrainian ports is initiated in Russia. It was in vain that the megalomaniac Nicolae Ceaușescu built what he dreamed to be the 'Eastern Rotterdam' in the Constanța complex: only an insignificant level of transit could be attracted here from the hinterland. After the disintegration of the Soviet Union, the collapse of mining and heavy industry and partial economic restructuring resulted in a drastic drop in the traffic of Black Sea ports. The economic crisis of 2008 posed many challenges, which were only effectively defied by the seaside terminal points of oil pipelines. The oil terminals on the shores near Odessa and Novorossiysk led Constanța in level of traffic. *The ports of the Black Sea (despite their ambitions) cannot become logistical hubs* (distribution centres) because these functions have been monopolised by Piraeus (which is near the intense-traffic Suez Canal–Western Europe route) and other node ports.

Of the Adriatic ports of the post-Yugoslav region, fierce competition has developed for the landlocked markets behind them. All three shoreline successor states have come up with development plans and executed them (modestly in Montenegro, at a grandiose scale in Croatia and realistically in Slovenia) to grow port capacity, and constructed infrastructure connecting them to the hinterland in the interest of making the ports more attractive (Great Potential …, 2013; Rathman et al., 2014).

The four most significant ports in the competition have various characteristics:

- *Bar* in Montenegro was established in the 1970s on Yugoslav military strategy grounds. Its disadvantage lies in the fact that its land connection routes to the hinterland are not part of the TEN-T network; it has weak development potential.
- *Ploče* in Croatia is in reality an 'outpost port' of Bosnia and Hercegovina and at the same time a terminal point of the V/c corridor that provides Central European connectivity. However, the construction of its railways and carriageways is in the mid-term unviable.

These two 'political ports' are contrasted by the characteristics of the two most northerly ports, which are considerably better, although the gap in their performance abilities is quite large, mostly because of the divergent quality in transportation infrastructure connecting them to the hinterland (corridors V and V/b). Acclimatising to globalisation is evident in the fact that development plans for both Rijeka and Koper are centred on the growth in international container transit traffic.

- The traditional market of the port of *Rijeka*, which has a long history, is the Carpathian Basin. The port is multifunctional and has many sections (bulk, oil, general cargo, timber, etc.).
- *Koper* has been operating since the 1960s/1970s and has preferred container handling. Its market cuts across that of Trieste, i.e. the Eastern Alps and Western Pannonia market areas (Çertalic, 2015).

The railway connecting Rijeka to Hungary is completely out of date and its container transport capacity is inadequate. Granted, Rijeka has a motorway that connects it to the hinterland, but it is unable to replace the railway (Mladen, 2014). For this reason Croatia plans to construct a railway that will connect to the Sava waterway by 2020, and by building a Sava–Danube canal it hopes to establish a high-capacity combined transport chain that would make it possible for Rijeka to be directly connected to Upper Austria (Twrdy et al., 2014). The budget for this project, however, is alarmingly high. But as long as the railway-capacity gap is not closed, Koper will maintain its competitive advantage. This advantage is illustrated in the fact that Koper's field of business includes the Carpathian Basin (Perkovic et al., 2013).

Compared to the above, Russia's ports underwent a larger-scale change in the 'geography' in port traffic, where the change in the centre of gravity took on a north-west orientation with the spatial transformation of the country's gates to the oceans – which included a will to disconnect the Baltic States. *The Baltic States lie at the intersection of German and Russian spheres of interest, and were able to profit from their 'threshold' or 'gateway' situation* during independence and during membership in the Soviet Union, given that Russia, which has few ports close to Western Europe, could not do without the almost ice-free Lithuanian, Latvian and Estonian ports (Erdősi, 2005b). In the 1970s and 1980s, 30–40 per cent of Soviet–Russian sea transport moved through the three Baltic States. For this reason Soviet transportation policy paid particular attention to the construction of railways near Baltic ports, and of pipelines that were a necessary condition for oil exports.

From a geopolitical and foreign trade point of view, in a paradoxical manner, it was Russia that defined one-time imperial policy that became one of the biggest losers of the collapse of the Soviet Union. The country had to face the fact that its geographic position in terms of the world economy power centres became peripheral at the time when its choice to adopt a market economy had not yet become a qualitatively measurable factor. As a result of the new independence of the Soviet successor states in the Baltics a significant lack of capacity emerged in the port stock of the Russian Federation – especially on the Baltic Sea. It was the climatically best situated, best placed for reaching foreign markets, and large capacity ports in the Baltic Sea and Black Sea regions fell outside the borders of the Russian Federation.

The Baltic States' policy relationship to Russian transit is rather two-faced. While from an economic point of view the traffic generated from Russia was seen not only as bearable, but as a factor that could increase national income through increased transit and port fees (as well as the employment of thousands), certain nationalist forces threatened blockades on occasion. (For example, after EU accession Lithuania drew attention to itself when it wanted to renege on its transit commitments as a corridor between mainland Russia and Kaliningrad.)

To this day Russia's economic situation is defined by the mass sale of natural resources. Foreign sea commerce serves this process. To eliminate political risks and to increase profits, in 1993 Russia launched a grand-scale port construction programme on its own shore of the Gulf of Finland. It defined a long-term goal of having 95 per cent of Russian oil exports move through its own ports (Deeg, 2004). By 2013 a weak result of 35 per cent was achieved through Baltic Sea ports (Table 16.2).

Most new Russian ports of export are in the area of St Petersburg. Given that the country's second largest city's port, which was modernised through western investments, cannot be expanded due to a lack of space, large specific ports have been (and are being) constructed beyond the city's agglomeration zone.

The competitiveness of the future use of domestic ports in comparison to those of the Baltic States (from economic viability and security of operation aspects) is threatened by several circumstances: for example, ports on the Gulf of Finland incur extra costs in weather

Table 16.2 The distribution of Russian sea foreign trade across Russian and foreign ports

Ports	1992		2000		2007		2009		2013	
	M t	%	M t	%	M t	%	M t	%	M t	%
Russian ports	228	71.2	166	68.6	412	76.3	410	76.6	533	85.4
[among these Baltic Sea ports]*	[68]	[21.3]	[53]	[21.9]	[184]	[34.0]	[162]	[30.3]	[218]	[34.9]
Foreign ports	92	28.8	76	31.4	128	23.7	125	23.4	91	14.6
[among these those of the Baltic States]**	[89]	[27.8]	[70]	[28.9]	[119]	[22.0]	[120]	[22.4]	[88]	[14.1]
Total Russian sea foreign trade	320	100.0	242	100.0	540	100.0	535	100.0	624	100.0

Source: The web pages of given ports and various studies, as well as data from the Shipping Statistics Yearbook, compiled by the author. The proportion of transit in foreign ports differs from the percentage proportions reported in Wenske and Breitzmann (2014).

Notes: * In the Gulf of Finland and the Kaliningrad port complex.
** Ports of the Baltic States and Finland.

where ice forms; the dependence of the Kaliningrad exclave on unstable transit through Lithuania; the situation of sea straits in Turkey, along with numerous other national security and extra expense risks.

The effect of half a dozen ports handling fuel and other raw material in the wide vicinity of St Petersburg on the European half of the Russian economic spatial structure is questionable and depends on the future development of the traffic situation in icy waters and the Black Sea, but there is *no question that the centre of gravity of Russian hydrocarbon exports has moved north*. However, it is possible that in the future this will be balanced by a growth of traffic in the Sea of Azov shore area.

No matter how large the new port capacity built by Russia in the Gulf of Finland, it *must* make further use of the ports of the Baltic States, whose total traffic in 2009 exceeded the level of the late 1980s and dropped significantly only after the launch of new Russian port capacities in the 2010s (Table 16.2). Transit continues to generate the largest portion of traffic for these ports, given that the export–import traffic of these low-population states is modest and that they barely produce mass products for export (Gopkalo and Goloviziniv, 2013). However, Russian exporters hardly have any means to replace the oil and natural gas pipelines that function here.

Conclusions

In CEE there is no close relation between transportation potential/*quality of transportation–geographic situation* and *economic development* (measured in GDP), or the economic situation (taking into account employment). In theory it can be assumed that countries with seaside ports have a large material advantage given opportunities to directly access the world market and from international transit mediation, but countries with severe economic problems like Croatia, Ukraine, Albania and Bulgaria, as well as those that have not significantly closed the gap like Romania, Latvia and Lithuania, serve to disprove this. To a degree Poland and Estonia, which have increased measurable growth in the last few decades, are exceptions. (In a paradoxical manner they are more distant from the definitive intercontinental sea shipping routes and from the Suez Canal than the less peripheral ports of the southern inland seas!)

At the same time it is a fact that the economic development of landlocked countries is mixed. The most developed include Czechia and Slovakia through Hungary, sloping down to the least developed Belarus and Moldova.

No matter how much rapid transportation infrastructure has contributed to dynamic growth in the number of foreign tourists, the significant income from tourism has not been enough to improve the state of 'holiday destination' countries in the Adriatic and Black Sea regions. The surprisingly high level of airport traffic in given countries (Montenegro, Bosnia and Hercegovina), much like the density of motorway networks (Croatia, Hungary) has not led to a faster pace in closing development gaps. That is to say that no matter how up-to-date transportation development is, it in itself is not a guarantee that there will be a dynamic effect on a macro-economic scale. Transportation can only become an effective force for development when combined with other necessary factors. (Hence the importance of a modern economic structure, a highly trained and creative, goal-oriented workforce, supportive public policy and state and local authority efforts to attract foreign capital.) The establishment of missing transportation routes is a much larger task than that of line-less communications networks. In terms of the technical characteristics of communications, thanks to multinational corporations there is now no significant difference among the countries of CEE. Further, differences in the intensity of utilising IC services between these countries are smaller than specific GDP indicators.

With the transformation of the global economy, the pull effect of the third power centre of (East) Asia is expected to be greater, but the quantity of interactions with Africa and the Middle East is also expected to rise. The South-East direction will gain importance in Europe's foreign trade intercontinental relations system. In what way will Eastern Europe be able to operate the logistical connections necessary to reach the new sales and acquisitions markets? It is not likely that the Western European mega-ports (with their top technology and special services offered to regular routes) will be omissible in the future.

For CEE the port groupings in the Adriatic Sea and the Black Sea can both be taken into account for developing eastern relations. (In terms of bulk, we may consider Constanţa, Burgas and Odessa, and Koper/Trieste and perhaps Rijeka for container and general cargo handling.) It is possible, however, that Istanbul will be able to play a hub role in both sea transport and air transportation for the East Balkans and other countries as well – there are already signs pointing to this. Among East Central European capitals Prague, Warsaw and Budapest compete with one another for air transport hub functions – but their services are generally restricted to European, North African and Middle Eastern markets. Only Istanbul offers wide services to reach distant world regions.

Acknowledgement

The publication of this research has been supported by project #104985 of the Hungarian National Research, Development and Innovation Office.

References

Autobahn Dresden–Prag, 2000. LKW Fahrverbot Bedingung für Baugenenmigung. *DVZ* (*Deutsche Verkehrzeitung*), 2 June.

Butkewicius, J. and Musteikis, P., 2005. Möglichkeiten zur Schaffung eines Hochgeschwindigkeitsnetzes in Litauen. *Zeitschrift der OSShD*, 48(6), pp. 4–6.

Çertalic, M., 2015. Luka Koper – Port of Koper TEN-T and MOS Opportunities. Liverpool, 21 May. http://superport.iprogress.netdna-cdn.com/wp-content/uploads/2015/05/masa-kertalic-port-of-koper-development.pdf

Croatian Transport Strategy, 2014. Transport Development Strategy of the Republic of Croatia 2014–2030. Ministry of Maritime Affairs, Transport and Infrastructure. Brussels, 9 October. Jaspers Network. www.jaspersnetwork.org/download/attachments/16711761/JASPERS%20transport%20strategies%20workshop%20-%20Croatian%20Transport%20Strategy.pdf?version=1&modificationDate=1413814912000&api=v2

Deeg, L., 2004. Neue Häfen braucht das Land. *DVZ*, 20 March.

Erdősi, F., 2001. Retesz vagy átjáró? A balkáni tranzitközlekedés [Latch or portal? Balkan transit transportation]. In: Z. Hajdú, N. Pap and J. Tóth, eds. *Az átalakuló Balkán politikai földrajzi kérdései. II. Magyar Politikai Földrajzi Konferencia.* Pécs: PTE TTK Földrajzi Intézet Kelet-Mediterrán és Balkán Tanulmányok Központja, pp. 102–120.

Erdősi, F., 2004. *Európa közlekedése és a regionális fejlődés. Második bővített, átdolgozott kiadás* [Europe's transportation and regional development. Second expanded and revised edition]. Pécs–Budapest: Dialóg Campus.

Erdősi, F., 2005a. *A Balkán közlekedésének főbb jellemzői* [The main characteristics of Balkan transportation]. Pécs: Pécsi Tudományegyetem TTK Földrajzi Intézet. (Balkán Füzetek, 3).

Erdősi, F., 2005b. *A Baltikum közlekedése és a kapuszerep* [Transportation and the portal role in the Baltics]. Pécs: MTA Regionális Kutatások Központja.

Erdősi, F., 2005c. *Magyarország közlekedési és távközlési földrajza* [Hungary's transportation and communications geography]. Budapest–Pécs: Dialóg Campus.

Erdősi, F., 2007. *Kelet-Európa országainak légi közlekedése* [Air transportation in the countries of Eastern Europe]. Pécs: MTA Regionális Kutatások Központja.

Erdősi, F., 2008. *Kelet-Európa országainak vízi közlekedése* [Water transportation in the countries of Eastern Europe]. Pécs: MTA Regionális Kutatások Központja.

Erdősi, F., 2009. *Kelet-Európa közlekedése* [Eastern Europe's transportation]. Pécs–Budapest: Dialóg Campus.

Erdősi, F., 2010. *A visegrádi országok vasúti közlekedése* [Rail transportation in the Visegrad countries]. Budapest: MÁV Zrt.

EU Strategy for the Danube Region, 2013. www.danube-region.eu/

Eurostat Freight transport statistics – modal split. http://ec.europa.eu/eurostat/statistics-explained/index.php/Freight_transport_statistics_-_modal_split

EU's Pie in the Sky: New Analysis Questions Further Funding Support for Aviation, 2012. *Bankwatch Mail*, 53. http://bankwatch.org/bwmail/53/eus-pie-sky-new-analysis-questions-further-funding-support-aviation

Fekete, D., ed., 2015. *Nyugat- és kelet-közép-európai járműipari térségek működési modelljei* [The operating models of vehicle industry regions in Western and East Central Europe]. Győr: Universitas-Győr Alapítvány.

Gopkalo, O. and Goloviziniv, A., 2013. The Russian Market Keeps Growing. *Baltic Transport Journal*, 8(5), pp. 6–10.

Great Potential for Further Development at Port of Bar, Montenegro, 2013. *World Maritime News*, 27 May. http://worldmaritimenews.com/archives/84362/great-potential-for-further-development-at-port-of-bar-montenegro/

Groß, S. and Schröder, A., eds., 2007. *Handbook of Low Cost Airlines: Strategies, Business Processes and Market Environment*. Berlin: Erich Schmidt Verlag.

Gudmundsson, H., 2003. Making Concepts Matter: Sustainable Mobility and Indicator Systems in Transport Policy. *International Social Science Journal*, 55(2), pp. 199–217.

Howkins, T.T., 2005. Changing Hegemonies and New External Pressures: South East European Railway Networks in Transition. *Journal of Transport Geography*, 13(2), pp. 187–197.

Jane's World Railways yearbooks, 2000–2014.

Komlós, A., 2001. Bosznia-Hercegovina vasúti közlekedése a balkáni háború után [Railway transport in Bosnia-Hercegovina after the Balkan war]. *Sínek Világa*, 44(2–3), pp. 82–86.

Mladen, J., 2014. Importance and Role of the Port of Rijeka in Transport and Economic Development of the Republic of Croatia. *Annuals of Maritime Studies*, 4, pp. 87–93.

Molnár L.A., 2007. Kelet-Közép-Európa úthálózata mint a felzárkózás eszköze [East Central Europe's road network as a tool for closing the gap]. *Közúti és Mélyépítési Szemle*, 11, pp. 1–9.

Perkovic, M., Twrdy, E. and Batista, M., 2013. The Increase in Container Capacity at Slovenia's Port of Koper. *TransNav*, 7(3), pp. 441–448.

Radloff, M., 2007. Fokas auf den Raum Berlin. Häfen Stättin und Swinemünde werben von deutsche Kunden. *DVZ*, 3 July.

Rathman, D., Debelić, B. and Stumpf, G., 2014. Structural Analysis of Development Capabilities of the Port of Ploče as a Potential Container Port within MoS Services. *Scientific Journal of Maritime Research*, pp. 145–150. [University of Rijeka Faculty of Maritime Studies].

Reggiani, A., Lampugnani, G., Nijkamp, P. and Pepping, G., 1995. Towards a Typology of European Inter-urban Transport Corridors for Advanced Transport Telematics Applications. *Journal of Transport Geography*, 3(1), pp. 53–67.

Robinson, F. and Kängsepp, L., 2013. Borders Raise Hurdles for Baltic High Speed Rail Link. *Wall Street Journal*, 19 December.

Statistical Yearbook of Croatia, 2013.

Statistical Yearbook of Poland, 2000–2014.

Taylor, Z., 1998. Polish Transport Policy: An Evaluation of the 1994/5 White Paper. *Journal of Transport Geography*, 6(3), pp. 227–236.

Taylor, Z., 2004. Recent Changes in Polish Transport Policy. *Transport Reviews*, 24(1), pp. 19–32.

Twrdy, E., Batista, M. and Stojaković, M., 2014. Competition among Container Ports in the North Adriatic. http://imet.gr/Portals/0/Intranet/Proceedings/SIGA2/twrdy_batista_stojakovic%5B1%5D.pdf

Voivodic, R., 1997. Höhere Leistungsfähigkeit durch Harmonisierung der Entwicklung. *Schienen der Welt*, 12, pp. 4–5.

Warsaw Modlin Airport. http://modlinairport.pl/

Wenske, C. and Breitzmann, K-H., 2014. Russlands Seetransport- und Hafenentwickung im Ostseeraum. *Internationales Verkehrswesen*, 66(2), pp. 48–52.

White Paper, 2011. Roadmap to a Single European Transport Area. Brussels: European Commission.

World Bank Transport Data & Statistics. http://researchguides.worldbankimflib.org/content.php?pid=31784&sid=232350

17
Nature and spatial planning in Central and Eastern European countries

Andrea Suvák

Introduction

Concerns about the natural environment have always been essential building blocks of what is called spatial planning today in Europe. In effect, environmental aspects were important drivers of the very formation of modern urban and spatial planning.

Their importance was reinforced by the Fifth Environmental Action Programme of the European Union that laid down the bottom lines of environmental policy integration, consideration of environmental costs and benefits, monitoring of environmental effects, cooperation with environmental authorities, and the public availability of environmental information. Environmental Impact Assessments and later Strategic Environmental Assessments have been tools dedicated to this integration.

The environmental policy of the Central and Eastern European (CEE) countries has largely been impacted by the environmental policy of the European Union, which is itself a blend of various policies, practices and even philosophies of Western European countries. The transformation of the CEE countries into market economies in the early 1990s gave an overdue start for policy formation, institutionalisation and awareness raising for environmental concerns. The EU has been an urging factor in this development through its legislation, and even more due to its funding schemes that channel EU environmental norms in the form of provisions. Amidst rushed marketisation and privatisation, the development of national economies and facing the crises of old industries, CEE countries did not build up their own norms and forms of environmental protection and policy integration; the methods supported by the EU were adopted, often resulting in incoherent systems with overlapping layers of conflicting approaches.

Environmental issues after the Second World War

After the Second World War, the centrally planned and coordinated economic system in the CEE socialist states, based on public ownership, was also peculiar in its relation to the natural environment. In early stages of the 'building of the socialist system' and industrial evolution, environmental problems were shown by socialist propaganda as a phenomenon exclusive to capitalist countries, the occurrence of which was impossible in communism. This fed the long-prevailing general assumption that environmental problems can only persist in capitalist states

– since in socialism the interest of the people and the interest of corporations (and in general that of the rich people at the top of the capital concentration pyramid) do not diverge from each other, thus problems affecting the whole population could not be overridden by special interests (Lange et al., 1938).

However, the reality was different. Forced industrialisation meant that industry was prioritised in all socialist countries over other sectors, which had a strong impact on land use as well. The location and building process of industrial plants undoubtedly effaced the needs of other sectors, not to speak of environmental protection. Agriculture was especially hit by this process, and large, quality agricultural areas were turned into industrial sites (Hall, 1987, cf. Dawson, 1987).

Besides their similarity in many aspects, all socialist countries had their peculiarities regarding their specific environmental problems and their approach to these issues (Pavlínek-Pickles, 2005). Local practices and results varied from region to region, so any standardised conclusion on environment-related operations within the region would be inappropriate. Nevertheless, state socialist CEE countries were home to some of the most notoriously polluted industrial areas in the world, such as Katowice and Krakow voivodeships in Poland, the coal-mining region of Northern Moravia, Czechia and many others all across the macro-region.

The uneven spatial development of industry resulted in the creation of hot-spots of severe environmental pollution. For example, the metallurgy industry in the Copşa Mică region of Romania released '12,000–67,000 tonnes of sulphur dioxide (SO_2), more than 500 tonnes of lead, 400 tonnes of zinc and 4 tonnes of cadmium, 3000 tonnes of carbon monoxide (CO), 200 tonnes of carbon dioxide (CO_2), and 3000 tonnes of particulate matter' every year. 'Air pollution by heavy metals exceeds legally permissible maximum levels by 600 times. Acid rain is devastating local forests, crops and soils. The total area affected by air pollution exceeds 180,000 hectares (ha), including 150,000 ha of polluted agricultural land, 31,000 ha of polluted forests, and 22,000 ha that are most severely damaged by air pollution' (Pavlínek and Pickles, 2005, p. 40).

The location of industry required labour to be settled close to industrial zones. This, similar to the late nineteenth-century Western European and English practice, was accomplished by building residential zones in the proximity of industrial plants, exposing inhabitants to severe health risks. In the newly built residential areas of the CEE countries, green areas were scarce. Heating was often provided by polluting central heating plants, and these areas were also lacking in appropriate sewage and waste management systems (Klarer and Francis, 1997; Klarer and Moldan, 1997).

Socialist states in general employed industrial and municipal management systems of much lower energy and material efficiency, which resulted in a higher pollution to economic performance ratio than the systems applied by Western European countries. Environmental problems occurred mainly due to the air pollution of large industrial concentrations, the soil and water pollution of industrial agriculture, and acid rain – the latter was partly generated by Western European pollution that, because of the climatic attributes of the continent, precipitated above Central European territories. Ownership structure was another barrier to the spread of responsible environmental thinking, since public ownership entailed a complicated prosecution process for cases of pollution.

Environmental conditions and the attitude towards nature were portrayed as 'flagrantly bad' by Western Europe and the USA. Anti-communist propaganda was certainly a component of this negative picture. On the other hand, there was an actual reason for overgeneralising environmental conditions based on the notorious CEE hot-spots: these sites were significantly over-represented in environmental quality statistics since the majority of monitoring stations were located in these areas. At the same time, it is often unmentioned in international analytical work that a very large share of the land in these countries was either a natural protection area (approximately 30 per cent according to a REC (1994) study), or untouched by heavy industrial development – although in this case, industrial agriculture was similarly harmful to the natural environment.

Despite the lack of reliable, consistent and comparable statistics for the environmental conditions of CEE countries, some investigations came to the conclusion that the environmental standards, the pollution of the natural elements and the industrial practices of post-war state socialist countries were quite similar to those of Western Europe and the USA some decades before. Hughes stated: 'apart from a number of severely damaged areas, the general level of exposure to major pollutants in Eastern Europe is not high in comparison with the OECD countries' (1991, p. 106). According to Juhüsz and Ragno (1993), air quality standards in CEE countries in the 1970s were stricter than those applied in Western European countries, and if a comparison was made based on EC standards of the time, it would turn out that the air quality of CEE cities was not necessarily worse than that of their Western European counterparts.

Rather than the general level of pollution, the problem lay in the efficiency of industrial production and the pollution to GDP ratio. Utilisation of poor quality coal for energy supply created very high pollution levels with low efficiency (Juhüsz and Ragno, 1993).

From the 1970s, almost all CEE countries adopted their own environmental legislation and institutions. This era can be characterised by protection-centred (conservationist) environmental interventions: natural protection areas, national parks and protected species were nominated. The regulations might have been seen as strict; however, the ongoing polluting practices were not halted due to the lack of means for control and prosecution, allowing them to continue without consequences.

Finally, the most important catalyst for Western European environmental progress was missing in the CEE macro-region: i.e., civil society. In a totalitarian system, public participation based on the freedom of expression was inconceivable. On the other hand, barely any information was available to the public about the actual state of the environment, since environmental problems did not officially exist. The social consciousness and sensitivity about the natural environment had developed significantly by the end of the 1980s, and the environmental issue became a platform for expressing dissatisfaction towards the workings of the entire system (Klarer and Francis, 1997).

During the transition

Transition itself was not a single act and it was not consistent (Stark, 1992). As environmental degradation was portrayed by communist and socialist states as the exclusive problem of capitalist states, those states positioned the industrial and agricultural practices of the socialist bloc as extremely wasteful of energy and materials (Podkorytova and Raskina, 2014). After the regime change, the socialist market was split up, leaving the production–consumption system unsustainable in many of these countries. The industrial and agricultural collapse naturally entailed the drastic decline of production, thus significantly decreasing certain activities that were environmentally harmful.

The regime change in CEE brought about some change in the perception of and attitudes towards environmental issues. In fact, environmental issues became indirect domains of criticism towards the system, and they were very strong elements of the political rhetoric adopted by the opponents of the communist regime. Leaders of the evolving environmental movement (like Ecoglasnost in Bulgaria) later became influential actors of political transition (French, 1990). However, after the political change, environmental movements and the issue itself lost much of their prominence. Deepening economic and social crisis came to the forefront of political debates, leaving environmental concerns of the economic transition in the background (Klarer and Francis, 1997). Socio-economic struggles affected not only the political importance of environmental issues, but also eroded the civic structures of democratic environmental policies and practice.

The fact that CEE countries managed to develop weak independence in many policy fields (including nature-related issues) might also be due to the above-mentioned socio-economic problems of the transition. As a result, the newly evolving environmental policy in the rapidly changing economic conditions was very much influenced by foreign governmental bodies and international agencies, and this influence was in fact warmly welcomed by CEE governments (Pavlínek, 2002). This new current of environmental policy blueprints and legislation was not in all cases stronger than the former practices of the socialist era.

Prior to accession, there were some concerns that the EU's environmental policy, which had been considered progressive to that point, would be attenuated by the much more permissive practice of the former socialist countries (Jehlicka and Tickle, 2004).

Changing conceptions of nature and spatial planning

Do CEE countries have their own specific conceptions about nature? Petr Pavlínek recalls the words of David Harvey: 'An enquiry into environmental history as well as into changing conceptions of nature provided a privileged and powerful way to enquire and understand social and cultural change' (1996, p. 166). This thought is in line with English studies from the same time. Howard Newby, for example, stated: 'beneath the conflict over "the environment" there is a much deeper conflict involving fundamental political principles and the kind of society we wish to create for the future' (1990, p. 4.). The environmental problem is not merely technological and scientific, it is deeply social and political, reflecting the way the given society (or civilisation) relates to the natural environment. A line in this dispute is concerned about spatial planning and its relation to and effect on the natural environment. Spatial planning, its process, actors, institutions and documents are a mirror of not only the different wills and interests of stakeholders, but also of the value systems and ideologies that are behind those wills. Thus, the approach of spatial planning to the natural environment is crucial – not only to effectuate more successful environmental protection interventions but also to uncover and help understand underlying complex value judgements. Myerson and Rydin (1994) see spatial planning as a forum that is capable of linking actual problems with otherwise hardly comprehensible cultural processes. According to Davoudi (2012), spatial planning in England has been the main scene of discourses about the natural environment. According to Whatmore and Boucher, 'planning represents an important institutional terrain for the contestation of the meaning and relations of the "natural environment"' (1993, p. 168). Spatial planning, beyond providing insight into the diverging conceptions about the human–nature relationship, is also a powerful tool in shaping these relations (Briassoulis, 2001; Williams, 2012).

Spatial planning and the very concept of planning have undergone significant changes after the regime change. In many of the CEE countries spatial planning has been understood as one of the sectors of development, rather than a unifying, integrating sphere. A strong emphasis on physical nature in spatial planning heritage, coupled with the markedly negative associations with former central command and government control, are challenges still to be overcome.

A solid Europeanisation of spatial planning has been taking place since the early 1990s. However, as Stead and Nadin (2011) argue, this is not synonymous with the convergence of planning systems or planning approaches. CEE conditions are fundamentally different, but notions and practices of spatial planning vary greatly among other European countries as well. In CEE, planning has a much weaker position in governments, and a much shorter contemporary history compared to Western European countries. The unsteady position of spatial planning in CEE countries is coupled with strong pressure from the European Union to adopt and follow EU legislation and guidelines regarding both spatial planning and environmental concerns encountered in the course of spatial planning. Changes are not only due to the regulatory framework that needs to be tailored to meet EU norms, but also to financial support from EU funds that are, in many cases, the only sources of development in these countries. The necessity of complying with EU guidelines to access funds (see Chapter 10 for the general dynamics) proved to be a stronger motivator than the intention to build up genuine, consistent, needs-based systems of environmental policy integration that would also resonate with national specificities in these countries. However, Pallagst and Mercier (2007) argued a few years ago that the most likely outcome of reforms in the CEE area would be a merger of styles of spatial planning with those applied in Western Europe, rather than a retention or reinforcement of difference. A 'merger of styles' can be more pessimistically viewed as the presence of a plethora of expressions and terms related to 'environmentally conscious' spatial planning, whereas it is extremely hard to find a systemic approach in the utilisation of these expressions. All in all they would remain environmentally conscious rhetoric lines that reuse and mix expressions of former and present spatial planning discourses. These include for example:

- sustainable development (sustainable urban development, sustainable economic development, etc.)
- green economy
- smart city
- green city
- compact city
- climate change
- resilience.

The meaning of analysis of the 'environment' in spatial planning has deep roots in the UK and especially in England. Summarising and completing the works of Newby (1990), Healey and Shaw (1993, 1994) and Blühdorn and Welsh (2007), Davoudi (2012) distinguished eight meanings of the natural environment in spatial planning. These treat nature as a(n):

- amenity
- heritage and landscape (canvas, background)
- nature reserve
- stock of resources and functions
- tradeable asset
- problem
- sustainability
- risk.

These categories are also applicable if we want to trace the different meanings of the environment in CEE spatial planning. A position of preceding studies was that the conception of nature and natural risk is different in CEE countries compared to the Western European view. Pavlínek (2002) claims there have been divergences in the Western and Eastern construction of the environment. Risk, a powerful view of Western discourses, was absent in Eastern environmental thinking, due to Marxist optimism about technological and scientific development. A 2012 study by Sandra Marquart-Pyatt found 'publics in CEE as a group simultaneously display heightened concern about environmental threats and lower willingness to sacrifice personally for the environment' (2012, p. 461). While the content of EU spatial and environmental policy guidelines possibly has the largest effect on the spatial planning system and on plans themselves in CEE, it is equally important to understand the value-systems behind the visible (or virtual?) nature-oriented approach of these spatial plans. For this, the above categorisation by Davoudi (2012) can serve as an appropriate starting point.

Environmental content of spatial planning documents in CEE

To gain insight into the concepts and value-sets regarding the natural environment in spatial planning, a brief examination was carried out in the Visegrad Four countries (Poland, Slovakia, Czechia, Hungary). The examination was based on the main national-level spatial planning document, since these documents are easily accessible in English and as expected, reflect the nature-related conceptions in lower level spatial plans as well (Table 17.1).

The framework for analysis consisted of the following questions:

- Is the natural environment included in the plans as an individual sector or are nature-related aspects integrated into all of the topics (sectors) of the plan?
- What parts of the plan address environmental problems? Are the situation analysis parts written in accordance with development proposals regarding environmental issues?
- Which meanings – based on Davoudi's categorisation – can be traced in the plan?
- What synonyms can be found for nature?
- Which 'environmentally conscious' rhetoric lines are followed in the plan?

The analysed documents were created at approximately the same time, preparing for the 2007–2013 Structural Funds period, but some of them also target long-term goals well beyond this time frame.

Table 17.1 Analysed national-level spatial planning documents of the Visegrad Four countries

Country	Title	Year published	Publisher	Planning time frame
Poland	National Development Strategy 2007–2015	2006	Ministry of Regional Development	2007–2015
Slovakia	National Regional Development Strategy of the Slovak Republic	2008	Government of the Slovak Republic	2007–2013
Czechia	Spatial Development Policy of the Czech Republic	2006	Ministry for Regional Development of the Czech Republic; Institute for Spatial Development	
Hungary	National Spatial Development Concept	2005	Government of Hungary	2007–2020 (2013)

Source: Author's construction.

In the *Polish* document, the situation analysis chapter dedicates a column to nature-related assets and opportunities. Here, three lines of thought appear: natural environment in less developed, mountainous areas are endowed with high biological diversity, and these sites are also attractive and important assets for nature-based tourism (the most important economic sector within these regions). It is also mentioned that the protection of the natural elements has improved in the last decades, and that the development of car transport poses new challenges. Later in the strategy, support for alternative tourism resonates with this problem; however, the document does not give any further details about its realisation. A third strong line is directly related to the resource narrative: the energy production of Poland is largely reliant on domestic resources, mainly coal, which is available in the country. The possible adverse effects of coal mining itself (which is mainly surface mining in Poland) or the use of this low quality coal for heating are not mentioned. In fact, the possible increase in energy prices is planned to be bridged over by alternative sites of import, and excavation of further domestic energy sources: coal remaining the main resource for domestic energy production. Hydropower and nuclear energy are listed as alternative energy sources.

In the vision statement part of the document, environmental protection, biodiversity and sustainable development are mentioned in the last paragraph (two sentences), but this mention is not related to any of the economic and social goals listed before. Later, environment is also mentioned, the cleanliness and healthiness of which is important for the quality of life of residents. The document gives the following interpretation for sustainable development: 'The state will pursue a policy of balanced development by integrating activities in the economic, social and environmental sphere in the interest of future generations' (p. 28.).

Eco-industry and eco-innovations are listed among the progressive sectors of the industry, which can result in economic growth and cost-reduction in production. Later the idea that the low efficiency and high pollution ratio of economic production can also be improved with the help of eco-industry is mentioned as well. A hint for more compact land use can also be found here.

Environment-related infrastructure was emphasised in the strategy; the development of a sewage treatment system, potable water and waste management were important aspects for the limitation of harmful substances entering uncontrolled into nature.

The development objective regarding rural areas underlines the significance of agricultural modernisation, setting aside and extending agriculture-related activities; however, the large impact of agricultural activity on the environment is not addressed. Similarly, the vulnerable yet still natural mountainous areas are not taken into consideration. The Polish document deals very little with the natural environment and, when it is mentioned, it is understood mainly as a stock of resources and an amenity, whereas fast economic growth is portrayed as inevitably damaging to the natural environment.

The National Regional Development Strategy of the *Slovak Republic* deals with the natural environment as a separate priority area. The relevant chapter proffers a detailed view about the state and pollution of natural elements within the country, specifying the most endangered areas. It discovers a 'paradoxical' coincidence between high potential for economic development and urbanisation and the most severe environmental problems, while marginalised territories (often not provided with potable water and sewage systems) remain the least stressed areas.

The most severe problems mentioned above include the inefficient implementation of environmental policies and the neglect of the environmental landscape. The phrase 'ecosystem services' appears here. Further issues range between insufficient material and energy intensity, a potential threat to imports, outdated and environmentally harmful technologies, and climate change as a threat to agricultural and other economic activities. The reduction of green areas and open space is a mostly urban problem, whereas it is admitted that the development and utilisation of alternative modes of transportation is in an initial phase. The document stresses that thematic areas for environmental interventions should be handled in a comprehensive manner together with other problems on a territorial basis. It also recommends that coordination and communication between different levels of decision-making should be improved.

The connection between environmental and other problems is addressed in other priority areas as well. On the one hand, the environment is still considered as something separate from other development issues; on the other hand, there is an intention to discover the synergies between environmental and other types of problems and solutions. These parts give detailed listings about certain environmental problems and proposed solutions. The strongest among these issues is the pollution of natural elements, especially land erosion due to flooding. Energy and material efficiency, natural risks and disasters, brownfield development and waste management are all listed among the potential intervention areas. Meanwhile, sustainable economic growth rhetoric is used when discussing economic development priorities.

The growing share of individual car-based transport is also mentioned in this document, with the insufficiency of environmentally conscious solutions in the realisation of infrastructure projects.

The Spatial Development Policy of *Czechia* states in its introduction that it would realise spatial development 'with regard to sustainable development of the territory above all' (p. 8) – however, it does not give any explanation of the perception of this notion. Among the proposals, sustainable spatial development is phrased as 'balanced relation of territorial conditions for favourable environment, for economic growth, and for cohesion of population living in the territory' (p. 12). Using Davoudi's categories, favourable environment presumably refers to the environment as an amenity. Ecological stability, greenbelts, the recreational and reproductive capacity of open areas are also mentioned. For transportation, environmentally friendly public mass transportation is proposed. The risk of natural disasters also appears in the initial chapter of the document.

Development pathways are distinguished in this document on a territorial basis. There are development areas, development axes and specific areas. It is clear that transportation infrastructure and the development of public transportation is the most important development goal for the regions of Czechia. Environmental issues are only mentioned in two development areas, and in certain development axes where environmental pollution is the highest. Otherwise the natural protection narrative is applied in specific areas with high natural value with the aim of protecting the economic basis of these tourism-related regions. Recreational functions and the importance of resource capacity (timber processing, mining industry) are also emphasised, pointing to a potential conflict in the use of these territories. Specific areas are proposed as rather identical development priorities that call for 'sustainable development in coordination with the nature and landscape protection' (p. 32), with no further details on what meaning, realisation and potential conflicts will result from this goal.

Transport corridors fall under the category of another type of territorial intervention and planning, where potential conflicts with Natura 2000 areas are mentioned. The development of rail corridors is emphasised as a more environmentally friendly means of transportation.

Among the long-term objectives of the *Hungarian* strategic document, we find 'sustainable territorial development and protection of heritage' (p. 4). However, only increasing the renewable energy share is listed among the medium-term goals. Later, an explanation of sustainable development reflects a resource-based understanding of nature: 'The development and resource management that is taking place today does not jeopardise future generations' ability to securely fulfil their needs. The development process does not increase the threat to the local natural- and built environment, cannot lead to the depletion of resources or the disappearance of cultures rich in value, and at the same time ensures the conditions for a high standard of living for society' (p. 6). Later a land-use approach integrates the resource narrative and extends it with the conservation narrative, which affects mainly rural areas.

Among the medium-term goals, natural environment is mentioned related to the health of citizens, and the *green areas* within the Budapest metropolitan region. The prevention of sprawl seems to be a goal concerning the capital city area only. Ecological balance as a priority is present only in the goals of the Balaton area, strongly related to the economic base the lake provides through tourism. The topics of protection against nature (floods) and protection of nature are mixed in the case of the two large river areas (Tisza and the Danube), which again relates to the economic benefits of the environmental and cultural heritage. Biodiversity is mentioned here, unrelated to the other aspects. Renewable energy appears in relation to agricultural production (biomass utilisation for energy production) and with micro-regional energy supply. The Hungarian plan is the only one of the four that would promote cross-border environmental protection and hazard mitigation.

Natural landscape value is given priority in the development of specific rural areas where there is a large share of protected natural areas. At the same time, the concept that beyond aesthetic aspects, landscape ecology, the harmonisation of agricultural and forest management activities as well as the conjoining of forests are equally important is emphasised.

From the analysis of the four documents above it turns out that all of them use a variety of expressions related to the inclusion of environmental aspects in spatial planning. The most frequently used terms are: sustainable (economic) development, biodiversity, landscape, ecological systems or stability, pollution of environmental elements, and heritage. The Slovak document is the most integrating of them, one that goes beyond having a separate chapter for environmental aspects. It is also the most detailed concerning the possible solutions for the problems detected. The resource narrative is very strong in the case of the Polish and the Czech documents, and the nature-as-amenity approach is markedly present in all four. It is interesting that both the 'amenity' and 'biodiversity' aspects of nature are delegated to economically marginalised areas. Nature as a risk is also visible in all of the four documents. The passages listing problems to be addressed are often not in line with the strategic parts coming later in the documents: environment-related problems are mentioned many times, but the way they should be addressed and, more importantly, integrated with the solution of other problems, is not or only superficially detailed.

The interpretation of 'sustainable development' is widely divergent across the documents, with none of them trying to address the inherent conflict of the expression itself. The other terms of rhetoric (e.g. smart city, resilience, etc.) were not used however, some references to compact structure can be found in the promotion of brownfield development and the alleviation of sprawl.

Conclusions

Despite the assumed similarities, there were large spatial and territorial differences in environmental pollution in state socialist Central and Eastern European countries. Pollution was presumably larger compared to the economic output than that of Western European countries, but it is not at all proven that pollution in general was more severe. The environmentally polluting hot-spots of heavy industries were accompanied by untouched, marginalised areas of high natural value. Environmental regulation of the individual countries might have been even stricter than that of certain Western European countries, however, the enforcement of these regulations was highly sporadic. The regime change brought about an unprecedented economic disaster as industry and agriculture in all of the countries collapsed, which resulted in the drastic decrease of environmentally harmful activities as well.

The economic reconstruction period of these countries was parallel to preparation for accession to the European Union. The severe socio-economic problems and the entrance into the market economy bore much higher importance than the construction of a new economic base of high environmental standards, and the countries rather relied on the guidance of foreign bodies. This did not lead to consistent and tailor-made environmental policy integration.

Spatial planning serves as an integrating platform for many activities that have an impact on the natural environment. Thus spatial planning and plans have a significant role in handling conflicts between environmental concerns and sectoral interests; at the same time, they are capable of harmonising needs and finding synergies between these interests. In former socialist countries, spatial planning itself needed to be reformed, and it needed to be ready to leap towards more conceptual levels and integrate environmental goals as well. The main spatial planning documents of the Visegrad Four countries show that this was achieved only partially. The most popular expressions of environmental policy integration can be found in the plans, but they lack a consistent and meaningful content, only serving as slogans. The conflicts between environmental needs and economic interests are not addressed in the plans. Their wording, prepared only a few years after EU accession, still mirror a conception of nature (mostly the 'resource' narrative) that was characteristic of Western European countries in the 1980s, before the introduction of the sustainable development narrative. The natural protection (conservationist) line is also powerful in these plans; this intention is channelled mainly to remote, marginalised areas, where the state of natural elements is good anyway. In the case of urban areas, alternative transportation, green areas and conscious land use are among the environment-related objectives. The documents do not handle nature-related problems and solutions consistently, except for the Slovak document which mentions the environmental consequences of all of the proposed development objectives. The environmental issue is seemingly one of the sectors or development goals/principles at best, but a more in-depth integration of these concerns is yet to come.

Acknowledgement

This research has been supported by project #104985 of the Hungarian National Research, Development and Innovation Office.

References

Blühdorn, I. and Welsh, I., 2007. Eco-politics beyond the Paradigm of Sustainability: A Conceptual Framework and Research Agenda. *Environmental Politics*, 16(2), pp. 185–205.

Briassoulis, H., 2001. Sustainable Development and Its Indicators: Through a (Planner's) Glass Darkly. *Journal of Environmental Planning and Management*, 44(3), pp. 409–427.

Davoudi, S., 2012. Climate Risk Security: New Meanings of the 'Environment' in the English Planning System. *European Planning Studies*, 20(1), pp. 49–69.

Dawson, A.H., ed., 1987. *Planning in Eastern Europe*. London: Croom Helm.

French, H.F., 1990. *Green Revolutions: Environmental Reconstructions in Eastern Europe and the Soviet Union*. Washington, DC: Worldwatch Institute. (Worldwatch Paper, 99).

Hall, D.R., 1987. Albania. In: A.H. Dawson, ed. *Planning in Eastern Europe*. London: Croom Helm.

Harvey, D., 1996. *Justice, Nature and the Geography of Difference*. Oxford: Blackwell.

Healey, P. and Shaw, T., 1993. *The Treatment of Environment by Planners: Evolving Concepts and Policies in Development Plans*. Centre for Research in European Urban Environments, Department of Town and Country Planning, University of Newcastle upon Tyne. (Electronic Working Paper, 4).

Healey, P. and Shaw, T., 1994. Changing Meanings of 'Environment' in the British Planning System. *Transactions of the Institute of British Geographers*, New Series, 19(4), pp. 425–438.

Hughes, G., 1991. Are the Costs of Cleaning up Eastern Europe Exaggerated? Economic Reform and the Environment. *Oxford Review of Economic Policy*, 7(4), pp. 106–136.

Jehlicka, P. and Tickle, A., 2004. Environmental Implications of Eastern Enlargement: The End of Progressive EU Environmental Policy? *Environmental Politics*, 13(1), pp. 77–95.

Juhüsz, F. and Ragno, A., 1993. The Environment in Eastern Europe: From Red to Green? *OECD Observer*, 181, pp. 33–36.

Klarer, J. and Francis, P., 1997. Regional Overview. In: J. Klarer and B. Moldan, eds. *The Environmental Challenge for Central European Economies*. Chichester: Wiley, pp. 1–66.

Klarer, J. and Moldan, B., 1997. *The Environmental Challenge for Central European Economies*. Chichester: Wiley.

Lange, O., Taylor, F.M. and Lippincott, B.E., 1938. *On the Economic Theory of Socialism*. Minneapolis: University of Minnesota.

Marquart-Pyatt, S.T., 2012. Environmental Concerns in Cross-National Context: How Do Mass Publics in Central and Eastern Europe Compare with Other Regions of the World? *Czech Sociological Review*, 48(3), pp. 441–466.

Myerson, G. and Rydin, Y., 1994. 'Environment' and Planning: A Tale of the Mundane and the Sublime. *Environment and Planning D*, 12(4), pp. 437–452.

Newby, H., 1990. Ecology, Amenity and Society: Social Science and Environmental Change. *Town Planning Review*, 16(1), pp. 3–20.

Pallagst, K.M. and Mercier, G., 2007. Urban and Regional Planning in Central and Eastern European Countries: From EU Requirements to Innovative Practices. In: K. Stanilov, ed. *The Post-socialist City. Urban Form and Space Transformations in Central and Eastern Europe after Socialism*. Dordrecht: Springer.

Pavlínek, P., 2002. Ost- und Ostmitteleuropa: Transitionen im Bereich der Umwelt. *Ost-West Gegenwaltinformationen*, 14(3), pp. 3–12.

Pavlínek, P. and Pickles, J., 2005. *Environmental Transitions. Transformation and Ecological Defence in Central and Eastern Europe*. London: Routledge.

Podkorytova, O. and Raskina, Y., 2014. Former Soviet Union Countries and European Union: Overcoming the Energy Efficiency Gap. *EUI Working Paper*, Robert Schuman Centre for Advanced Studies (RSCAS, 2014/03).

REC, 1994. *Strategic Environmental Issues in Central and Eastern Europe. Environmental Needs Assessment in Ten Countries*. Budapest: Regional Environmental Center for Central and Eastern Europe.

Stead, D. and Nadin, V., 2011. Shifts in Territorial Governance and the Europeanisation of Spatial Planning in Central and Eastern Europe. In: N. Adams, G. Cotella and R. Nunes, eds. *Territorial Development, Cohesion and Spatial Planning*. London: Routledge, pp. 154–178.

Stark, D., 1992. Path Dependence and Privatization Strategies in East Central Europe. *East European Politics and Societies*, 6(1), pp. 17–51.

Whatmore, S. and Boucher, S., 1993. Bargaining with Nature: The Discourse and Practice of 'Environmental Planning Gain'. *Transactions of the Institute of British Geographers*, New Series, 18(2), pp. 166–178.

Williams, J., 2012. Regulative, Facilitative and Strategic Contributions of Planning to Achieving Low Carbon Development. *Planning Theory & Practice*, 13(1), pp. 131–144.

Analysed documents

Ministry of Regional Development, Poland, 2006. National Development Strategy 2007–2015.

Government of the Slovak Republic, 2008. National Regional Development Strategy of the Slovak Republic.

Ministry for Regional Development of the Czech Republic, 2006. Spatial Development Policy of the Czech Republic (Decree of the Government of the Czech Republic as of May 17th, 2006, No. 561 on the Spatial Development Policy of the Czech Republic).

Government of Hungary, 2005. National Spatial Development Concept.

18
Spatial researches in Central and Eastern Europe[1]

Gyula Horváth

Introduction

The twentieth century marked an era of specialisation of the sciences and the birth of new scientific disciplines. The investigation of increasingly complex social processes and economic phenomena required the development of new scientific branches. In the background of this evolution was the need for a conscious organisation of society and the economy: the organisation of a country, the operation of an economy, the development of human relations requires substantial knowledge. Various branches of sociology, political science, psychology and economics already played a dominant role in the practice of modern states at a relatively early period.

In the focus of interest of the early classical economists were new space shaping forces that resulted in the transition from a rural society to the industrial age in Western Europe. The mercantilists of the golden age of commercial capitalism highlighted the role of spatial relations in market expansion and the reduction of production costs. Almost every economic current of thought during industrial capitalism integrated space, local, regional, national and international dimensions into its system of thinking. Economics joined the traditional discipline of geography in the investigation of territorial processes; spatial economics even gained a predominant role in theory building and the formulation of spatial policy. Due largely to the methodological development of the social sciences, the expansion of empirical analysis and the interest of social management, spatial relations of socio-economic phenomena entered the horizon of other social sciences as well (Benko, 1999; Egyed, 2012).

During the second half of the twentieth century a large number of research programmes were launched in Western Europe and the United States to investigate the spatial structure of the economy and society; new theories, instruments and institutions enriched the science and practice of social management. Institutions and departments were organised to facilitate research and training in the area of scientific problems related to space; monographs, book series and journals were published (Isard, 1975). Just as economists adopted the spatial approach, so did the powerful need for the modernisation of traditional regional geography emerge. Despite the numerous disciplinary results, international scientific public opinion associates the notion and research methodology of the new economic geography with the work of economist Paul Krugman (1991a, 1991b, 2000). The theses that contain the counter-opinions of prominent figures of economic geography reject several aspects of the spatial grounding of the tentative renewal of economics (Martin, 1999).

The leading figures of economic geography sometimes commit the natural mistake of neglecting the recent results of other scientific disciplines while being immersed in studying the new development perspectives of their own scientific discipline. The rapid development of the integrative spatial science, 'regional science' occurred simultaneously with the modernisation of traditional disciplines. There are still debates concerning the nature of its 'autonomy' as a discipline; nevertheless, social scientists engaged in spatial research have arrived at a consensus concerning its definition. According to this definition regional science provides a unified system of the common basic notions, theories and methods of social scientific disciplines engaged in the study of space, through the utilisation of which they investigate economic phenomena and processes (Boyce, 2004; Boyce et al., 1991; Nemes Nagy, 1998; Mészáros, 2006).

Starting from the 1970s, the advanced industrialised countries have experienced periods of grave crisis: geographical disparities have intensified, and the decline of Fordist-type industrial production together with the emergence of new space shaping forces gave rise to countless questions regarding spatial transformation. Massive structural changes have shaped the post-Fordist economy and these have formed the core of research on social and settlement change during the past two decades. The number of regional scientific publications multiplied; half of over a dozen journals currently published were established during the boom phase. The spectrum of research topics became broader, new trends emerged, e.g. the investigation of space structuring effects of innovation and technological development and the network economy.

The results of the development of regional science in Western Europe and the USA were summarised in several studies and books (Florax and Plane, 2004; Isard, 2003; Isserman, 1993, 1995). New works were published about the publication forums of regional science and the activities of its international organisations during the past decade. In these works we only find a couple of references to Eastern and Central European spatial research. The modest references can be explained by the fact that the examination of the spatial evolution of the economy and society and the organisation of spatial research into an autonomous discipline were not reflected in CEE research programmes. On the other hand, we may recall that neither the results achieved in determining the regularities of national social and economic spaces, nor the attempts at the organisation of science caught the attention of international professional public opinion. It is likely that both suppositions contain an element of truth.

A specific feature of the investigation of CEE regional development is that several Western European researchers, such as the British scholars Ian Hamilton (1974, 1982), Michael Bradshaw (1993; Bradshaw and Stenning, 2004) and David Turnock (1978, 1989), and French researchers Marie-Claude Maurel (1982, 2002, 2004) and Violette Rey (Rey and Gerbaud, 1996), were specialised in the examination of this area and studied the spatial processes of one or several countries, while in CEE an extremely narrow circle of researchers performed systematic research on this macro-region. György Enyedi (1981, 1989, 2003) was the only scientist to perform comparative analyses about this area on a regular basis. Around the turn of the millennium, the scientific activities of Iván Illés, coordinator of the CADSES-programme for the German Federal Office for Construction and Spatial Development were closely related to this area (Illés, 2002). Polish researcher Gregorz Gorzelak published a book in 1996 about the Eastern European regional aspects of the transformation (Gorzelak, 1996). However, the investigation of the area did not constitute a permanent theme in his scientific portfolio. At the turn of the year 2000, preparations for the EU accession of CEE countries were accompanied by several scientific publications; these, however, were collections of studies prepared by researchers of the various countries.

The present study evaluates the regional research capacities of the countries of the former socialist bloc. It provides a picture of the historical antecedents of spatial research, the specifics of regional tasks to be resolved, the characteristics of the institutionalisation of regional science and its publication forums. As a conclusion, it summarises the presence of various criteria of regional science in the individual countries. The summary and evaluation of results in spatial research and the contribution of 'European added value' of the results may be the topic of a separate study. This future study may choose from a wide spectrum of results of the past two decades. Significant results were achieved in the development of European regional science, e.g. in studying the regional effects of the CEE market economic transition, the new democratic public administrational spatial organisation, regionalisation and regional decentralisation, the system of objectives, instruments and institutions of EU-conforming regional policy, the competitiveness of urban networks and cross-border cooperation.

Recent publications with a special thematic focus were published about the state of spatial research in some countries, yet no comparative analysis has been performed as yet about CEE regional science. A summary about the state of social geography was prepared in the framework of a research investigating the state and results of CEE social sciences. We find ample references to spatial research in these (Maurel, 2002). The evaluation prepared by György Enyedi for the International Geographical Union contains valuable information about the results of Central European applied geography (Enyedi, 2003). The methodology of the study is based on the evaluation of professional literature, analysis of Internet databases and the personal experiences of the author.

Regional development and spatial research in the twentieth century

There is hardly any period in the millennium-long history of Europe when Eastern territories followed an 'outstanding' development path. The long-term trends of marginalisation were only interrupted by short periods of recovery during the prosperous centuries of the Middle Ages, the Enlightenment, and later on the unfolding of the industrial revolution. General European development had a decisive impact on the metropolitan elements of the CEE system of settlements. Core regions did emerge in the spatial structure, but their expansion was limited, they did not exert a decisive influence on regional development outside their immediate catchment areas.

As a result of the state organisational and socio-political tasks related to the territorial shifts that took place during the short period between the two world wars, several scientific disciplines had to place a growing emphasis on the analysis of territorial economic processes, the organisational system of territorial public administration and governance models, the settlement system and population redistribution. During this era, the activities of several acknowledged social scientists left their mark on the functioning of the state. To cite a few examples, we can mention the rural sociological works of the Romanian Dimitrie Gusti, the political scientific research of the Hungarian András Rónay, the economic and socio-geographical analyses of the Czech Viktor Dvorský. The research results of significant figures of the generally prominent Russian (later Soviet) applied economic geography, Ivan G. Aleksandrov, Nikolay N. Baransky and Nikolay N. Kolosovsky contributed to the creation of economic districts and provided the scientific groundwork for spatial planning.

During World War II, new institutions were established in several parts of Central Europe which regarded the identification of regional assets to be their main task in order to provide scientific bases for post-war reconstruction. Among these, the Baltic Institute in Toruń, the Silesian Institute in Wrocław, the Western Institute in Poznań, the Silesian Institute in Opava,

Moravia, and the Transdanubian Research Institute in Hungary are worth mentioning. Two of these institutes are still functioning at present. Currently, the main profile of the Institute of Poznań is the research of Polish–German relations. The institute founded in Pécs in 1943 became the centre of basic research in Hungarian spatial development, and maintained its functioning as the seat of the Centre for Regional Studies of the Hungarian Academy of Sciences (HAS) from 1984.

Socialist planned economies desired to achieve the moderation of spatial disparities primarily through industrial development programmes. Infrastructural investments linked to industrialisation and the ever-growing speed of urbanisation resulted in the reduction of income disparities of macro-regions. Spatial planning fulfilled an important role in the system of the planned economy: in Poland, an act on spatial planning was adopted in 1961. Long-term spatial plans contained the key elements of regional development strategies; settlement network development concepts were elaborated in several countries (Bulgaria, Hungary, Romania).

The ideology of the regional and settlement policy of state socialism (classical Marxist theory, urbanist utopias, planning theory) and the objectives derived thereof (balanced development, the moderation of disparities in the civilisation of villages and towns, the spatially balanced distribution of free or highly subsidised social transfers) posed serious barriers to scientific disciplines engaged in the study of spatial processes. The notion of an ideally homogeneous society conquered scientific thought in the various countries in a differentiated way. Due to their intensive links with Western science, Polish and Hungarian social science provided significant research results about spatially unequal development, the anomalies of the transformation of the settlement structure at a relatively early period, the middle years of the 1960s; they questioned 'expressis verbis' the efficiency of central planned economy, an economic policy neglecting local–regional assets (Domański, 1983; Dziewoński, 1967; Enyedi, 1981).

The results of the analyses about the spatial transformation of the planned economies pointed to the fact that the economic structure and type of urbanisation characteristics in Eastern and Central Europe did not represent an autonomous model, but a repeat of Western-type urbanisation and development cycles with a significant delay. The disparities of spatial development could be attributed to late industrialisation, on one hand, and the functioning of the system of state socialism, on the other.

Due to the specific development paths of Eastern and Central Europe, research in the field of social and economic space shows quite a few unique features as well. *Socialist science policy, following the guiding principles of the power structure, did not consider spatial research as a priority issue.* This was mainly because not one tier of the strictly centralised state administration was interested in the analysis of local–regional specifics. Political practice aimed at homogenisation and considered spatial aspects only to the extent that central planning required.

Even though the era between 1948 and 1990 was characterised by challenges related to the research of socio-economic space for different scientific disciplines, the demands of the commissioners were neither complex nor demanding of thematic cooperation between scientific disciplines from the aspect of social management. The traditional scientific disciplines investigating spatial relations (economic geography, settlement and public administration sciences, economics to a certain degree) could all pursue independently of each other their activities in academies or universities. The scientific bases of spatial research were established primarily in public institutions, national planning offices and urban planning institutes.

The catalytic effect of the investigation of spatial processes can also be detected in the process of the differentiation of CEE social sciences. The research results relating to the detection of inter-settlement disparities in the structure of society served as an important force behind the greater autonomy of sociology in terms of theory and methodology, whilst the investigations of spatial-settlement components of public administrative-power relations contributed to the legitimisation of political science (Bihari, 1983; Kulcsár, 1986; Musil, 1977).

The institutional coordination of territorial research was achieved in two countries. The Council for the Study of Productive Forces maintained its functioning in the Soviet Union. In Poland, the Polish Academy of Sciences (PAS) established the Committee of Space Economy and Regional Planning in 1958 (Kukliński, 1966). These two institutions disposed of autonomous financial resources facilitating the validation of their general coordination competences. In a third country, Hungary, the governmental decree on the national spatial development concept delegated the organisation of basic research in spatial development to the HAS in 1971; however, this did not imply state level coordination.

The major scientific branch involved in the examination of spatial processes was social and economic geography. Almost every scientific academy had their own geographic institutions whose results in applied geographical research had a significant impact on spatial development decisions of the era. Poland was also prominent in the area of institutional innovations: the name of the geographical institute of the PAS was changed to Institute of Geography and Spatial Economics at the beginning of the 1970s. In Hungary, the Centre for Regional Studies functioning in the form of a network was established in 1984 whose base was the Transdanubian Research Institute of the HAS performing territorial basic research. The research centre, in collaboration with the Faculty of Economics of the University of Pécs, launched a postgraduate training programme in spatial development in 1988. The scientific capacities of the geographical departments and institutes of the university were quite significant despite the fact that this discipline was rather more oriented towards teacher training in socialist higher education. Departments of urbanism in polytechnical universities were also acknowledged research groups in several countries.

A significant factor in the long-term development of Polish and Hungarian spatial research was the reformist spirit in the political systems of the two countries. Consequently, the relations between scientific workshops of the two countries were maintained with Western European research units, joint research programmes were launched, and the national, regional and local political elite expressed interest in their research results. In fact, it is not too bold to state that regional research played an active role in preparing for the regime change (Maurel, 2002). Research results called attention to the fact that the modernisation of the economy required a substantial transformation of the spatial structure and, as a result, the reconceptualisation of objectives, principles and institutions of spatial development policy was inevitable. The cooperation and development coalition between the central state, local-territorial communities, public and private sectors was to become the basis of the new model of social management. Hungarian research analysing the spatial structural transformation of planned economies highlighted the fact that CEE economic structure and urbanisation did not constitute an independent model but an example of delayed Western-type urbanisation and development cycles. The disparities of spatial development are attributed to the lateness of development on one hand and the functioning of the system of state socialism on the other (Enyedi, 1989).

Regional transformation after systemic change

New processes could be observed in Eastern and Central European spatial development in the 1990s. Their impacts were as dramatic as the changes linked to forced industrialisation. Demographic, labour market, economic and environmental processes showed significant disparities in Eastern and Central Europe during the transition period to market economy. Western European experts tend to treat this territory as a homogeneous entity. However, the heritage of state socialism, the regional effects of transformation and the economic policy instruments and institutional solutions utilised in the management of post-socialist change have produced quite heterogeneous results. The radical transformation of the economic structure has not affected the various regions in the same way. The losers of the transformation – similarly to other European countries – were heavy and extractive industrial regions and, as a CEE specificity, cohesive agricultural areas. The building of market economy resulted in an aggravation of regional disparities, most spectacularly in Russia.

One of the characteristic deficiencies of the activities of the first democratic governments in Eastern and Central Europe following regime change was the lack of attention given to the spectacular and rapid spatial restructuring of the economy. Hungary was the only exception, where, following the democratic elections of 1990, a Ministry of Environmental Protection and Spatial Development was established in the governmental structure, and governmental programmes were elaborated for the restructuring of heavy industrial regions experiencing acute crisis symptoms. None of the governments elaborated a coherent regional policy strategy covering the entire area of the country, nor were former spatial development programmes adjusted to the new development objectives – rather, they were eradicated. The political elite did not comprehend the essence of spatial development; there were quite a few who identified it with the dated instruments of the planned economy, and regarded it as a remnant of national economic planning. This political atmosphere was not favourable for the thematic and institutional modernisation of spatial research in the 1990s.

EU membership, institutional changes and expanding financial opportunities have created more favourable conditions from the aspect of spatial research as well. New knowledge about the practice of Western European spatial development policy and the economic and human resources potential of regions mentioned in national development plans was required for the application of the Structural Policy of the European Union and the elaboration of regional development programmes and concepts. The new requirements generated scientific demand; regional development aspects have also been present in the training of professionals in several countries. In the largest successor state of the former USSR, the Federation of Russia, regional scientific knowledge gained importance due to the reorganisation of interregional relations and the widening competences of regional authorities.

The institutional frameworks of spatial research at the beginning of the twenty-first century

Demand for a better comprehension of spatial processes significantly increased after the change of regime. The institutional structure of spatial research has also undergone major transformations. Academic research institutions have found themselves in a difficult financial situation in several countries. The Czech Institute of Geography was closed, while a research centre of earth sciences was established in Bulgaria where the role of social geography is quite peripheral. There have been institutional integrations in Hungary as well: the Centre for Regional Studies has been deprived of its managing functions, the national network has become weaker. The new research centre's seat is in Budapest; the positive experiences in the decentralised management of science have presumably gone to waste. Large public urban planning institutes with remarkable intellectual capacities which had played a significant role in the elaboration and execution of spatial and settlement development tasks of the socialist era were closed.

On the other hand, the weight of regional scientific capacities of universities has increased. Research has once more become a priority in universities; the structure of training has also been transformed. In geography training, applied geography masters programmes have been launched, which also specialise in the training of spatial and settlement development experts. A significant result of the comprehensive reform of the economics curriculum was the organisation of a masters programme in spatial economics and regional policy.

According to calculations based on Internet data collection in the research institutes and university workshops of the six countries of Eastern and Central Europe, the number of employees engaged in spatial research exceeds 900. The distribution of student numbers is quite uneven within and also between the respective countries (Table 18.1, Figure 18.1).

Among the countries investigated in depth, Poland has the largest capacity in regional scientific research and training. Poznań, Łódź, Warsaw, Krakow and Wrocław are the country's most significant centres of regional scientific research. Hungary ranks second (the most important workshop centres being Pécs and Budapest), with its spatial distribution of research units in nine cities and towns, which is more even than in Romania, where regional scientific workshops can be found in four cities. In Czechia, only the three largest cities can be regarded as centres of regional scientific research. Slovakia is tri-polar from the aspect of regional science, and in Bulgaria only the academic and university geographical institutes of the capital city are engaged in regional scientific research. Approximately 60 scientific workshops with regional research as their main profile have been organised in 30 cities of CEE since the beginning of the 2000s. These workshops have multi-annual research programmes; they publish their results on a regular basis, their colleagues frequently attend international scientific forums, publish their works and participate in conferences.

In the following, the author cannot refrain from evoking some features of the institutional background of Russian regional science. The analysis of this country is not possible in the framework of the present study, since the collection and processing of the massive volume of information would require a longer time. The leading institutions with a century-long tradition of Russian regional scientific research are still functioning, as has been demonstrated, 'regional'naya nauka' is an acknowledged scientific discipline in Russia. The discipline has two dominant intellectual centres: Moscow and Novosibirsk. Regional topics can be found in the research plans of dozens of the academic and federal sectoral research institutions. Two institutions deserve special attention. Several colleagues of the Institute of Geography of the Russian Academy of Sciences (RAS) Russia regularly publish high quality works with Western scientific publishers (Artobolevsky, 1997; Ioffe and Nefedova, 2000; Lappo and Hönsch, 2000). The other significant workshop is the Council for the Study of Productive Forces already mentioned, which exerts its coordinating functions through its several research programmes and the publication of books and journals. The scientific centre of Novosibirsk is the Institute of

Table 18.1 The number of regional science researchers in Eastern and Central European countries, 2012

	The number of scientific researchers, person	Distribution, %	The rate of researchers employed in research units in capital cities, %
Bulgaria	30	3.3	100.0
Czechia	115	12.6	34.8
Poland	425	46.7	17.5
Hungary	150	16.5	20.0
Romania	130	14.3	31.9
Slovakia	60	6.6	50.0
Total	910	100.0	21.4

Source: Author's estimations based on online data collection. Contains university and research institute workshops whose name, research programmes and publications contain reference to regional science topics.

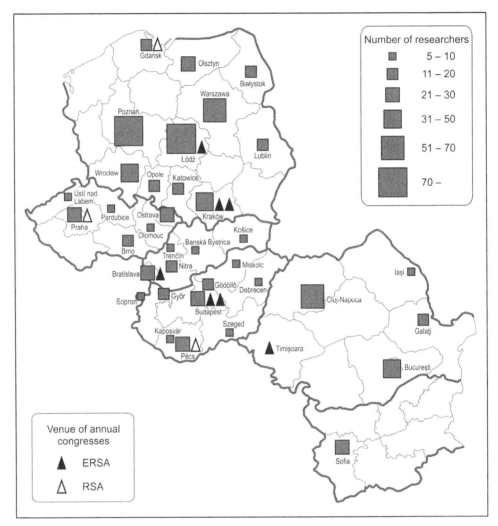

Figure 18.1 Spatial research workshops in Central and Eastern Europe, 2012

Source: Author's construction based on online data sources.

Economics and Industrial Production Organisation of the Siberian Division of the Russian Academy. One of the reform institutions of Soviet Perestrojka can boast of three scientific schools in regional science. One is responsible for laying the scientific groundwork for the spatial development strategy of Siberia and the further development of spatial planning, the other is the leading workshop of Russian settlement sociology, the third is a scientific community that functions on the basis of the most advanced Russian traditions of mathematical–statistical analysis methods and modelling. The two institutions in Moscow and the one in Novosibirsk constitute the scientific basis of the federal research programme titled 'The interdisciplinary synthesis of the spatial development of the Federation of Russia' coordinated by the Russian Academy (Kuleshov et al., 2012; Kotlyakov et al., 2012).

Apart from research institutions, scientific associations constitute the other important base of spatial research. Besides researchers engaged in the field, a scientific association assembles practising professionals interested in the application of scientific results and intellectuals interested in regional development. These forums for intellectuals function as autonomous institutions or national divisions of international regional science associations. The first group contains the Hungarian and Romanian Regional Science Associations. The Romanian Regional Science Association was founded in 2000. Currently it has 140 members. The results of Romanian spatial research are presented during its annual thematic conferences. It publishes a journal with two issues annually, titled the *Romanian Journal of Regional Science*. The Hungarian Regional Science Association was established in 2002, and has a current membership of 300. Its annual general assemblies are joined by thematic conferences. The organisations of regional scientific researchers in the rest of the countries are the national divisions of either the European Regional Science Association or the Regional Studies Association. In Poland, the Committee for Spatial Economy of the Polish Academy of Sciences (Komitet Przestrzennego Zagospodarowania Kraju PAN) can be regarded as the integration centre of regional scientific research. The committee operates six working groups and publishes three publication series annually. The 115 members of the Regional Scientific Committee of the Hungarian Academy of Sciences are employed in five working committees.

The development of publication forums

As was the case more globally, the publication of spatial research results in Central and Eastern Europe was only possible in the scientific journals of other disciplines during the first half of the twentieth century. Journals in geography, economics, sociology, public administration published the results of spatial research. Regional science did not have its own publication forums in any of these countries until the middle years of the 1980s, apart from the publication series of the Polish Academic Committee, the public administration journal of the Academic Institution of Economics of Novosibirsk or the Hungarian Regional Statistics. Before regime change, the publication of the Hungarian *Tér és Társadalom* in 1987 was considered a scientific novelty, and was followed with interest among international professional circles as well.

The first decade of the 2000s was the main period of the foundation of journals and over two dozen series were established by institutions, publishers and institutional consortiums engaged in regional research. The data concerning the major journals of regional science in Eastern and Central Europe are summarised in Table 18.2.

The level of institutionalisation, the traditions and scientific capacities of regional research have a decisive impact on publication activity. The Polish and Hungarian publication forums reveal a complex picture. The number of regional scientific monographs is the highest in these two countries. In the framework of the series titled 'Spatial and Settlement Research' (Területi és Települési Kutatások) under the care of the Hungarian Academic Press and later on under Dialóg Campus Publishing with the subtitle 'Studia Regionum', over 40 scientific monographs were published summarising the results of Hungarian regional scientists until the end of the last decade. Between 2010 and 2015, it was the 'Modern Regional Science' series under the care of the Akadémiai Kiadó which promoted the results of Hungarian researchers. However, this series was discontinued by Wolters Kluwer, the publisher's international owner, in 2016 for failing to meet expected sales targets. High quality Polish publishers which operate in regional centres publish a large number of regional scientific works as well.

Table 18.2 Some characteristics of the major regional scientific journals

Name of journal	Year of foundation	Publisher	Annual frequency of publication	Language
Czechia				
Regionalní Studia	2007	Faculty of Economics of the University of Prague	4	English
Poland				
Biuletyn KPZP PAN	1958	KPZK PAN	Occasional, 4–5 volumes a year. 250 published volumes.	Polish
Studia KPZK PAN	1958	KPZK PAN	Occasional, 2 volumes. 115 published volumes	Polish
Studia Regionalia	1986	KPZK PAN	Occasional, 1–2 volumes. 40 published volumes.	English
European Spatial Research and Policy	1994	University of Lodz, University of Groningen, Comenius University, Tbilisi State University, The Federal Office for Building and Regional Planning (Germany)	2	English
Studia Regionalne i Lokalne	2000	University of Warsaw Centre for Regional and Local Studies	4	Polish
Hungary				
Területi Statisztika	1960	Central Statistical Office	6	Hungarian
Tér és Társadalom	1987	HAS Centre for Regional Studies	4	Hungarian
Falu, Város, Régió	1999	Hungarian Nonprofit Ltd. for Regional Development and Town Planning	3	Hungarian
Discussion Papers	1986	HAS Centre for Regional Studies	Occasional, 105 published volumes.	English
Russia				
Region: Ekonomika i sociologiya	1963	RAS Siberian Branch, Institute of Economics and Industrial Production Organisation	4	Russian
Prostranstvennaya ekonomika	2004	RAS Far-East Branch, Institute of Economics	4	Russian
Region: sistemy, ekonomika, upravlenie	2007	Nauchnaya Zhizn'	4	Russian
Ekonomika regiona	2011	RAS Ural Branch, Institute of Economics	4	Russian
Sovremennye proizvoditel'nye sily	2012	Council for the Study of Productive Forces	4	Russian
Regional Science of Russia	2010	Pleiades Publishing, Springer Verlag distribution	2	English

Name of journal	Year of foundation	Publisher	Annual frequency of publication	Language
Romania				
Romanian Journal of Regional Science, Online	2007	Romanian Regional Science Association	2	English
Romanian Review of Regional Studies	2007	Babeș–Bolyai University Centre for Regional Geography	2	English
Journal of Urban and Regional Analysis	2011	Interdisciplinary Centre for Advanced Research on Territorial Dynamics, University of Bucharest	4	English

Source: Data collected by the author based on Internet resources.

Conclusions

The positive and negative effects of processes shaping socio-economic spaces can be observed in the twentieth-century development of Eastern and Central Europe, just as in other parts of the continent. Spatial aspects were also represented in the policies of past eras characterised by heterogeneous forms of state organisation. Research results were useful for decision-makers in terms of their ramifications for specific regions. The research results of various social scientific disciplines were incorporated into spatially related decision-making processes during the last years of the twentieth century as well. Nevertheless, the ruling elite of the communist era required only superficial knowledge about the evolution of spatial processes. Spatial research was conducted within national borders; international professional cooperation – with the exception of Poland and Hungary – remained weak and occasional.

Profound regional transformation was experienced due to the introduction of the market economy after 1989 and the collapse of the Soviet Union. The manageability of these changes naturally called for the thematic and organisational development of spatial research. The preparations for EU accession provided a further impulse for the research, and regional studies research groups flourished in all Central and Eastern European countries at the beginning of the twenty-first century. Disparities can be detected regarding the volume, the institutional system and the spatial distribution of research work. The ample availability of factors that contribute to the identity of regional science as an autonomous discipline can be demonstrated in the two EU member countries, Poland and Hungary, and in Russia as well (Table 18.3). Disciplinary criteria are partly lacking or show a weak level of development in the remaining countries.

The spatial distribution of regional scientific research units is somewhat more decentralised than in the case of other scientific disciplines. Research and development capacities of Eastern and Central European countries show a high degree of concentration in the capital cities; this may be regarded as an unfavourable phenomenon from the aspect of scientific and regional development (Horváth, 2009). In Poland, Hungary and Romania, the weight of capital cities in terms of the number of employees in regional science is one-half to one-third compared to other scientific disciplines. Regional science is a symbol and role model of the decentralisation of social activities. This discipline has accumulated valuable experiences in the operation of its decentralised and network-based organisational system; its methods may be efficiently transmitted to other economic and social sectors as well.

The examination of the development history of regional research demonstrates that outstanding scientists play a decisive role in the upswing of the scientific discipline and the broad

Table 18.3 The development level of the disciplinary criteria of regional science

	Bulgaria	Czechia	Poland	Hungary	Russia	Romania	Slovakia
Research units	■	■■	■■	■■	■■	■	■
University masters training	■■	■■	■■	■	■	■	
Doctoral schools			■■	■■	■■		
Journals		■	■■	■■	■■	■■	
Book series			■■	■■	■■	■	
Scientific association and scientific academic coordinating organisation			■■	■■	■■	■	
International regional scientific congress		■	■■	■■	■	■	■

Source: Author's construction.

Key ■ – Weakly developed; ■■ – Developed.

utilisation of research innovations. In Poland, Antoni Kukliński (1927–), professor of the University of Warsaw, contributed to the foundation of several organs and institutions of regional science. During the past two decades in Russia it was Aleksandr Granberg (1936–1910) – former director of the Novosibirsk Institute of Economics and Organisation of Industrial Production of the Russian Academy of Sciences and president of the Council for the Study of Productive Forces of RAS – who contributed with his work to the development of Russian regional science. In Hungary, the scholar György Enyedi (1930–2012) was the founder of this scientific discipline. All three of them made significant efforts towards the integration of Eastern and Central European research results into the international system of regional science.

Acknowledgement

The publication of this research has been supported by project #104985 of the Hungarian National Research, Development and Innovation Office.

Notes

1 The author of this chapter is deceased. Editing and minor updates were undertaken by the editor.

References

Artobolevsky, S.S., 1997. *Regional Policy in Europe*. London: Jessica Kingsley.
Benko, G., 1999. *Regionális tudomány*. Budapest–Pécs: Dialóg Campus Kiadó.
Bihari, O., 1983. *Korszerű tendenciák az államhatalom gyakorlásában* [Recent tendencies in the practice of state power]. Budapest, Közgazdasági és Jogi Könyvkiadó.
Boyce, D., 2004. A Short History of the Field of Regional Science. *Papers in Regional Science*, 83(1), pp. 31–57
Boyce, D., Nijkamp, P. and Shefer, D., 1991. *Regional Science: Retrospect and Prospect*. Berlin: Springer-Verlag.
Bradshaw, M.J., 1993. *The Economic Effects of Soviet Dissolution*. London: Royal Institute of International Affairs.
Bradshaw, M.J. and Stenning, A.C., eds., 2004. *The Post Socialist Economies of East Central Europe and the Former Soviet Union*. Harlow: Pearson/DARG.

Domański, R., 1983. *Gospodarka przestrzenna*. Warszawa: PWN.
Dziewoński, K., 1967. Teoria regionu ekonomicznego. *Przegląd Geograficzny*, 39(1), pp. 33–50.
Egyed, I., 2012. A regionális tudomány az elmélet és a a gyakorlat között [Regional science on the crossroads between theory and practice]. *Tér és Társadalom*, 26(4), pp. 17–36.
Enyedi, Gy., 1981. *Földrajz és társadalom* [Geography and society]. Budapest: Magvető Könyvkiadó.
Enyedi, Gy., 1989. Településpolitikák Kelet-Közép-Európában [Settlement policies in Eastern Central Europe]. *Társadalmi Szemle*, 44(10), pp. 20–31.
Enyedi Gy., 2003. Alkalmazott földrajz Közép-Európában [Applied geography in Central Europe]. *Földrajzi Értesítő*, 52(3–4), pp. 145–160.
Florax, R.J.G.M. and Plane, D.A., eds., 2004. *Fifty Years of Regional Science*. Heidelberg, NY: Springer-Verlag.
Gorzelak, G., 1996. *The Regional Dimension of Transformation in Central Europe*. London: Jessica Kingsley.
Hamilton, I.F.E., 1974. *Poland's Western and Northern Territories*. Oxford: Oxford University Press.
Hamilton, I.F.E., 1982. Regional Policy in Poland. A Search for Equity. *Geoforum*, 13(2), pp. 121–132.
Horváth, Gy., 2009. *Cohesion Deficiencies in Eastern and Central Europe – Inequalities of Regional Research Area*. Discussion Papers 72. Pécs: Centre for Regional Studies of the Hungarian Academy of Sciences.
Illés, I., 2002. *Közép- és Délkelet-Európa az ezredfordulón. Átalakulás, integráció, régiók* [South and Southeast Europe at the turn of the millennium. Transformation, integration, regions]. Budapest–Pécs: Dialóg Campus Kiadó.
Ioffe, G.V. and Nefedova, T.G., 2000. *The Environs of Russian Cities. A Case Study of Moscow. Russian Views of Transition in the Rural Sector*. Washington, DC: World Bank.
Isard, W., 1975. *Introduction to Regional Science*. Englewood Cliffs, NJ: Prentice-Hall.
Isard, W., ed., 2003. *History of Regional Science and the Regional Science Association International*. Berlin: Springer-Verlag.
Isserman, A.M., 1993. Lost in Space. On the History, Status, and Future of Regional Science. *Review of Regional Studies*, 23(1), pp. 1–50.
Isserman, A.M., 1995. The History, Status and Future of Regional Science: An American Perspective. *International Regional Science Review*, 17(3), pp. 249–296.
Kotlyakov, V.M., Glezer, O.B., Treivish, A.I. and Shvetsov, A.N., 2012. Novaya programma fundamental'nyh issledovaniy prosranstvennogo razvitiya Rossii. *Region: Ekonomika i sociologiya*, 59(2), pp. 24–44.
Krugman, P., 1991a. *Geography and Trade*. Cambridge, MA: MIT Press.
Krugman, P., 1991b. Increasing Returns and Economic Geography. *Journal of Political Economy*, 99(3). pp. 483–499.
Krugman, P., 2000. A földrajz szerepe a fejlődésben [The role of geography in development]. *Tér és Társadalom*, 14(4), pp. 1–21.
Kukliński, A.R., 1966. Research Activity of the Committee for Space Economy and Regional Planning. In: J.C. Fisher, ed., *City and Regional Planning in Poland*. Ithaca, NY: Cornell University Press, pp. 389–405.
Kulcsár, K., 1986. *A modernizáció és a magyar társadalom* [Modernisation and Hungarian society]. Budapest: Magvető Könyvkiadó.
Kuleshov, V., Seliverstov, V., Suslov, V. and Suspicyn, S., 2012. Sibirskaya shkola regional'nyh issledovaniy v programme Prezidiuma RAN 'Fundamental'nye problemy prostranstvennogo razvitiya Rossiyskoy Federacii: mezhdisciplinarny sintez'. *Region: Ekonomika i sociologiya*, 2, pp. 3–23.
Lappo, G.M. and Hönsch, F.W., 2000. *Urbanisierung Russlands. Urbanisierung der Erde*. Berlin: Gebrüder Borntraeger.
Martin, R., 1999. The New 'Geographical Turn' in Economics: Some Critical Reflections. *Cambridge Journal of Economics*, 23(1), pp. 65–91.
Maurel, M.-C., 1982. *Territoire et stratégies soviétiques*. Paris: Economica.
Maurel, M.-C., 2002. Central-European Geography and the Post-socialist Transformation. A Western Point of View. In: M. Kaase, V. Sparschuh and A. Wenninger, eds. *The Social Science Disciplines in Central and Eastern Europe. Handbook on Economics, Political Science and Sociology*. Budapest: GESIS/Social Science Information Centre, Collegium Budapest Institute for Advanced Study, pp. 578–587.
Maurel, M.-C. 2004. Différenciation et reconfiguration des territoires en Europe centrale. *Annales de géographie*, 113(2), pp. 124–144.
Mészáros, R., 2006. A társadalomföldrajz és a regionális tudomány Magyarországon [Social geography and regional science in Hungary]. *Magyar Tudomány*, 167(1), pp. 21–28.

Musil, J., 1977. *Urbanization in Socialist Countries*. Praha: Svoboda.
Nemes Nagy, J., 1998. *A tér a társadalomkutatásban*. Budapest: Hilscher Rezső Szociálpolitikai Egyesület.
Rey, V. and Gerbaud, F., 1996. Nouvelles campagnes de L'Europe Centre Orientale. *Espaces & Milieux*. Paris: CNRS.
Turnock, D., 1978. *Eastern Europe*. London: Dawson Publishing.
Turnock, D., 1989. *The Human Geography of East Central Europe*. London: Routledge.

19
Conclusion
An evolutionary look at new development paths
Gábor Lux and László Faragó

Introduction

With an eye towards the most important development paths, this chapter aims to present both the current situation and the upcoming challenges of CEE regional development. This development undoubtably rests on strong path-dependence, but while some trajectories are old mainstays in socio-economic development, others are getting exhausted or are already on their way out. Which is which and what makes the difference? As the contributions of this volume attest, the post-crisis period increasingly appears to be a significant turning point in many respects.

When dealing with the long-term development paths of the CEE macro-region, we inevitably have to face questions about the *longue durée* of regional development and the weight of history. Indeed, where does post-socialism end? This term has been in common use since 1990 (although some transformation processes were already under way by the 1980s), and it has served well to describe a period of transition and European (re-)integration. But even if the inherited structures are prominent and the socio-economic phenomena path-dependent, can we describe a macro-region with the legacies of a political system that has been defunct for more than 25 years? Judging by the popularity of 'ruin porn', the photographic exploitation of decaying post-industrial landscapes, there is a certain fascination with this wreckage in the public eye. Although they enjoy worldwide popularity, these images are indelibly linked to the perception of Central and Eastern Europe, and they are enduring symbols of a 'historical failure' (Herrschel, 2007) or 'the Eastern wilderness' (Domanski, 2004a). They also play a role in the 'reinscription of otherness' in Europe's eastern enlargement (Kuus, 2004), something that is just as often rejected by the macro-region's own citizens as it is embraced (Sucháček and Herot, 2014).

Even if the weight of history remains too important to deny or reject, the findings of this book suggest that we might speak of post-socialism coming to its end, or as something that has already become a thing of the past. It is hard to draw a definite dividing line. The 2004 EU enlargement seems to be too early. The onset of the 2008 crisis and the new landscape of post-crisis Europe appear to be a point of departure for something new. The existing development paths produce diminishing returns, and hopefully other opportunities will emerge to take their place. The post-transition era is still greatly influenced by historical legacies, among them the surviving patterns of the socialist period are just one, while the sustainability of the post-socialist model is becoming a primary question.

Social sciences in CEE have always had to struggle with the issues of delayed or under-development, and after a few exuberant years following 1990 these questions returned as a series of apparent paradoxes. How to develop efficient and prosperous economies from badly decayed foundations (Chapters 3, 4 and 5)? How to build entrepreneurship without entrepreneurs (Chapter 6), or effective regional governance with persistent legacies of centralisation (Chapter 9)? What about the openness of borders in a macro-region where the former borders have so often been painfully redrawn and contested (Chapter 11), and where ethnic minorities have never had it good – but there were important distinctions between 'bad' and 'much worse' (Chapter 12)? Since the new financial and economic crisis, a spirit of defeatism and cultural determinism has come to haunt the discussions about socio-economic issues (Pogátsa, 2014, 2016), and it has been fashionable in CEE discourse to treat the macro-region's prospects with dismissal or blatant contempt. Yet contempt for the self is not more noble than contempt for the other.

Indeed, many of the aforementioned issues have shown gradual improvement, if they have not been completely or even permanently resolved. Good practices have emerged and gained traction. Time may heal many wounds – in which case the questions to be asked are perhaps: 'But do we have that time?' and 'Where should we be going next?' In lieu of a recapitulation of the preceding chapters, such questions – and there are more questions than answers – will occupy this chapter. Two sections will deal with a discussion of current development processes and future challenges *within* Central and Eastern Europe, while the other two will extend the scope of the debate to focus on the issues of *broader* European integration from a CEE perspective.

Scale: an issue of staying power or one of increasing importance?

Regional development in our time favours the centres rather than the peripheries and foments concentration on all spatial scales, while abandoning, or at least diminishing, the relevance of the regulative barriers and redistributive policies which had previously kept these concentration processes under control. The *global interface* – formed by myriads of linkages in communication, transport and trade – emerges as a highly permeable medium which does not give rise to 'the death of distance', but serves to greatly reduce its significance. Globally integrated spaces become 'closer' than the ones separated by less physical distance, but they are not deeply embedded within the world system, which results in 'subverted proximities' (ESPON ET 2050, 2015). Globalisation is not merely a top-down process, since it is also a product of localities and individuals engaging in everyday trade and consumption. However, the overall process has its dominant drivers and unambiguous winners. The worldwide rescaling process allows the inter-connected space of global integration to wedge itself between the smaller local, regional and national spaces, reconfiguring them according to its own logic. The success of metropolitan city regions is a potent argument in favour of the policies that provide further privilege for their development model. In adapting to the conditions set by 'limitless globalisation' and a 'metro-politan world', Europe itself reframes the debate about its strengths, values and identity – in fact, it reframes the questions before it could even consider the answers.

Several chapters in this book deal with the various ways in which the rescaling process affects Central and Eastern Europe, a macro-region with less dense networks and less prominent centres. Scale and density are the most path-dependent variables in regional development; they are the results of long-term accumulation processes which are revealed most clearly in urbanisation. The long-term urbanisation deficit as compared to Western Europe has historical roots and is a key issue of future development. As 'carriers of history', urban networks can be more stable than states; while the state formation processes in CEE entered a new phase after 1990, the towns and cities remained relatively stable. Even where the cities themselves have declined due to the late arrival of suburbanisation, city regions have gained importance.

In examining this 'foundation' underpinning many other networks, Chapter 8 has shown how *post-socialist restructuring was city-led in many ways*. Transformation was mostly about the affairs of cities, and groups of large cities and metropolitan areas have increasingly emerged as the most competitive actors of the post-transition period. Corridors and transport systems mainly serve to connect these regional hubs and to further their integration and competitiveness at both European and global levels (Chapter 16). The present era of globalisation does not offer minor cities and small towns a correspondingly important role; they find it much harder to develop competitive functions, or even to maintain their former positions in structuring and integrating the space economy. It remains a question whether large cities can assume the role of integrating the surrounding territories, or whether they will integrate into European and global networks, thus undergoing disembedding, reducing their attachment to their hinterlands. The lack of successful territorial integration will undoubtedly result in increasing territorial and social polarisation.

The relevance of city networks and the rescaling process can be found in many aspects of regional development. The uneven geographies of the post-industrial service economy affect the economic structure (Chapter 2), manufacturing (Chapter 3), business and financial services (Chapter 4), entrepreneurship (Chapter 6), and also the spread of the knowledge economy (Chapter 14). Successful service-based economies thrive in the central regions and at selected points in space, while manufacturing remains a powerful dynamising force in the provinces, delineating the spaces of integration and disintegration. It is not simply the level of urbanisation that matters: there are also other dividing lines, most prominently the one between the Visegrad countries and Slovenia on the one side, and South-Eastern Europe on the other. The most advanced functions of the modern post-industrial age seem to be restricted to the metropolitan or submetropolitan settlement tier, as is the case with KIBS, ICT activities or R&D and innovation – factors which in fact reinforce territorial concentration in the globalised economy (Audretsch, 1998; McCann, 2008; McCann and Ortega-Argilés, 2015). This 'geographic determinism' of our time (Sucháček, 2010) sometimes appears to be insurmountable.

In many CEE countries (except for Poland, and partially Czechia and Romania), strong urban counterpoles are missing, and thus the benefits of modern knowledge economy may remain restricted to national capitals and their surroundings, while the costs may be spread out over the peripheries. As Chapters 4 and 14 have evidenced (findings reinforced by the national data from the World Bank's Knowledge Assessment Methodology, 2012), the CEE macro-region does not have a fully-fledged knowledge economy yet, only some of its developed 'islands' have these characteristics. The missing tier of large cities is a development challenge emphasising the importance of polycentric development scenarios and the role of territorial cohesion, something which might represent an alternative path to the current mode of unlimited global/metropolitan competition. We should learn more from the good examples of those German, Italian and French regions which have relied on mid-tier, non-metropolitan urban centres and achieved strong social cohesion and network integration. Although this growth model has been slightly 'out of fashion' since the rise of metropolitanism, its study should be revived and its lessons be applied broadly.

Regional development patterns and emerging structures in a system of dependencies

The relationships in which the CEE economies and social milieus have found themselves after transition are characterised by multiple dependencies in terms of financial capital, the source of know-how and even policy transfer. These dependencies are notable because while they are gradually becoming mutual, they remain both asymmetric (the periphery is more dependent on the core than vice versa) and unilateral (the periphery is dependent on one core, but the core has a set of multilateral linkages all of which are relevant but none critical). Integration can offer mutual benefits, but it can also transmit the results of shocks which get magnified on the periphery or become unequally distributed in a centre–periphery relationship – as exemplified by the Eurozone crisis. This relationship has far-reaching consequences for the macro-region's development outlook and its political bargaining power; and it has led to a unique brand of modern capitalism that simultaneously resembles the core's model and is distinct from it.

Indeed, the whole CEE development path fits into what the 'varieties of capitalism' debate in comparative economics dubs the 'dependent market economy' (DME) model (Bohle and Greskovits, 2004, 2006; Rugraff, 2008; Nölke and Vliegenthart, 2009). In contrast to the less regulated liberal market economies (LMEs, e.g. the USA), as well as their coordinated counterparts (CMEs, e.g. Germany), DMEs' competitiveness is tied to 'a specific type of comparative advantage that is not based on radical innovation (LMEs) or incremental innovation (CMEs), but rather on an assembly platform for semistandardised industrial goods' (Nölke and Vliegenthart, 2009, p. 679). The mutually dependent relationship has led to unquestionable advantages – rapid integration into global networks, capital inflows, knowledge and policy transfer – but the 'fast-track' development path has also entailed trade-offs and increased vulnerabilities.

- TNCs exhibit different firm behaviour on home markets and near their subsidiaries: the most valuable segments of the value chain are kept close to corporate centres in developed economies, while the associated production functions on the peripheries receive much less attention. There are further differences in the local features of value chain management, the use of business services, local connectivity, attitudes towards the social net, etc.
- There are notable risks associated with capital movements, as production sites engage in intense competition for the reinvestment of company profits which can be easily repatriated or moved to other sites according to corporate strategies.
- Even in successful regions, over-reliance on FDI can result in crowding-out and congestion effects, targeting the product and labour markets of local companies and taking over their development niches.
- Dependent financialisation leads to different lending activity and development priorities in the centres and on the peripheries; risk-taking and consumer protection are both lower, while profit repatriation is higher.
- Most importantly, low-road competitiveness and external dependency pose long-term problems in the accumulation of financial, human and maybe even social capital. Low-road, low-income competitiveness leads to a development trap: it hinders the formation of new, well-capitalised domestic enterprises, while encouraging skilled workers to move westwards in pursuit of higher wages. This leads to long-term human capital loss in CEE and undermines the potential sources of qualitative improvement.

The structures of dependency are self-reinforcing and can lock regions into static development paths, eventually making them succumb to economic crises and low-cost competitors.

The FDI-driven restructuring of manufacturing (Chapter 3) has reused the salvageable production factors of declining socialist industries, but has not yet produced a domestic milieu of high value-added production with strongly integrated networks. In many respects, the same applies to the service sector (Chapter 4) and agriculture (Chapter 5). In the early years of transformation there had been too much faith placed in the benefits of 'creative destruction', which entailed significant social and opportunity costs. Today we can see that over-reliance on FDI-based development was not only the easy path of European integration, but it has also been a source of vulnerabilities. The lack of 'national champions' – important actors in the rise of East Asian economies, and also in many EU states – and the weakness of domestic entrepreneurship are opportunity costs of transition. While the former problem cannot be remedied easily, the second should receive more attention. This is becoming increasingly a question of quality: high value-added economies with higher factor intensity and knowledge content, together with localisation, local embeddedness and network development (even network design). While the vertical logic and the authoritarian philosophy of central planning were antithetical to the development of horizontal socio-economic networks, the FDI-driven spaces of modern production also lack strong networks, because the main actors are *simply not interested* in them. Strong localisation is still more the exception than the rule. It is therefore a challenge of the next decades to proceed towards greater embeddedness and network development in production, and taking the 'high road' is the surest way to achieve that.

The functional decline of rural areas (Chapter 5) particularly raises the issue of reconsidering the current development model. Rural areas and small towns experienced further marginalisation in the post-socialist period, losing much of the small employment role they could muster (this process was especially marked in Hungary, causing the decline of a slowly emerging, broad and well-educated rural middle class). Neither agriculture, nor the labour-intensive food and light industries of small towns escaped without heavy long-term losses. Rural development has been a rather contested field of transition, shaped not only by the values and interests of local society, but also by the economic and social concerns of the non-rural elite groups. The national policy approaches, themselves the products of long-term historical development paths, both 'channel' and 'filter' the EU's policy goals, and this strongly influences their effectiveness.

Conflicts between agriculture and its different models (small- and large-scale farming, private and cooperative farms, crop production and complex product chains), as well as questions about recreational use, landscape preservation, rural tourism and the quality of life, are more often determined by decision-makers in the core regions than by the real needs of rural society. Therefore, issues of employment and social cohesion have been sidelined by both business and, paradoxically, the 'post-modern rural economy' (Póla, 2016). Although this raises serious European concerns, the deep socio-economic problems of underdeveloped rural areas are not treated properly. One of the consequences of provincial hollowing-out has been rural poverty. However, the long-term result is depopulation and migration to both the national capitals and the European core regions. Population losses in Poland, Romania and, since the economic crisis, in Hungary have been intrinsically tied to these migration flows. Emigration might serve as a partial solution to underemployment issues, but the long-term losses of human capital, knowledge and working-age population can be expected to further worsen the prospects of rural CEE in terms of the economy, society, local politics, land use patterns and above all resilience.

Which way leads forward in the post-crisis, post-transition era? One key to the development of the CEE regions is capital accumulation, the other is greater investment into human capital. Path-dependent development can be seen in both successful and unsuccessful regions, in virtuous as well as vicious accumulation processes. The mutual dependencies and the co-evolution of local industries, knowledge creation and governance take shape in different forms. Successes tend to imply some sort of collective capital accumulation: in the form of valuable, localised knowledge, regional re-specialisation or examples of good governance. All of these can become the seeds of new development paths, and they hold the promise of strengthening endogenous development capacity. The future will provide greater opportunities, but also greater responsibility for the local actors: effective governance (Chapter 9) will be just as important in the success or stagnation of a region as localised knowledge – all around the space economy.

Human capital is still a neglected field of regional development. While human potential is the key to unlocking new development paths, to the modern knowledge economy, and is an important aspect of resilience (Chapter 15), its relevance has not been sufficiently considered in the CEE countries. Paradoxically, the conditions of post-socialism led to the chronic underfunding of education, healthcare and research and development just as they gained critical importance in development policy. Despite the rising number of higher education graduates, the regional knowledge creating and disseminating institutions have always been the first to suffer funding cuts. Meanwhile, the regional knowledge transfer role of universities, even the mid-range ones (Chapter 14), has become ever more important. Human capital is a major cornerstone of regional resilience, whether we perceive it as shock absorbing capacity, the ability to adapt to the future or to develop efficient and democratic institutions. The findings of Chapter 15 suggest that national capitals show the strongest resilience and peripheries the weakest. This clearly calls for investments in the peripheries to strengthen their own knowledge-creation, diffusion and adaptation capacities. It is a great challenge to change the trend of metropolitan knowledge concentration and to solve the absorption problems in the less favoured regions.

This necessitates rethinking the development philosophies, including the presently used policy instruments. The CEE countries haphazardly adopted the EU policies based on knowledge and human capital, and we may often speak of a kind of 'mimetic' development, where the development policy rhetoric was adopted, but the genuine content is missing. It is always easy to find exceptions, but we should generally be more concerned with the average. We are mindful of the good examples and good practices, but we are suspicious of the idealised case studies originating from the most advanced European or US regions, stripped of their contradictions and complexity. Some policies, such as the EU's Smart Specialisation Strategy (S3) platform or the LEADER approach, are particularly suited to serve local needs, but what really matters is to empower the local actors and communities to reintegrate the socio-economic space (Chapter 2).

Questions for European regional policy

The enlargement of the European Union to 28 member states (2004, 2007, 2014) raised new challenges for regional policy, too. While the EU's territory grew by 36 per cent and its population by 30 per cent, its GDP growth barely exceeded 10 per cent. Almost all of the newly joined regions are 'less developed', below 75 per cent of the EU's average GDP at purchasing power parity. The EU's 20 least developed regions are all in the new member states. In spite of development funding, it is now clear that cohesion and catching-up are much slower than many had expected. The most developed member states do not necessarily have an interest in solving the problematic centre–periphery relationship, rather they are interested in developing a more accommodating and adaptable periphery.

In the last 15 years, two important shifts have taken place in EUropean regional policy:

- Regionalism fell out of fashion as the member states' role in treating regional differences increased and attention gradually shifted from regions to cities, particularly the metropolitan areas.
- Regional policy based on redistribution was replaced with the narrative of 'encouraging growth and jobs' in *all regions*. In practice, support previously reserved for the less developed regions has been extended to their developed peers.[1]

Previously, the EU's structural funds and the Cohesion Fund had supported the catching-up of selected (underdeveloped, restructuring, sparsely populated, etc.) regions, focusing on the 'cohesion countries' where most of these were to be found. The regional approach was based on the idea that functions taken over from national governments (e.g. restructuring, the mitigation of regional differences) would be easier to implement in a regional framework and that the regions can better formulate their own economic problems and cultural needs, and implement their own development programmes. The political and economic 'federalist movements' and the decentralisation efforts based on subsidiarity (see e.g. the works of Leopold Kohr, Denis de Rougemont, Jean-François Gravier and Guy Héraud; or in Hungary, Gyula Horváth, Ilona Pálné Kovács and János Rechnitzer) had brought regionalism to the forefront by the last decade of the previous century, well expressed in the term 'Europe of the Regions'. Cross-border cooperation (the Interreg or the European Territorial Cooperation programme) contributed to filling interregional cooperation with content.

European new regionalism was mainly built on the idea that the new territorial networks and the clustering of the post-Fordist economy would take place in regional frameworks. At the same time, there was neo-liberal consensus in questioning the role of central governments in economic development and regulation. The EU's 'four freedoms' have partially eroded the influence of national governments, but the 2008 crisis has shown that the operation of the free market needs more (local and national) community control. Companies, clusters and networks operating on the global markets exist within their national framework and function by using local resources. Consequently, they are heavily influenced not only by international competition, but also by national regulations and local circumstances (milieu, atmosphere, territorial capital). If Europe is built merely on the free flow of products, capital, services and persons, then concentration processes arising from international competitiveness and increasing returns to scale will become the primary force behind economic localisation, and regions will be unable to provide better (more flexible or accommodating), framework conditions than the nation state.

The meaning of 'bottom-up', 'endogenous', 'place-based', 'smart' or 'tailor-made' development is self-evident: every state/region/settlement can grow through the most effective exploitation of its territorial capital. This is one of the undeniable truths of spatially aware development approaches. Yet the same principle receives a peculiar spin in contemporary EUropean regional policy. These strategies start from the assumption that unfavourable, 'less developed' territories are able to utilise only smaller amounts for development effectively (cf. the lowering of the absorption ceiling), whereas support for developed regions is utilised more efficiently. Every region should fully use its capabilities, and whether they are more or less developed, all of them should be supported in this effort. This approach can be found in various documents and declarations made by EU leaders supporting implicitly or openly 'multiple-speed Europe'.[2] In setting the mission for the Directorate-General for Regional and Urban Policy (the name itself is a telling shift from DG REGIO) in 2014, the mission statement set forth a plan 'where people in all our regions and cities can realise their full potential' (EC, 2014, p. 6).

The problem is that neither the priorities of the EU2020 strategy, nor the thematic objectives breaking it down (Community Strategic Framework) contain explicit goals relating to territorial cohesion; they suffer from 'spatial blindness'. None of the five thematic goals outlined there are 'spatially aware'. They focus on employment, R&D investments, climate protection and sustainable energy, education, and the fight against poverty and social exclusion. The specific issues and aspirations of countries only appear in connection with their implementing and localising these objectives, while regions are entirely omitted.

As Chapter 10 – and also some others in this volume – implies, territorially differentiated regional policies focusing on redistribution can best serve the objectives of reducing European territorial differences, and it is the principles of subsidiarity and multilevel governance which can help 'translate' them onto the regional and local levels. It is not sufficient for the EU's political and administrative centres to occupy themselves with supra-national concerns; effective national and subnational solutions are also needed. The Mediterranean as well as the Central and East European countries are all interested in supporting the principle of solidarity and in continuing or expanding cohesion funding. We argue that this interest is mutual with the EU's core countries because:

- A further increase in regional development gaps will lead to significant political and socio-economic costs (*especially* opportunity costs) that are higher than the costs of the amelioration of regional differences would be.
- The expansion of the internal market, the improvement of socio-economic conditions and new cooperation opportunities between core and non-core regions are all mutually beneficial.
- Further European integration is unlikely to succeed with the EU's current development differences, whereas increasing community-level territorial cohesion would provide a further impetus for this project.
- An internally strong EU will also find it easier to deal with external challenges which can be expected to increase with climate change, the waning influence of US policy on European development and security, and the growing uncertainty in this multipolar world.
- Most importantly, a return to the European model of shared responsibility and solidarity will contribute to the strength of the Union, which it will not find if playing by the rules of unmitigated global (metropolitan) competition.

Concluding thoughts: Europeanisation and the challenge of decentralisation

In the last two decades, many had expected *the further unification of Europe* by transferring functions from national governments to supra-national institutions. Transfer processes were also expected from national government level to regional institutions being between the national and the local levels. Through the creation of the Committee of Regions by the Maastricht Treaty (Art. 198) and the preparation of the Community Support Framework (CSF) plans, regions had become part of the EU's planning and decision-making mechanism. At the same time, Commission and Brussels bureaucracy measures manifested themselves at the regional level. Cooperation between the Union and the regions served one another's legitimacy, 'capturing' national governments in a 'vise', 'sandwich' or 'nutcracker' (Borrás-Alomar et al., 1994). Today we already know that this has brought no fundamental structural changes in the operational logic of either the EU or the national governments. Regions are only weakly involved in EU-level decisions, while the other two tiers are locked in a power struggle. The notion of 'federalist Europe' was not underpinned by real development processes.

The EU started to see itself not only as an organisation representing European unity and striving for some 'sum total' or 'average' of diverse European interests, but also as the prime facilitator of further integration, having its own goals and supra-national or post-national interests (Chapter 10). These interests sometimes overruled the concerns of both the member states and EU citizens. Fearing to lose their sovereignty not only because of the EU, but of the apparently 'unlimited' globalisation as well, and also due to the series of crises sweeping the continent in 2008, 2010 and 2015, member states have responded in a similar way. This contradiction, always present beneath the surface of European integration, has come to the fore in recent years.

This conflict may be the result of a phase of development, or it might foreshadow a process of European disintegration. At the core of the matter we find that in abandoning their mutually set goals, the main shapers of integration – the nation states and the EUropean centre – have subverted the concept of regionalism and replaced its operational logic with their own centralist ambitions. We can speak of subverted regionalism in multiple respects:

- The EU–member-state balance and the spirit of cooperation have been replaced by the mutual antagonism of competing centralisms: one in Brussels and several others in the national capitals, neither of which serve the needs and interests of their respective peripheries.
- In lieu of subsidiarity, the EU has developed its own 'civilising mission', to be achieved by any means if it serves the common interest; meanwhile nation states have returned to nation-building agendas at the cost of the community. Both of these are the products of top-down philosophies.
- Instead of serving the interests of territorial cohesion and the development of the peripheries, regional policy has been repurposed to fit central agendas: supra-national and pro-globalisation interests at community level, and national goals at the nation-state level.
- Development policy has become increasingly homogenised, abandoning for uniform solutions its erstwhile flexibility and the idea of local needs, and imposing central ideas about growth and progress on a differentiated socio-economic landscape.
- The prevailing development model – as discussed previously – hinges on privileging the metropolitan (global cities and national capitals) over the non-metropolitan, and subverts the interests and values of the latter in favour of the former.

Subverted regionalism does not only lead to losing an opportunity for growth and a potential European compromise (*Ausgleich*), but it indisputably abandons a valuable achievement (*acquis communautaire* in the original sense) of the European model. Substituting the competing centralisms of nation states and EUropean institutions for subsidiarity, we sacrifice ourselves on the altar of globalisation.

Yet distrust of the EU in the CEE member states has been less motivated by integration fatigue than by the consistent power asymmetries and the lingering fears both of external control originating from unpleasant historical experience and of becoming 'voiceless' or 'unheard'. The EU institutions' reaction to dissent on the peripheries has not served to build bridges and increase trust, instead, it often seemed to confirm these countries' wildest fears. The breakdown of good faith has affected both sides adversely: in Brussels it has become tempting to use blunt force to achieve a desirable 'European' solution, while on the peripheries it has resulted in governments strengthening their grip on 'their territory' to defend vital interests. This is a conflict without winners, and the related negative prophecies may be self-fulfilling.

There has also been an important misunderstanding in the West–East relationship, explored a decade ago by Domański (2004b) as 'the pitfalls of paternalism and a claimant attitude'. CEE has a long, deeply embedded tradition of criticising the European core. The new notion of 'Central Europe' itself emerged as a criticism of the great powers' callousness in their treatment of the macro-region. This topic was discussed most prominently in Kundera (1984), and was expressed by the political movements demanding to end state socialism, restore sovereignty and return to European normality lost in the world wars. However, *criticism does not imply rejection*. Ideas and views challenging the European status quo are articulated with the intent to *better* European civilisation, not *best* it (EUrope's CEE critics like to raise the point that they like the continent better than the post-1968 West does[3]). Questions striking to the heart of European identity – such as those raised about the financial and economic crisis or the ongoing migration challenge – do not ask 'Do we want an integrated Europe?', but rather 'What kind of Europe do we want to integrate into?' Such critical views are similarly common in the European core, but their legitimacy is less often – and less *easily* – contested and/or dismissed. (In fact, the notion that the beneficiaries of EU funding should not criticise it is a disturbingly common attitude, even in CEE professional circles.)

However, the misunderstanding strikes both ways. CEE's notion of dependency and victimhood can also become a convenient excuse for rejecting responsibility and substituting the criticism of the other for self-examination. Increased national sovereignty can just as easily subordinate the interests of individuals and communities; the only difference is that instead of a nebulous 'European project', the name of the concept is 'the national interest'. Centralists of every stripe are not fond of autonomy and independent thought: they seek to marginalise these and reassert their control by one means or another.

The position of the CEE regions and cities – and of course their citizens – in these processes is a key question. Are they simply recipients (subjects) of regional development and Europeanisation, or can they emerge as autonomous actors whose input can shape these processes or even initiate new ones? *Dependent relationships disempower individuals and communities alike. To be achieved through the principle of subsidiarity and a consistent agenda of decentralisation on multiple territorial scales (i.e. at European, national and regional levels), it is only autonomy that ensures that the actors take their fate into their own hands and become equal in exploring new European futures.*

Acknowledgements

The publication of this research has been supported by project #104985 of the Hungarian National Research, Development and Innovation Office. While writing this paper, Gábor Lux was supported by the János Bolyai Research Scholarship of the Hungarian Academy of Sciences.

Notes

1. In the 2014–2020 period, 'transition regions' receive €35.4 bn, while the 'more developed regions' get €54.4 bn. Under the old scheme, disadvantaged regions could have received all €89.8 bn.
2. See Juncker (2014, p. 12): 'My firm conviction is that we must move forward as a Union. *We do not necessarily all have to move at the same speed* – the Treaties provide for that and we have seen that we can work with different arrangements. Those who want to move further, faster, should be able to do so.'
3. Kundera (1984), again, writing in the 1980s: 'By virtue of its political system, Central Europe is the East; by virtue of its cultural history, it is the West. But since Europe itself is in the process of losing its own cultural identity, it perceives in Central Europe nothing but a political regime; put another way, it sees in Central Europe only Eastern Europe.'

References

Audretsch, D.B., 1998. Agglomeration and the Location of Innovative Activity. *Oxford Review of Economic Policy* 14(2), pp. 18–29.

Bohle, D. and Greskovits, B., 2004. Capital, Labor, and the Prospects of the European Social Model in the East. *CES Central & Eastern Working Paper* 58.

Bohle, D. and Greskovits, B., 2006. Capitalism without Compromise: Strong Business and Weak Labor in Eastern Europe's Weak Transnational Industries. *Studies in Comparative International Development* 41(1), pp. 3–25.

Borrás-Alomar, S., Christiansen, T. and Rodríguez-Pose, A., 1994. Towards a 'Europe of the Regions'? Visions and Reality from a Critical Perspective. *Regional Politics and Policy* 4(2), pp. 1–27.

Domański, B., 2004a. Moral Problems of Eastern Wilderness: European Core and Periphery. In: R. Lee and D.M. Smith, eds. *Geographies and Moralities: International Perspectives on Development, Justice and Peace*. London: Blackwell, pp. 47–61.

Domański, B., 2004b. West and East in 'New Europe': The Pitfalls of Paternalism and a Claimant Attitude. *European Urban and Regional Studies* 11(4), pp. 377–381.

ESPON ET 2050, 2015. *Territorial Scenarios and Visions for Europe.* Luxembourg: ESPON and MCRIT.
European Commission (EC), 2014. *Regio Management Plan.* Ref. Ares(2014)160839 – 24/01/2014.
Herrschel, T., 2007. Between Difference and Adjustment – The Re-/presentation and Implementation of Post-socialist (Communist) Transformation. *Geoforum* 38(3), pp. 439–444.
Juncker, J-C., 2014. *A New Start for Europe: My Agenda for Jobs, Growth, Fairness and Democratic Change. Political Guidelines for the Next Commission. Opening Statement in the European Parliament Plenary Session.* Strasbourg.
Kundera, M., 1984. The Tragedy of Central Europe. *New York Review of Books* 31(7), pp. 33–38.
Kuus, M., 2004. Europe's Eastern Expansion and the Reinscription of Otherness in East-Central Europe. *Progress in Human Geography* 28(4), pp. 472–489.
McCann, P., 2008. Globalization and Economic Geography: The World Is Curved, Not Flat. *Cambridge Journal of Regions, Economy and Society* 1(3), pp. 351–370.
McCann, P. and Ortega-Argilés, R., 2015. Smart Specialization, Regional Growth and Applications to European Union Cohesion Policy. *Regional Studies* 49(8), pp. 1291–1302.
Nölke, A. and Vliegenthart, A., 2009. Enlarging the Varieties of Capitalism: The Emergence of Dependent Market Economies in East Central Europe. *World Politics* 61(4), pp. 670–702.
Pogátsa, Z., 2014. Cultural Defeatism in Central Europe. *Visegrad Revue* 27 January.
Pogátsa, Z., 2016. *Magyarország politikai gazdaságtana: Az északi modell esélyei* [The political economy of Hungary: The chances of the Northern model]. Budapest: Osiris.
Póla, P., 2016. *Helyi erőforrásokra alapozó vidékfejlesztés Magyarországon* [Local resource-based rural development in Hungary]. Pécs: Publikon Kiadó.
Rugraff, E., 2008. Are the FDI Policies of the Central European Countries Efficient? *Post-Communist Economies* 20(3), pp. 303–316.
Sucháček, J., 2010. On the Emergence of Minor Cities. In: J. Sucháček and J.J. Petersen, eds. *Developments in Minor Cities: Institutions Matter.* Ostrava: VŠB – Technical University of Ostrava, pp. 13–28.
Sucháček, Jan and Herot, P., 2014. The City of Ostrava – From Industrial Image to Industrial Image 2.0. In: F.M. Go, A. Lemmetyinen and U. Hakala, eds. *Harnessing Place Branding through Cultural Entrepreneurship.* Basingstoke: Palgrave, pp. 191–210.
World Bank Knowledge Assessment Methodology, 2012. http://web.worldbank.org

Index

accession 143, 160, 170, 179; transitional restrictions 212
acquis communaitaire 160, 288, 317
agglomeration 15, 169–70, 230; *see also* concentration
agriculture; developmental role 75, 136, 313; output **77**; state socialist 71–2
airports 273–6, *275*
aviation *see* transport
autonomy 149, 151, 189, 192–4

Balkans 130
Baltic states 5, 84n1, transport role 265, 268–9, 276, 279–80
banking 36; network 55, 57; ownership *56*, 61; transition of 54–5
borders; crossings 263–5, *264*; de-bordering 178, 180; rebordering 178–9, 265

capital; accumulation 24, 107–11, 312–3; human 24, 313–4; movements 312 ; territorial 160, 169–70
cathedrals in the desert *see* embeddedness
centralisation 8, 142; EU-level 161–3, 317
Central and Eastern Europe; image 3, 309; scope 5
centre–periphery relationship 159–62, 220, 314, 317
cities; capital 51, 125, 136–7, 160, 219; gateway 135; main data **127**; minor 16; regional centres 51, 125, 128, 170, 219
clusters 16, 36, 38, 108, 227, 315
co-evolution 30
cohesion 169–71, 311, 313
cohesion policy *see* regional policy
common agricultural policy (CAP) 70, 76
companies *see* firms; *see also* trans-national corporations
concentration 163; of capital 56, 315; of creativity 107–8; of knowledge 227, 229; metropolitan 15, 21, 51–2, 136–7, 160–1, 227
Council of Mutual Economic Assistance (COMECON) 31, 276
creative economy 106, 110, 114

cross-border cooperation 135, 176, 178, 263, 292; institutions 183
crisis 40, 56, 58–61, 143, 244, 312
cultural determinism 310
culture; culture-based development 107–10, 113; heritage 110

de-industrialisation 17–19, *18*, 107; *see also* tertiarisation
decentralisation 141, 143, 153–4, 172, 253, 299–300, 315, 318
delocalisation 23, 33, 41
demographics 216, 219; of the peripheries 81, 135–6, 313, 219
dependent market economy 4, 41, 52–3, *61*, 107, 160–1, 226, 312, 318
depopulation *see* demographics
development; endogenous 16, 24, 34, 83, 313; high and low-road 16, 23–4; paths 31, 36, 313; place-based 169–70; polycentric 129, 136, 171–2, 242, 311; sustainable 39–42; 290–1
developmental states 4
disparities; between wealthiest and least developed regions *164*; in innovation *234*, **236–7**; macro-regional 5, 21–2, 34, 72, 311; regional 40–1, 94, 96, 136–7, 160, 311
diversification 32
dual economies 34, 76, 220, 161

education 97; higher education figures **228**
embeddedness 17, 232, 313
employment; in agriculture *69*; in industry *22*; in services *22*, *49*
energy 72, 244, 252, 257, 286, 290; renewable 85n2; 291–2
enlargement *see* accession
entrepreneurship 24, 42, 87–90, 100, 313; enterprise formation 88; regional disparities 94, *96*
environment 252–3; planning models 289–92; policy 287; state socialist legacy 284–6
European Capital of Culture (ECoC) 111, 113–14; budgets *113*

Index

European Grouping for Territorial Cooperation (EGTCs) 177, 181, 183
europeanisation 160, 175, 177, 179, 317
euroregions 177, 181, 265
evolutionary economic geography 3, 29

financialisation 58, 312
firms; medium-sized 42; density 98
foreign direct investment (FDI) 19, 33, 160–1; by country **20**; effects on innovation 226; effects on settlement network 124, 136; financial services 52–4; inward 53; sectoral breakdown **48**; as source of competitiveness 23, 312; in South-Eastern Europe 36
fragmentation 130, 133, 141, 153, 265, 269
freight *see* transportation
functional urban areas 128–9; main data **129**

gentrification 109–10, 114–15
geographic determinism 4, 24
global cities *see* metropolitan areas
global financial and economic crisis *see* crisis
global integration 34, 163, 310
global–local paradox 36, 80
global production network 23, 50, 107
global value chains 33–4, 36, 41, 107, 312
governance 253; European 162–3; medium-level 142, 144–52; rescaling 111; transition 141

highways *see* motorways

industrial districts *see* clusters
industrialisation 30–1, 285; *see also* reindustrialisation
industry; core area 23, 34–5, *35*; evolutionary processes in 29–30; food and light 34–5, 313; localisation 36–9; location factors 36, 38–9, **39**; spatial structures 34–5; upgrading 36, 39–41
innovation 33; performance 232–7, **236–7**; policy 232, 298; transfer 230, 232
institutions 142–3; evolutionary role 3, 242; informal 242, 255; institutional sclerosis 32; regional policy 165, 167
International Monetary Fund (IMF) 163, 226

knowledge economy 41, 87, 225, 311, 314; effects on entrepreneurship 98

labour; markets 218–21, 273; unskilled 23, 215; skilled 8, 52, 100, 172
land; ownership 73, 77, **79**; use **69**, 292
LEADER programmes 83–4, 314
local governments 141, 153; national models 144–52
localisation 24, 36–9, 80, 313, 310
lock-in 31–2, 71, 107, 312

manufacturing *see* industry
metropolitan areas 15, 129, 160–1, 163, 220, 256–7, 310–11; concentration in *see* concentration; in South-Eastern Europe 133
migration; balance 215–16; crisis 216–18; determinants 214–15; to EU-15 23, 100, 172, 211–15, 313; mobility 218, 263, 273; source countries 216
minorities 176, 180, 182, 189; legal status 194–5, 201n4
modal split 266–8, **268**
motorways 266, 272–3
movements; against cultural redevelopment 114–6; environmental 286; ethno-regionalist *see* regionalism, ethno-
municipalities *see* local governments

national champions 32, 42, 313
nation-building 133–6, 142, 175–6, 189, 317; and transport 265–6
networks 310; industrial 30; railway 265; and regional integration 25; in state socialism 31; trans-European *see* trans-European networks

offshoring 50–2, 107; main sites *53*
old industrial regions *see* regions

path-dependence 23, 30, 123, 130, 309–10; in environmental policy 288; in industrial transition 33–4; virtuous 32, 42
peripheries 23, 131, 169, 170; homogeneity of 36
planning 161–3, 163, 287, 289–92; central 255, 298, 313; contracts 167
policy transfer 112, 142–4, 267, 287–8
pollution *see* environment
polycentricity *see* development, polycentric
population density *68*
ports 272, 276–80; main nodes and flows *277*
post-Fordism *see* post-industrial society
post-industrial society 47, 311
post-socialism *see* transition
post-transition 1, 55, 309
pre-accession 160, 178, 181, 267; funds 145, 180, 263, 76
privatisation 32–3, 53–5, 73, 109–10

railways *see* transportation
refugee crisis *see* migration
regional centres *see* cities
regional disparities *see* disparities
regionalisation 7, 142–3, 254; national models 144–52
regionalism 255; ethno- 189–91; ethno-regionalist organisations **190**, **198**; new 5, 142, 315; subverted 317

regional policy 142–3, 153, 170; challenges 314–6; institutional system 165; instruments **168**; state socialist 298; target regions *166*
regional science 295–6; institutions 299–303, *302*; origins in CEE 297–9; publication forums 303–5, **304–5**; researchers **301**
regions; border 180; central 20–1, 137, 160, 169; hollowing-out 19, 23–5, 33, 36, 67, 312; intermediate 23; least and most developed 164, **167**; meso- 142, 144–52; old industrial 30–1, 33, *35*, 37, 108, 220
reindustrialisation 34, *40*
rescaling 15, 111, 310
research and development; expenditures 226, **231**; institutions 226–7; transition of 226; *see also* innovation
resilience 313–4; concept 240; determinants 241; economic 244–8, *245–7*; environmental 252–3; institutional-political 253–6; measuring 243–4; technological 248–51, *249–50*
restructuring x; sectoral 17–19, 47–9
rural areas 79–83, 130, 219; delineation 67, 79–80; functions of 81, 290, 313; in state socialism
rural development 82–4
ruralisation 17, 75, 125–6

separatism 194, 196, 200
services; business 21, 50–7, 136–7; concentration 48, 136–7; exports **51**; location advantages 50
settlement networks; hierarchy 128–30, 134; main data **125**; in South-Eastern Europe 130–6, *131*; transition 127; in Visegrad countries 124–30, *126*
shocks; exogenous 58, 240, 258, 312
South-Eastern Europe 21; specialisation patterns 34, 55; urban network 130–6

spatial justice 3, 5
specialisation 42; de-specialisation 19, 33, 36; overspecialisation 32; smart specialisation (S³) strategies 37, 42, 314
structural change *see* restructuring
subsidiarity 5, 7–8, 171, 315, 317–8
suburbanisation 127
sustainable development *see* development, sustainable

territorial cooperation *see* cross-border cooperation
tertiarisation 17, 47, 107–8; evolutionary nature 258; forms 19
towns 23, 125, 128, 130, 313
trans-European networks (TEN) 269
transition; concept of 3; recession 72, 74; studies 2
trans-national corporations (TNCs) 16, 312
transportation; aviation 273–6; corridors 269–70, 291; freight 271–2, 276–80; policy 266–9; railway 264, 270–1; regional development role 281; state socialist heritage 261–3
Triple Helix model 30

unemployment 220; informal 31
universities 300–3, 314; economic impact 231; mid-range 227–9, 232
urbanisation 127, 132, 136; deficit 310

varieties of capitalism 4, 160–1, 312
Visegrad countries 21; settlement system 124–30

war; consequences in former Yugoslavia 134
water transport 271–2
World Bank 226
world cities *see* metropolitan areas

Taylor & Francis eBooks

Helping you to choose the right eBooks for your Library

Add Routledge titles to your library's digital collection today. Taylor and Francis ebooks contains over 50,000 titles in the Humanities, Social Sciences, Behavioural Sciences, Built Environment and Law.

Choose from a range of subject packages or create your own!

Benefits for you
- Free MARC records
- COUNTER-compliant usage statistics
- Flexible purchase and pricing options
- All titles DRM-free.

Benefits for your user
- Off-site, anytime access via Athens or referring URL
- Print or copy pages or chapters
- Full content search
- Bookmark, highlight and annotate text
- Access to thousands of pages of quality research at the click of a button.

REQUEST YOUR FREE INSTITUTIONAL TRIAL TODAY

Free Trials Available
We offer free trials to qualifying academic, corporate and government customers.

eCollections – Choose from over 30 subject eCollections, including:

Archaeology	Language Learning
Architecture	Law
Asian Studies	Literature
Business & Management	Media & Communication
Classical Studies	Middle East Studies
Construction	Music
Creative & Media Arts	Philosophy
Criminology & Criminal Justice	Planning
Economics	Politics
Education	Psychology & Mental Health
Energy	Religion
Engineering	Security
English Language & Linguistics	Social Work
Environment & Sustainability	Sociology
Geography	Sport
Health Studies	Theatre & Performance
History	Tourism, Hospitality & Events

For more information, pricing enquiries or to order a free trial, please contact your local sales team:
www.tandfebooks.com/page/sales

 Routledge Taylor & Francis Group | The home of Routledge books

www.tandfebooks.com